T0143098

Cybersecurity Data Science

Scott Mongeau • Andrzej Hajdasinski

Cybersecurity Data Science

Best Practices in an Emerging Profession

Foreword by Timothy Shimeall

 Springer

Scott Mongeau
Nyenrode Business Universiteit
Breukelen, Netherlands

Andrzej Hajdasinski
Nyenrode Business Universiteit
Breukelen, Netherlands

ISBN 978-3-030-74898-2   ISBN 978-3-030-74896-8   (eBook)
https://doi.org/10.1007/978-3-030-74896-8

This Springer imprint is published by the registered company Springer Nature Switzerland AG
The registered company address is: Gewerbestrasse 11, 6330 Cham, Switzerland

Frontispiece design: Andreas Kallipolitis, iamtraum.com

"Ars longa, vita brevis, occasio praeceps, experimentum periculosum, iudicium difficile."

"Life is short, the art long, opportunity fleeting, experiment treacherous, judgment difficult."

—Hippocrates

Dedicated to Marloes, family, and friends

Foreword

While data science has been emerging as a profession since 2005, the professionalization of its application to cybersecurity is less mature. One reason for this relative immaturity is that both data science and cybersecurity have been undergoing extensive change and accepted practices are still evolving. Another reason is that, unlike many fields of data analysis, cybersecurity has intelligent opposition to its methods, specifically attackers who wish to intrude on computer systems and networks. To date, cybersecurity has been in a race against that opposition, and the state of data science for cybersecurity reflects that race. Under those conditions, accepted practices are rapidly challenged and modified.

Despite these challenges, the importance of cybersecurity data science is increased due to a number of pressures. The velocity of cybersecurity data is large, and increasing. A single, moderately busy, server or firewall generates gigabytes of log entries every day. A network traffic log for a large network generates tens of billions of entries per day. Security event analysis systems only deal with some of the more immediate and easily recognized issues. Data science approaches that can efficiently categorize and focus attention on the most impactful streams within this fire hose of data are urgently needed. At the same time, the activities of the attackers are increasingly diverse in subtlety, impact, and targeting. While some are easily recognized and of immediate effect on a recognizable target within the perception of the defenders, others mimic desirable traffic, lie latent within the target until desired by the attacker, or hit outside of the defender's perception, in unmonitored portions of their infrastructure or in the infrastructure of suppliers or vendors. By employing explicit feature engineering and sensitivity analysis, cybersecurity data science may focus on those features most revealing of even subtle activities and also provide the chance to secure on a community-wide basis. Federating and sharing data within even a tightly related community is often difficult due to the lack of common methods for data analysis and interpretation. Cybersecurity data science, with its explicit consideration of the characteristics of data and of analysis methods, offers an opportunity to bridge the federation and sharing difficulties.

This book thoroughly, if not exhaustively, documents the lack of maturity in data science applied to cybersecurity. More than identifying this lack of maturity, it uses a mixed-mode data collection, both qualitative and quantitative, to point to how the gaps in cybersecurity data science can be filled as it emerges as a full profession. Using a multifaceted mix of detailed literature review, survey of experts, and modeling, Dr. Mongeau has carefully delineated both where this data science profession is currently lacking and how those lacks could be addressed in future work. A wide range of factors are included, and clear recommendations are provided.

The reader who comes to this volume from an interest in cybersecurity will gain much in understanding how data science methods apply in this space. The book refers to various methods of analysis, and how those methods lend insight into cybersecurity objectives. This book serves as a broad and useful introduction to how data science contributes to cybersecurity, as that science is practiced by modern professionals.

The reader who comes to this volume from an interest in data science will find this book summarizes the state of data science as a profession (and the path forward) but then focuses directly on the specific needs of cybersecurity and how the profession would help to protect data in the modern world. The analysis problems (such as the chronic lack of ground truth) and methods to remediate those problems are covered thoroughly, and from a perspective that speaks to the data scientist.

The reader who comes to this volume from a managerial perspective, or one who seeks to understand the emergence of this field and the current capability of those practicing data science for cybersecurity, will find the clear description of the state of the field useful. This book offers a solid state of the field and is supportive to both realistic appraisal of what can be gained from practitioners and to what to look for as emerging capabilities in the near future. The interviews and quantitative analysis give in-depth understanding of what is, and what is coming.

Taken in total, this book offers an extremely useful clarification to the emergence of cybersecurity data science as a specific profession, borrowing from both cybersecurity and from general data science. Dr. Mongeau has provided a useful degree of clarity to these rapidly developing fields. From this basis, a variety of useful work will spring in the days to come.

Timothy Shimeall, Ph.D.
Senior Member of the Technical Staff
CERT Situational Awareness Team
Software Engineering Institute
Carnegie Mellon University
Pittsburgh, Pennsylvania

May 2021

Preface

During the concluding stages of this research effort in early 2020, the global Coronavirus (COVID-19) pandemic emerged and rapidly evolved. It would be remiss not to comment on this world-changing event regarding its relationship to and impact upon the central research topic.

It became increasingly evident that dramatic societal effects were occurring as a result of the coronavirus outbreak. For one, reactions to the pandemic brought about a rapid acceleration of the use of the Internet for communication and collaboration. There was a sudden mass shift towards people working from home in order to limit the spread of the contagion. Information workers, students, families, and friends moved quickly online to communicate and collaborate.

Quite suddenly, a large proportion of global work and social life has moved online. Although telecommuting was already a growing trend, the rapid movement to adopt remote working as the de facto approach has introduced a radical shift. Many popular press commentators suggest that this movement towards remote and virtual work will persist for many months and might possibly lead to more permanent shifts (Burr and Endicott 2020; Lichfield 2020).

The cybersecurity implications of this shift mirror those which were already developing before the crisis, albeit at a much larger scale. Defenders suddenly are confronted with safeguarding radically expanded, distributed, and defuse networks and devices. It is clear the future of corporate networks will be increasingly decentralized as a result. The "castle-and-moat" approach to defending organizational networks will be increasingly outmoded.

In the wake of this shift, the interruption of Internet carriage and digital devices will have far greater impacts on social, political, and economic life. Adversaries are presented with a range of enticing new targets to exploit and schemes to perpetrate (Gallagher and Bloomberg 2020; Jowitt 2020).

One need not look far to see that criminal and malicious actors have been quick to adapt and exploit current events. Phishing emails have adopted subject matter related to the coronavirus in an attempt to deliver malware and other malicious payloads (Reiner 2020). As well, state-level actors quickly adapted to the coronavirus crisis as an opportunity to sow disinformation (Beaumont et al. 2020) and to stage attacks to raid medical research and interrupt healthcare systems (Cohen and Marquardt 2020). These trends are disheartening and demand concerted responses.

This research profiles a range of mounting challenges afflicting the cybersecurity profession. A central challenge concerns the fact that rules-based methods to detect security events are increasingly ineffective due to the ever-increasing complexity and scale of digital infrastructure. To the degree that information professionals are increasingly working remotely *en masse*, approaches to network defense will be challenged to develop novel approaches.

An effort has been made herein to detail how cybersecurity data science (CSDS) can address a broad range of challenges to improve security assurance. The coronavirus pandemic has suddenly highlighted the importance of pursuing and enabling such improvements to cybersecurity practice. The stance here is that this dramatic event has increased the importance and visibility of CSDS.

CSDS provides refined and focused approaches to gain situational awareness. This includes the ability to refine understandings of normal versus abnormal behavior and to categorize and detect complex incidents. CSDS methods support complex statistical understandings of user behavior and an ability to monitor distributed events through dynamic pattern analysis. This highlights the rising importance of machine learning-driven approaches, such as applying unsupervised machine learning to extrapolate patterned understandings of user and device behaviors.

The advent of physically distributed and diffuse networks, not to mention the increase in cloud-based collaboration, conferencing, etc., demands the methods and treatments made feasible by CSDS. Growing dispersion and complexity will necessitate data analytics-based monitoring methods to assure and defend increasingly distributed organizational networks. In short, the coronavirus crisis has effectively made prescriptive security assurance approaches, of the type outlined in this work, more a pressing necessity than an aspirational work-in-progress.

A circumstantial similarity between the corona crisis and the subject matter herein concerns the broader challenge of undertaking scientific inquiry in complex environments beset by incomplete or unclear data. A major challenge in efforts to address and contain the outbreak revealed confusion, frustration, and conflict resulting from misunderstandings and disagreements surrounding the proper application of statistical and scientific methods. The crisis revealed how environments of uncertainty and complexity lead to confusion and challenges in marshaling focused and coordinated responses.

While it is not appropriate to compare like-to-like in terms of impact, cybersecurity faces a similar environment of complexity, uncertainty, confusion, and conflict related to data, methods, and responses. Confronting this topic, the aim of this work is to offer a potentially generalizable set of approaches to programmatically address complexity and uncertainty through the orchestration of data management, scientific inquiry, and organizational collaboration. The organizational implementation of such efforts is emphasized as requiring information sharing, communication, collaboration, and aligned incentives…

Moving beyond recent events, it is worthy to comment briefly on this research effort's broader origins and intentions. It should be mentioned that this research embeds what might be considered a conceptual Trojan Horse. A core topic addressed in the course of this manuscript is a detailed critique of the data science field in general, taking aim at its often-capricious sense of rigor. The root of this critique stems from the legacy origins of this research effort, going back more than a decade.

From 2009 through to 2020, I have been a part-time academic researcher alongside my full-time (and then some) practitioner work. Throughout this period, as a data analytics, and later, data science practitioner, my initial research efforts, starting during my MBA and Master of Financial Management degree programs at the Erasmus Rotterdam School of Management (RSM), and migrating into the Nyenrode Executive Doctorate Program (EDP), were focused on methodological and organizational challenges in the practice of data science (née data analytics). Early on, this included confronting process-based challenges of data diagnostics, analytics model validation, and model explainability.

A set of core critiques and views emerged over the last decade, integrating my research and field experience as a data science practitioner. I gradually built up a set of principles and best practices to address perceived shortcomings in the field. However, as an early research topic, a broad-based critique of the data science discipline lacked sufficient grounding and specificity. By 2015, my initial research and writing became conceptually unwieldy and suggested an overly ambitious purview.

Subsequently, focusing on a particular applied area of data science, the eponymous cybersecurity data science (CSDS), provided much-needed grounding and refinement. This was a natural fit as I had been practicing as a data scientist focused on the broad area of security analytics, encompassing cybersecurity, fraud, financial crime, security, and intelligence domains.

Focusing on CSDS allowed for practical examples to be framed, highlighting data science critiques and demonstrating best practices. However, the status of CSDS as a very new and rapidly emerging field created a fresh set of research challenges. For one, there was a need to qualify key terms such as "profession" and, thereafter, to distinguish emerging from mature professions systematically. This guided and informed the scope of the multifaceted literature analysis undertaken in *Phase I*, an opportunity to frame *as-is* CSDS practice in a systematic way.

For another, as the CSDS domain was and still is quite new, there was a need to understand what CSDS *ideally* intends to be in time. This led naturally into an investigation of the gaps between "practitioner idealized" visions of the domain versus limitations experienced in practice. My own experience as a practitioner reinforced a growing conviction that much of security analytics in practice was at quite a low level of maturity as compared to, say, applied data science in pharmaceutical, manufacturing, financial, or marketing domains. Part of this is that there is a core emphasis on exploration and discovery in security analytics. In contrast, more mature analytics sub-domains typically have a relatively clear set of target objectives supported by well-developed body of theory to guide practice.

That there were fundamental questions unanswered concerning the goals and methods of the CSDS domain resulted in a key decision to conduct interview research, as opposed to survey research (a more typical approach in organizational scholarship). Further, it was resolved to focus the research scope on diagnostic problem-solving, rather than hypothetical-deductive approaches. The results of this effort are discussed and analyzed in *Phase II*.

I felt strongly then, and still feel, that the interview approach was crucial to exploring and extrapolating hidden and subtle aspects of this new profession as an emerging phenomenon. Conducting interview research brought its own challenges associated with mechanics, methodological grounding, interpretation, and some discussions concerning rigor, etc.

Efforts to ensure rigor and completeness in qualitative and quantitative interpretations of the interviews challenged me to operate outside my comfort zone. I have since developed a new appreciation for interviews as an insightful organizational research method. I feel happy with the approach taken and the results achieved. I also think that the results are interesting and important to report.

Having provided both deductive and inductive perspectives as a foundation, the design science approach in *Phase III* took form organically. My prior research on methodological challenges and prescriptions in the data science field found new relevance once grounded in CSDS practice. Nonetheless, *Phase III* still bears broader observations and guidance of potential interest and value to data science generalists. Essentially, CSDS, being a new and complex domain, was an excellent "toy problem" to explore organizational challenges and approaches relevant to wider data science practice.

The result provides findings of interest to a broad set of stakeholders, from academic to practitioner to policy maker. Similar to a complex musical score, the write-up attempts to orchestrate various themes and voices into one. The ten research questions are a drumbeat for those intently focused on developing the CSDS domain. The vocal component addresses managers and organizational leaders through practical punctuations, and weaves throughout. The CSDS practitioner perspective is a high note resonating continually in the background. Along the path, the baseline, key theoretical concepts and contributions to academic research are highlighted. To satisfy the methodologist, a horn section drives the effort to ensure that the complex and novel methods are clear and resonant.

In the end, the intention was to satisfy an array of people, which can risk diffusion. Complex scores can be hit-or-miss. Sometimes, the syncopation falls flat. There may be uneven points where a stride is hit, but then suddenly ends. Music is a matter of taste, especially new music, and it is expected that not everyone will enjoy it. For those, I invite them to spend some time listening, and to recognize that several individual melodies are harmonizing as a polyphony.

With these "key notes," one on the very latest current events and one on the origins of this research spanning back over a decade, I will conclude my opening comments. I thank you sincerely for your attention and greatly appreciate your interest in this research effort. I hope that you may find what is covered of interest, import, and value.

Breukelen, Netherlands   Scott Mongeau
December 2020

Acknowledgments

This manuscript embodies my work at the Nyenrode Business Universiteit Executive Doctorate Program (EDP) during an 8-year period between 2011 and 2019. Based in Breukelen, Netherlands, Nyenrode was established in 1946 to build bridges between entrepreneurs and to connect public and private interests. As an international professional with ties to both the Netherlands and the USA, Nyenrode was an inspiring and unique place to undertake this research. Nyenrode facilitated my growth by exposing me to a diverse range of research methodologies and teaching opportunities.

Many have contributed along this journey. Foremost, my central advisor throughout the process, Prof. Dr. Ir. Andrzej Hajdasinski, has been an inspiring beacon and deserves my deepest appreciation and gratitude for his unwavering support and encouragement during this effort. Similarly, my co-advisor Prof. Dr. Henry Robben has been an invaluable and immensely appreciated supporter of this effort, especially in navigating the crucial final phase.

Deepest gratitude goes to the members of the reading committee, who selflessly contributed their valuable time and thought into reviewing and providing feedback to improve this research effort. I am greatly honored by their contribution and consideration. The honored members of the reading committee receive my deepest gratitude: Prof. Dr. Ir. A.C.P.M. Backx of Technische Universiteit Eindhoven, Prof. Dr. Ir. M.F.W.H.A. Janssen of Technische Universiteit Delft, Prof. Dr. B. Hoogenboom of Nyenrode Business Universiteit, and Prof. Dr. R. Blomme of Nyenrode Business Universiteit.

Similar great appreciation goes to Prof. Dr. Ad Kil, the founding director of the Nyenrode EDP. Deepest and most sincere appreciation goes to Els Sonneveld, Nyenrode-pedel and former EDP Program Manager, for her unwavering support and encouragement. Great appreciation goes to Dr. Melanie de Ruiter, Nyenrode PhD Director, and Marijke Hoogendoorn, PhD Support, for their excellent administrative stewardship.

Many others at Nyenrode contributed their thoughts, knowledge, and support during this process. Special thanks go to those who provided guidance and feedback through the EDP program, including Prof. Dr. Hans Doorewaard, Prof. Dr. Edward

Groenland, Prof. Dr. Robert Jan Blomme, Prof. Dr. Lineke Sneller, Prof. Dr. Dimitri Mortelmans, and Prof. Dr. Ivo de Loo.

A number of globally respected cybersecurity data scientists shared their wisdom and input. CSDS encompasses a small but growing community to which I am proud to offer this contribution. I would like to particularly thank Dr. Joshua Neil for his generous time and encouragement. As well, Gabriel Bassett was instrumental in connecting me to a number of CSDS professionals, for which I am very grateful. Others who contributed their knowledge and expertise to the project include Dr. Alexander Kent, Dr. Donald Peterson, Dr. Emmanuel Tsukerman, Dr. Rudy Guyonneau, Charles Givre, Clarence Chio, Joshua Saxe, Shawn Riley, and Chiheb Chebbi. Through LinkedIn, I benefitted from the contributions of Mark Powell, Scott Nester, and Thomas Speidel. I also wish to recognize my SAS Institute colleagues Allan Russell, Henrik Kiertzner, Alex Anglin, Kelly Chen, Liz Goldberg, Catherine Truxillo, and Gerhard Svolba.

A very special thanks goes to my Erasmus RSM OneMBA and Nyenrode EDP colleagues, Dr. Ali Fenwick and Ross Gardner, who have been on this extended journey since 2009. A special recognition goes to Dr. Fenwick for his encouragement and support during the final phase. He reminded me that great doubts stand testament to meaningful and important projects, and that if the doubts are missing, something is likely amiss. I would also like to recognize my fellow EDP classmates, particularly Tanja de Jonge, Marian Dragt, Wilbert Snoei, and Eric Mantelaers.

Deep and greatest appreciation goes to my lovely wife, Marloes Ligtenberg, for her boundless patience and support throughout this time-consuming and involved effort. Likewise, great appreciation goes to my family in the United States, Barbara Barrows, Beth Mongeau, Allen Mongeau, and Barbara McKenna, and in the Netherlands, Loes Vogel, Peter Krans, JP Ligtenberg, Mirjam Boon, Pieter Ligtenberg, and Elise Ligtenberg. With fond memories, we also remember Lucky Ligtenberg.

My friends have lent their support and patience, particularly in understanding my frequent absences over a not insubstantial period of time. Thanks go to my friends Peter Moorhouse, Lee Razo, James Decker, Willem Burgmans, James Musick, Deppe Tjandra, Leo van der Sluijs, Sebastiaan Ruinard, Simone Ipenburg, Marjan Wölcken, Marieke Bloemsma, Annelies Witteveen, Berdina de Boer, Maartje Kijne, Lidewijde de Jong, Willie Torres, Jennifer Kessler, Thomas Kessler, Ronald Dijk, Volkert Claassen, Lydia Koedam, Brandon Hartley, Marie Diamond, Joe Silber, Rachel Oakley, Marjits Vriesendorp, and Kemal Yerlikaya. Similarly, I wish to recognize my Erasmus Rotterdam School of Management 2009 MFM and 2011 OneMBA classmates. Many others contributed their valuable time and support, for which I am deeply grateful.

Breukelen, Netherlands   Scott Mongeau
December 2020

Contents

List of Figures

List of Tables

Chapter 1
Summary Introduction

1.1   Summary Overview

Modern enterprise is increasingly facilitated by a pervasive global network of hyperconnected digital devices and distributed services. For many organizations, choosing not to connect to and participate in this digital cybersphere is a self-limiting option. However, assuring the continuity and integrity of digital infrastructure is a growing challenge.

Along with participation, interconnectivity exposes participants to actors seeking to damage, exploit, or otherwise misuse digital infrastructures. The exponential growth and development of networks, data volumes, and digital devices have expanded risks and vulnerabilities. Professionals charged with assuring cybersecurity are increasingly challenged by a complex of factors including expanding vulnerabilities, evolving threats, overworked staff, false alert overload, challenges orchestrating remediation, and difficulties integrating security systems and data.

Even as risks multiply, many organizations pursue ambitious strategies for digital advancement. The sheer pace of digital expansion frequently bypasses rigorous security oversight. Security professionals face pressures to mediate between growing demands for connectivity and increasing exposure to a range of opportunistic threats.

A central managerial challenge in defending networked infrastructure is the growing imbalance between the capabilities of defenders and attackers. While threats and vulnerabilities expand, there is a growing lack of skilled practitioners to accommodate and monitor the risks manifested. Cybersecurity professionals increasingly struggle to keep pace with the security impacts of continual growth and technical innovation.

In aggregate, these factors have led to an interest in applying analytical, algorithmic, and machine-driven methods to semi-automate aspects of the growing security protection and assurance gap. Increasing threats, staff shortages, and an environment of growing complexity impel the need for more sophisticated, data-driven

© Springer Nature Switzerland AG 2021
S. Mongeau, A. Hajdasinski, *Cybersecurity Data Science*,
https://doi.org/10.1007/978-3-030-74896-8_1

approaches. Growing cybersecurity pressures thus organically frame requirements for cybersecurity data science (CSDS) practice.

CSDS is a rapidly evolving practitioner discipline at the intersection of two fields of intense public and commercial interest. CSDS professional practice emerges from the functional intersection of its two parent domains: applying *data science* methods to improve *cybersecurity* assurance. As an emergent, hybrid profession, CSDS offers a range of novel approaches to address growing challenges facing the cybersecurity field.

Data science presents an array of prescriptions to discover, detect, and prevent a rapidly shifting complex of security threats. CSDS methods thus derive from core data science activities, including analytics problem framing, exploratory data analysis, visualization, diagnostics, data preparation, data engineering, statistical analysis, feature engineering, machine learning, optimization, semantic analytics, and ensuring scientific rigor in data-focused inquiry (Davenport 2013; Kelleher and Tierney 2018; Sallam and Cearley 2012).

Data science offers methods to support cybersecurity assurance goals, spanning security threat identification, protection, detection, response, and recovery (NIST 2018). Focused techniques are capable of addressing a broad range of security activities, including, but not limited to, network monitoring, system monitoring, penetration (PEN) testing, forensics, security operations (SecOps), software security engineering, incident response, policy compliance, and security awareness (Carnegie Mellon University and John Hopkins University 2019; NIST 2017).

Analytics-based methods drive improved security threat detection, triage, and operational efficiencies, introducing avenues for partial automation. Improving intrusion, threat, and incident detection in security monitoring is currently the most common CSDS activity (Loaiza et al. 2019; Ponemon Institute 2017).

Hosting a growing body of titled, active professionals, CSDS is a discipline emerging from dynamic practice, as opposed to being a synthetic, theoretically derived program. It is, in this sense, an opportunistic field growing at the confluence of pressing practical need (the need to improve digital security assurance) and evolving analytical methods (the broad and fast-growing range of data science approaches and supporting technologies).

The emerging CSDS professional domain is characterized by newness, rapid change, and complexity, spanning technical, methodological, and organizational contexts. Efforts to individuate and advance the field are hindered by rapid change in the parent domains. Exacerbated by these factors, the CSDS field currently lacks a clearly articulated set of best practices.

As an emerging profession, CSDS lacks characteristics of mature professions. A lack of standards and codified best practices, *body of theory*, limits the ability to train staff, manage programmatic efforts, ensure rigor, and track effectiveness. As CSDS efforts expand, body of theory gaps will become self-limiting as practitioners increasingly encounter conflicting understandings concerning proper practice.

By improving an understanding of current challenges and best practices in the CSDS field, a foundation for advancement of the domain is framed. The systematic gap analysis undertaken compares where the CSDS discipline is currently (*as-is*), where it is evolving in terms of dominant trends and idealized practice (*to-be*), and where organizations need to foster development to address shortcomings (*gap-diagnosis*). Outcomes frame a set of requirements and prescriptions for iterating CSDS professional practice.

The resulting manuscript delivers an integrated practitioner problem-solving research report intended to provide support to the development of the new field. Grounded in the management of information systems (MIS) domain, the inquiry encompasses a systematic, mixed method exploration of CSDS as an emerging profession.

As staged problem-solving research, this effort encompasses three research phases, bridging diagnostic research (background, opinion, and gap analysis) and design science problem-solving:

1. *Phase I—Diagnostic background (problem) analysis*: literature analysis surfaces challenges related to CSDS professionalization and frames key CSDS methodological concepts.
2. *Phase II—Diagnostic opinion (interview) research and gap (mixed method) analysis*: interview research and quantitative analysis of results frames three key gaps in CSDS practice.
3. *Phase III—Design science problem-solving*: focused recommendations are provided to address CSDS gaps through a problem-solving design process.

In terms of the integration of the three phases, literature analysis (*Phase I*) frames CSDS sensitizing concepts to ground, motivate, and inform research. Interview research supported by quantitative analysis (*Phase II*) extrapolates a set of key gaps forestalling CSDS practice. Based on interviews with 50 CSDS practitioners, a set of three key gaps in practice are surfaced: data management, scientific processes, and cross-domain collaboration.

Combined, *Phase I* and *II* results establish a foundation for design problem-solving (*Phase III*). The gaps surfaced are framed as problem-solving requirements that are developed into CSDS best practices through a design science treatment (Gregor and Hevner 2013; Hevner et al. 2004; Wieringa 2014). Design prescriptions (nascent designs) are surfaced and advocated to address key gaps. Results are framed as practical processes to strengthen CSDS practice and professional body of theory. The prescriptions resulting are asserted as novel MIS research contributions of interest to advancing CSDS practice.

In pursuing problem-solving research, the scientific focus of this effort is on rigorous diagnostic pattern extrapolation and substantiation in a complex new domain. Interview research, comparative literature analysis, and text analytics compose the core empirical elements, supported by literature-based, deductively derived concepts to guide design recommendations. Deductive and inductive methods are integrated to abductively derive a set of recommended CSDS best practices.

A central point of advocacy resulting from this research is that for the cybersecurity domain to advance, cybersecurity must be framed not simply as engineered treatments, but as *scientific phenomenon*. The traditional instinct in cybersecurity is to respond to each new challenge with technical prescriptions, an approach which risks devolving into a game of so-called whack-a-mole, a circular battle with no end and no victor. A set of methods and processes for inculcating greater scientific rigor, and thus theory building, in cybersecurity efforts vis-à-vis CSDS are outlined.

1.2 Motivations and Audience

Assuring digital infrastructure integrity is essential to the continuity of modern economic, political, and social life. Global enterprise is increasingly facilitated by a pervasive and expanding network of hyperconnected digital devices and distributed services. Due to inherent vulnerabilities, safeguarding digital infrastructure is a persistent and growing challenge for organizations globally.

Security assurance efforts are challenged by a complex of factors, including expanding digital infrastructure, proliferating vulnerabilities, fast-evolving threats, shortages of security professionals, overwhelming volumes of security alerts, and labor-intensive investigation and remediation efforts. Cybersecurity professionals increasingly struggle to keep pace with the repercussions of continual technological advancement and constant digital infrastructure expansion.

The traditional perimeter-protection *castle-and-moat* security assurance paradigm has been rendered ineffective by expanding distributed networks, wireless devices, virtual services, cloud offerings, data proliferation, and distributed software services. Even as risks multiply amid growing complexity, many organizations continue to pursue a strategy of aggressive digital development. The sheer pace of digital expansion frequently bypasses prudential security oversight.

Data science methods such as predictive machine learning are already applied to a range of security challenges such as phishing detection in emails, malware and virus detection, and network anomaly detection (Bhattacharyya and Kalita 2013; Chio and Freeman 2018; Sarker et al. 2020; Saxe and Sanders 2018; Spring et al. 2019; Stamp 2017; Tsukerman 2019b). The focused application of statistical analysis to network and computer intrusion monitoring goes back at least two decades (Marchette 2001).

Primarily due to cost economics, entrenched implementations of data science for security have hitherto been relegated to large-scale institutions, principally nation-states, financial and telecommunications majors, transnational corporations, and global software and solutions providers. In the past decade, the combination of more robust and available computing resources, innovative analytics methods, increasing pressure from threats, and knowledge gained from institutional and research efforts has democratized and popularized interest from a broader set of stakeholders.

An increasingly diverse range of organizations and solution providers have expressed an interest in the field of CSDS. A growing set of security software and service startups have grounded their core offerings as packaged data science innovations. Beyond this, according to a US Institute for Defense Analyses report, "It appears that essentially all the vendors of cybersecurity products are working to adopt AI/ML components in their products" (Loaiza et al. 2019).

Whether commercially or prudentially motivated, interest in CSDS is impelled, at base, by growing challenges associated with assuring the security and integrity of digital infrastructure. While the cybersecurity profession has codified a set of best practices and approaches, unchecked network proliferation and technological advancement are outstripping the human-delimited capacity of many security operations centers (SOCs). There are simply too many alerts and events overwhelming traditional security monitoring functions.

One byproduct of the steady expansion of networks, digital access, and networked devices is that there is a growing amount of data available concerning the behavior of users, agents, and devices accessing and utilizing machines and networks. This data is of great potential use in efforts to monitor devices and networks for untoward behavior. However, the sheer proliferation of security-related data has led directly to a secondary challenge: how best to manage and analyze this data? The significant volume, speed, and diversity of security data present data engineering challenges. As well, methodological challenges arise concerning how best to disassociate normal and expected events from threats and incursions. The challenge is one of detecting minute contextual signals in an ocean of noise composed of massive sets of data.

These factors frame the practical motivations which have led to the organic emergence of the CSDS discipline. As will be explored in-depth, CSDS presently offers a range of practical prescriptions to improve security assurance efforts. However, such treatments are typically cast as focused engineering solutions. Strong, overarching theoretical foundations are mainly absent from CSDS practice at present.

A central assertion of this research is that for the CSDS profession to advance, there must be a tighter and clearer integration of practice with theory. This research posits that CSDS, as an emerging profession and field of inquiry, can only advance through a closer symbiosis between practice (*technê*) and logic-guided theory (*logos* driven *epistêmê*). This implies a more concerted focus on scientific methods and scientific rigor in practice.

Unfortunately, there is a poverty of theoretical guidance available from the CSDS parental domains. Both cybersecurity and data science lack strong theoretical foundations, especially regarding processes for ensuring scientific rigor. The cybersecurity profession, arising from pragmatic and opportunistic origins, embeds an inherent engineering focus and bias (Craigen 2014; Craigen et al. 2014; Schatz et al. 2017). The data science domain, while providing a vast and growing range of analytical approaches, ironically suffers from an overarching lack of guidance concerning processes for ensuring scientific rigor.

The CSDS parent domains are themselves undergoing rapid transformation, complicating efforts to codify CSDS standards of practice. As the progeny of these two domains, CSDS inherits a lack of rigorous theoretical foundations to draw upon. Although a rapidly developing field, professional and institutional consensus regarding systematic underpinnings for CSDS are lacking. As a result, managers, practitioners, and researchers interested in pursuing CSDS programs are challenged to engage in structured collaboration. As long as CSDS persists as a purely engineering-driven set of treatments, the field will remain a set of loose practices characterizing a proto-profession.

Being in the early stages of professionalization, there are gaps in CSDS theory and practice that impede effectiveness. This project's central motivating goal entails elucidating foundations for CSDS body of theory, both current and idealized, to promote the profession's development. Improving an understanding of CSDS body of theory establishes a foundation for a range of organizational activities, including general management, research, strategy, planning, design, implementation, coordination, and training.

A staged structural understanding of the process of professional emergence (professionalization) is explored in-depth in the literature review undertaken in *Phase I*. The inherent gaps between practitioner and academic perspectives of cybersecurity, data science, and CSDS domains are examined systematically. This frames the core of the research effort, which is an attempt to catalog and address CSDS practice-theory gaps in the interests of staged professionalization. This research's central benefit rests in surfacing, framing, and advocating a theoretical foundation for the CSDS field. A set of practical gap-prescriptions are profiled and explored as nascent design artifacts (design artifacts that require implementation and assessment).

In mapping CSDS disciplinary boundaries and gaps, this analysis offers guidance to a range of stakeholders interested in advancing CSDS professional practice and general effectiveness. Summarizing the intended benefits of the three phases of this research inquiry to a range of audiences:

- *Security practitioners* seeking professional development and insights will benefit from a focused perspective on how data science can be applied to improve cybersecurity assurance, with *Phase I* providing a primer in this regard.
- *CSDS professionals* will be interested in guidance offered concerning developing the CSDS profession. This includes a range of best practice proposals framed by CSDS colleagues in practitioner interviews covered in *Phase II*.
- *Data scientists* are offered insights into how data science methods can be fruitfully applied to the cybersecurity domain. Those seeking focused insights into refining CSDS methods will be interested in design science guidance in *Phase III*.
- *Security managers*, *planners*, and *strategists* will be interested in practical guidance concerning implementing CSDS programs and solutions. Design science prescriptions profiled in *Phase III* offer guidance on enabling CSDS in the organization.
- *Software and service firms* seeking to understand the market for CSDS products, solutions, and services, including consulting offerings and managed security services, will find this work of interest. Topics covered will interest management, business development, product development, marketing, sales, and professional services professionals alike.

- *Researchers* will benefit from the broad-based foundation provided to frame follow-on research efforts, including organizational surveys, case study implementations, and/or action research.
- *Instructors* and *students* may choose to utilize this work as a CSDS primer and reference. The work as a whole provides a comprehensive overview of the CSDS field. Those seeking to develop curricula and training programs will find both *Phases I* and *III* of particular interest.
- *Institutional stakeholders* seeking to facilitate the development of the CSDS domain will be served throughout the inquiry by the practical perspectives and mechanisms espoused for iterating the new profession.
- *Military, intelligence, and government cybersecurity stakeholders* will encounter many topics of interest to national cybersphere stewardship. This work addresses the evolution of the cybersecurity field in reaction to emerging CSDS capabilities and threats.

1.3   Research Questions and Management Problems Addressed

This work aims to analyze challenges in the emerging CSDS profession, diagnose central gaps, and prescribe design treatments to facilitate domain advancement. Problem-solving design guidance is framed based upon a systematic accounting of CSDS challenges and best practices. The diagnostic analysis produces derived gaps that inform problem-solving requirements and subsequent design prescriptions. Advocacy is provided to strengthen professional rigor and standards, body of theory, in the emerging CSDS field.

The central research question addressed herein:

RQ0: What treatment designs are prescribed from diagnostic research to address gaps impeding the development of cybersecurity data science (CSDS) professional practice?

The central research question informs theoretical (TQ), empirical (EQ), and analytical (AQ) questions (Verschuren and Doorewaard 2010) triangulated through three research phases:

TQ0: *Phase I*—Extrapolating from diagnostic *background research* of literature, what challenges do practitioners face in the CSDS domain?
EQ0: *Phase II*—Extrapolating from interview-based diagnostic *opinion research*, what challenges do practitioners face in the CSDS domain?
AQ0: *Phase III (triangulating Phases I and II)*—Extrapolating from mixed method diagnostic *gap analysis research*, what categorical treatment design prescriptions are recommended to address identified CSDS practice gaps?

The following three associated research objectives (ROs) are addressed sequentially through a series of ten associated research questions (RQs) addressed across the three phases:

RO1: *Phase I—Analyze* the as-is state of the CSDS field based on comparative literature

> **RQ1**: What is the basis for asserting CSDS as a nascent professional domain?
>
> **RQ2**: What are the disciplinary boundaries of CSDS as a hybrid professional domain?
>
> **RQ3**: Where is CSDS in a process of professionalization?
>
> **RQ4**: What challenges does the CSDS domain face on the path to professionalization?

RO2: *Phase II—Diagnose* gaps impeding CSDS professionalization based on qualitative research

> **RQ5**: What gaps can be diagnosed in the emerging CSDS field from practitioner perspectives?
>
> **RQ6**: What treatments are prescribed to address gaps based on practitioner input?

RO3: *Phase III—Prescribe* design treatments based on gap analysis and extrapolation from literature

> **RQ7**: How should CSDS methodological challenges be addressed?
>
> **RQ8**: Which data science methodological treatments are prescribed for the CSDS domain?
>
> **RQ9**: What processes are advocated to implement prescriptions?
>
> **RQ10**: How can CSDS prescriptions be implemented as organizational processes?

The following three practical management problems (MPs) are addressed via the three core phases:

MP1: What gaps challenge the effectiveness of the CSDS field and impede professionalization?

MP2: What prescriptive treatments address categorical CSDS gaps?

MP3: What guidance can be offered to orchestrate the implementation of CSDS programs?

The research objectives, research questions, and management problems are explored, addressed, and answered in sequence through the three main research phases. In terms of the staging of the phases, structured results from diagnostic literature analysis (*Phase I*), interview research of CSDS practitioners (*Phase II*), and quantitative analysis of interview results (*Phase II*) are used to frame requirements for design problem-solving (*Phase III*).

1.4   Research Process and Methods

In terms of an academic research locus, this effort encompasses *practitioner problem-solving research* in the management of information systems (MIS) domain. The inquiry spans a partial *problem-solving design cycle*; problem-solving is informed by multi-phased diagnostic analysis. Triangulated mixed methods are applied to explore, define, and surface gaps associated with CSDS professional practice.

The three-phase partial problem-solving design cycle encompasses diagnostic background analysis (*Phase I*), diagnostic opinion research and gap analysis (*Phase II*), and design problem-solving (*Phase III*). Results of the problem-solving effort are presented in *Phase III* as a set of advocated best practices for the emerging domain. Results are intended to offer practical guidance to CSDS practitioners, as well as to provide a foundation for future research efforts.

Given the novel and complex nature of the CSDS domain, *Phase I* begins with a systematic examination of the *as-is* discipline through a derived professional maturity model. This allows for a structured understanding of where the field is in a staged process of professionalization. One of the key assertions resulting is that stronger theoretical foundations are necessary for the CSDS domain to advance given the lack of a clearly articulated *body of theory*. Accompanying the assessment of CSDS professional maturity, a semi-structured thematic analysis of 30 manuscript-length works associated with the CSDS domain is undertaken. This results in an observation of key themes characterizing the CSDS field, along with an understanding of topical gaps.

Phase II undertakes and analyzes semi-grounded interviews with 50 global CSDS professionals. Quantitative approaches are applied to analyze themes extrapolated from interviews (interview code frequency, factor analysis of interview codes, factor-to-factor correlation, and regression analysis). Diagnostic efforts mix inductive methods (semi-structured literature thematic analysis, semi-grounded interviews with practitioners, thematic text analytics of interview transcripts) and deductive approaches (theoretical concept extrapolation from literature, coding of interviews supported by sensitizing concepts, memoing). *Phase II* results in the surfacing of key practitioner-perceived themes and gaps. The resulting structured gaps are presented as a basis for problem-solving requirements to drive *Phase III* design treatments.

Table 1.1 summarizes key research phases, methods, and results at a high level. Literature analysis conducted in *Phase I* serves multiple functions. Analysis grounds, motivates, and substantiates the research objectives. Literature-based themes and models are framed as sensitizing concepts to support semi-grounded interviews of CSDS professionals undertaken in *Phase II*. Extrapolations from literature also surface a series of key CSDS conceptual models to support design problem-solving in *Phase III*.

Figure 1.1 summarizes the interaction between the main research phases. *Phase I* literature-derived themes and models are framed as sensitizing concepts to support

Table 1.1 Summary of research and results—overall overview

Chapter	Phase	Methods	Results
2. CSDS as an Emerging Profession	I. Diagnostic background analysis	Multi-faceted literature analysis	CSDS maturity gaps / CSDS demand model / Literature corpus / Literature gaps / Sensitizing concepts / Analytical methods
3. CSDS Practitioner Interviews & Gaps	II. Diagnostic opinion research	Qualitative interview research	Key challenge & best practice themes
	II. Diagnostic gap analysis	Quantitative analysis of themes	Diagnosis of CSDS gap-prescriptions
4. CSDS Gap-Prescription Designs	III. Design problem-solving	Design science	Design requirements & prescriptions

Fig. 1.1 Interrelationship of key research phases

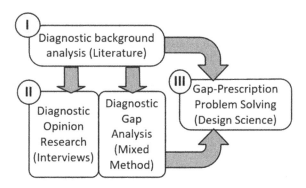

interviews of CSDS professionals undertaken in *Phase II*. *Phase I* also surfaces a series of key CSDS conceptual models to support design problem-solving in *Phase III*. Deductive concepts from *Phase I* and inductive findings in *Phase II* support design problem-solving in *Phase III*.

The newness and complexity of the emerging field motivated triangulated mixed methods, combining both inductive and deductive approaches. The often unruly reality of CSDS as a domain emerging from dynamic practice benefited from a mix of both top-down and bottom-up approaches to support abductive problem-solving. This reflects an approach more similar to medical diagnostics, befitting the underlying phenomenon's complexity and novelty.

As a practitioner problem-solving research effort supported by mixed diagnostic and design research, methods are focused on exploring and providing a practical understanding of a complex new domain. The goal is problem discovery, analysis, validation, and prescription, rather than hypothetical-deductive theory framing and testing. A partial problem-solving research cycle is pursued, with implementation and assessment framed for future research efforts.

Motivated by Gregor and Hevner's design science guidance in the information systems (IS) research domain (2013), the main outcome is an advocated set of structured gap-prescriptions (nascent design artifacts) intended to guide and advance the evolution of CSDS practice. Results achieved are intended to provide a foundation both for academics to undertake follow-on domain research and for practitioners to pursue follow-on implementations.

1.5   Theoretical Contributions

Research outcomes present an array of academic, theoretical, and methodological contributions:

1. Research extrapolates systematic understandings of challenges and best practices in the emerging field of CSDS from an MIS perspective, addressing a research lacuna.
2. Assessment of CSDS professional maturity through a structured model highlights gaps forestalling CSDS professionalization. This approach has potential broader promise for assessing other emerging IT-related professions.
3. A functional assessment of CSDS practice frames the field as serving a critical diagnostic function in prescribing data science treatments to address cybersecurity challenges. This assertion highlights the importance of body of theory in CSDS practice.
4. A CSDS literature corpus of 33 manuscript-length works is identified and assessed, resulting in a comparative analysis of key themes.
5. Key data science methodological concepts of particular interest to the CSDS domain are raised and framed.
6. The novel mix of quantitative methods applied in the interview research component suggest new approaches for analyzing interview data.
7. The outcome of interview gap analysis surfaces three central CSDS domain gaps of interest to practitioners and researchers: data management, scientific processes, and cross-domain collaboration.
8. Based on the CSDS domain gaps identified, a set of process-based prescriptions are extrapolated and recommended to support advancement of the emerging field.
9. Gap-prescriptions resulting from the problem-solving design process frame focused recommendations to advance CSDS body of theory.
10. Gap-prescriptions frame structured approaches to instill greater empirical, theoretical, and scientific rigor in cybersecurity research.

1.6 Contributions to Management Practice

Results of this research will be of interest to a range of parties spanning professional, commercial, and public domains. Key stakeholders served include the following: security, data science, and CSDS practitioners; cybersecurity stewards; organizational managers, planners, and strategists; software and service firms; governmental organizations and NGOs; and military and intelligence services.

In mapping CSDS challenges and prescribing practical treatments, this work offers guidance to those interested in advancing CSDS professional practice and general cybersecurity effectiveness, including stakeholders planning programmatic implementations. Discrete contributions in this regard are the following:

1. The integrated report offers a comprehensive analysis of gaps facing the CSDS profession leading to guidance for future development. Treatments are advocated as approaches to advance CSDS body of theory and thus to promote the process of professionalization.
2. CSDS is asserted as a hybrid domain and nascent profession. The field has emerged from the fusion of cybersecurity needs and data science methods in response to a growing economic and organizational pressures associated with the effects of rapid digital expansion and evolution. From a managerial perspective, a structured understanding of the broad dynamics shaping the emergence of CSDS provides an ability to control and foster this transformation.
3. Guidance based upon interviews with active CSDS practitioners advocates the need to integrate and align human and technical resources through organizational processes. Results suggest that resource coordination, data management, and scientific methods must together conjoin to drive CSDS as a functional organizational process. This presents CSDS as necessitating multiple, overlapping processes, particularly cyclical data discovery and scientific inquiry.
4. The three CSDS gaps surfaced, data management, scientific processes, and cross-domain collaboration, roughly align with the MIS focus areas of technology, process, and organization. A key observation raised concerns the degree to which these gaps overlap. Design prescriptions presented in the conclusion frame data management and scientific processes as mutually reinforcing.
5. CSDS programs ideally facilitate and incentivize social processes associated with knowledge exchange, creation, and validation. Managerial supervision is necessitated to frame data science programmatically, an effort that is as much organizational as it is methodological and technical. Suitable recommendations to facilitate such programmatic implementations are provided.
6. CSDS best practices profiled outline a range of methods to measure phenomena, make better decisions, identify threats, control risks, and optimize processes. This aggregation of best practices frames a vision concerning ideal CSDS practice. Collectively, prescriptions suggest a basis for a CSDS maturity model. Maturity indicators are useful to managers in benchmarking capabilities, providing a structured basis for gap analysis and strategic planning.

Chapter 2
Phase I: CSDS as an Emerging Profession—Diagnostic Literature Analysis

2.1 Research Objectives

Due to the nature of cybersecurity data science (CSDS) as a novel field emerging in the midst of rapid technological change, there is a gap in CSDS-focused organizational research. Challenges operationalizing CSDS solutions lead to a call for an increased theoretical focus on organizational problem-solving research. To address this gap, CSDS fits the profile of an organizational problem that is "relatively new or fairly complex," necessitating an effort to "clarify the relevant background and the reasons for the problem" (Verschuren and Doorewaard 2010).

To improve an understanding of the emerging domain, this phase undertakes *diagnostic background analysis* in the form of multifaceted literature analysis. The effort serves multiple goals: (1) profile structured gaps in the process of CSDS professionalization, (2) frame the discrete forces propelling growth of the domain, (3) identify a corpus of key literature in the domain, (4) substantiate the assertion of gaps in CSDS literature, (5) summarize key concepts to support interview research conducted in *Phase II*, and (6) position key data science methodological concepts to drive design problem-solving in *Phase III*.

Literature focused on CSDS is analyzed and compared, integrating academic, practitioner, and industry sources. Professional press and industry sources are referenced alongside scholarly sources to systematically frame the nature of and drives impelling the new field. Efforts result in a series of six focused outcomes that serve to substantiate, motivate, and iterate the larger research effort.

CSDS, as a nascent profession, lacks several characteristics of mature occupations. To understand discrete gaps, a structured maturity assessment is undertaken based on a professional maturity model (Greenwood 1957). The outcome, *Result 1*, profiles key professionalization gaps in the emerging domain. *Result 2* frames a CSDS supply-demand model generated from Looijen and Delen's information systems management model (Looijen and Delen 1992). As CSDS is a de facto

© Springer Nature Switzerland AG 2021

S. Mongeau, A. Hajdasinski, *Cybersecurity Data Science*,
https://doi.org/10.1007/978-3-030-74896-8_2

profession hosting a growing number of active practitioners, this highlights the forces propelling domain growth in terms of the discrete functional benefits offered.

As a gap in CSDS organizational research has been asserted, a thematic literature gap analysis is conducted. *Results 3* identifies and summarizes a key corpus of manuscript-length CSDS works. Based upon this corpus, *Result 4* substantiates the organizational theory thematic gap asserted.

In order to establish a foundation for diagnostic interview research of CSDS practitioners carried out in *Phase II*, *Result 5* frames a set of CSDS *sensitizing concepts*, primary concepts derived to orient qualitative inquiry (van den Hoonaard 1996). Given the newness and complexity of the emerging field, sensitizing concepts provide a basis for framing and interpreting practitioner interviews. Finally, an array of CSDS-relevant data science methodological concepts are framed in *Result 6* which provide guidance to problem-solving design efforts in *Phase III*.

The *Phase I* literature analysis and resulting six outcomes, profiled in Sects. 2.6.1–2.6.6, are offered as novel organizational research contributions addressing a research lacuna from the management of information systems (MIS) perspective. A summary of key focused research methods and outcomes in this phase are presented in Table 2.1. The focused problems, objectives, and questions addressed in this phase are presented in Table 2.2.

Table 2.1 Summary of research and results—literature review

Chapter	Phase	Method	Result
2. CSDS as an Emerging Profession	I. Diagnostic background analysis	Multi-faceted literature analysis	CSDS maturity gaps
			CSDS demand model
			Literature corpus
			Literature gaps
			Sensitizing concepts
			Analytical methods

Table 2.2 Summary of literature review research questions

Management problem 1	What gaps challenge the effectiveness of the CSDS field and impede professionalization?
Research objective 1	*Analyze* the as-is state of the CSDS field based on comparative literature
Research question 1	What is the basis for asserting CSDS as a nascent professional domain?
Research question 2	What are the disciplinary boundaries of CSDS as a hybrid professional domain?
Research question 3	Where is CSDS in a process of professionalization?
Research question 4	What challenges face the CSDS domain on the path to professionalization?

2.2   CSDS in a Management Research Context

2.2.1   Working Definition

Assuring the continuity and integrity of digital information and infrastructure is a persistent and growing challenge for global organizations. Modern enterprise is increasingly facilitated by a pervasive global network of hyperconnected digital devices and distributed services. For many organizations, opting out of connecting to and participating in this evolving cybersphere, the realm of network-mediated information technology and electronic communication (Oxford Dictionary 2019), is a self-limiting option.

However, along with participation, interconnectivity intrinsically exposes participants to a broad range of fast-evolving, sophisticated threats (Ramirez 2017). Interconnective complexity in digital infrastructure has resulted in increasingly brittle and vulnerable systems as multiple layers evidence prolific interdependencies, from hardware and carriage through to data, software, and applications (Forrester 2020; Jang-Jaccard and Nepal 2014; Schneier 2018).

Organizations increasingly desire to optimize the economic upside of prolific digital access while controlling and minimizing downside risks (World Economic Forum 2015). This conflation of practical necessity with exposure has equated to the growing acceptance that cybersecurity has emerged as a general enterprise risk (Beaupérin 2019; Bevan et al. 2018; Brockett et al. 2012; Poppensieker and Riemenschnitter 2018).

The assertion "if you can't measure it, you can't manage it" is often attributed to Peter Drucker (Coffeen 2009). In keeping, control and assurance of cybersecurity risk depends upon the ability to measure vulnerabilities and exposure in environments of rapid change and uncertainty (Cheng et al. 2014; Hubbard and Seiersen 2016; Lehto 2015; Refsdal et al. 2015).

The field of data science hosts a range of practical methods and processes for discovering, validating, forecasting, and optimizing opportunities and risks in dynamic digital environments (Skiena 2017; Voulgaris 2017). The combination of the need for improved cybersecurity control and assurance, supported by analytical methods to measure and assess risk, frames the emerging field of cybersecurity data science (CSDS) (Fisk 2019; Sahinoglu 2016).

A proposed working definition for CSDS will be explored and extrapolated in the literature-based diagnostic background analysis that follows:

the practice of data science to assure the continuity of digital devices, systems, services, software, and agents in pursuit of the stewardship of systemic cybersphere stability, spanning technical, operational, organizational, economic, social, and political contexts.

2.2.2   Expanding Scope, Scale, and Risk

For firms, connecting to the cybersphere means access to the opportunities provided by participating in diverse global markets. In terms of scale, by 2022, it is anticipated six billion people will be using the Internet, 75% of the then-projected global population of eight billion (Cisco Systems Inc. 2018; Morgan 2017a). The benefits and reach of such access are expanding and diversifying as networked digital device technologies rapidly evolve in lockstep (Ramirez 2017).

Traffic in the expanding cybersphere increasingly encompasses emerging forms of media, such as virtual reality, high-definition video, and interactive entertainment. Data traffic is increasingly produced and consumed by powerful new computerized devices such as smart sensors, autonomous vehicles, drones, and interactive AI-powered agents (Schneier 2018). Mobile technology continues to proliferate. By 2022, smartphone and mobile device traffic is projected to account for more than 60% of global data traffic (Cisco Systems Inc. 2019).

Exponential growth trends in global data traffic are well-documented and sustained, with a 40% annual growth rate being projected through to 2030. From 32.6 billion zettabytes (32.6 billion terabytes) total data generated in 2018, 2023 volume is expected to more than triple to 102.6 zettabytes annually (for scale: 1 zettabyte = 1,000,000 (1 million) petabytes = 1,000,000,000 (1 billion) terabytes = 1,000,000,000,000 (1 trillion) gigabytes). By this time, aggregate global data storage capacity is projected to reach 111.7 zettabytes (Reinsel and Rydning 2018).

By 2023, the number of devices connected to networks is projected to reach 27.1 billion (Cisco Systems Inc. 2018), nearly 3.5 devices for every human. The most significant growth in traffic over the next decade will occur from autonomous machine-driven systems and devices, including increasing numbers of sensors, smart devices, industrial devices, and automated systems generating and consuming traffic (Barabási 2014; IDC 2014; Lohr 2011; Morgan 2017a).

Cybersecurity professionals increasingly struggle to keep pace with the security impacts of continual exponential growth and technical innovation. The traditional perimeter-protection *castle-and-moat* paradigm has been rendered ineffective by the expansion of distributed networks, devices, services, and data (Leuprecht et al. 2016).

Castle-and-moat-oriented defense represents an increasingly anachronistic model for security architectures in which organizational information and devices are viewed as being protected behind sets of "hard" restricting and retaining mechanisms such as firewalls and linear access control. Detection of security events in this paradigm is driven by rule- or signature-based mechanisms, which are increasingly regarded as insufficiently focused and result in many more alerts than investigators can operationally review.

This castle-and-moat security stance is premised on two-dimensional models of network participation and access. In contrast, emerging systems and devices are multimodal and multidimensional, connecting promiscuously across and underneath traditional defenses. The emerging reality of modern network-mediated access to computing resources and applications is virtual, decentralized, and hybridized (Lehto 2015; Leuprecht et al. 2016).

Along with cloud-based, virtualized, containerized, and microservice-based composite applications, the proliferation of wireless access and mobile devices continues to challenge and invalidate "hard" two-dimensional models for securing access, infrastructure, and information (Wolff 2014).

The dramatic effects of the 2020 Coronavirus (COVID-19) pandemic have brought about sudden and potentially long-term shifts towards mass teleworking (Burr and Endicott 2020; Lichfield 2020). As distributed online telework becomes the de facto norm, organizational networks will become increasingly diffuse and geographically distributed, further puncturing the feasibility of "hard barrier" approaches to cyber defense.

Newer frames and paradigms for accommodating multimodal and multidimensional networks espouse situational awareness and operational readiness stances. There is an assumption that incursions may already be underway on the network and that adversarial methods are quite likely to be actively evading traditional blocking mechanisms and rules-based detection approaches.

The volume, scope, and scale of cybersecurity risks and incidents are growing in lockstep with technological advancement (World Economic Forum 2015). The continued expansion of network traffic and the proliferation of networked digital devices for storing, processing, and generating data has led inexorably to growing vulnerabilities and risks (Greengard 2016; Shostack 2012).

From a risk management and assurance perspective, expanding opportunities for digital proliferation leads to greater exposure: data which can be stolen or altered, devices which can be compromised or misused, and increasingly physical infrastructure which can be interfered with or damaged. In cybersecurity terms, access and accessible resources expand attack surfaces and targets (Hubbard and Seiersen 2016; Lehto 2015; Poppensieker and Riemenschnitter 2018).

As digital devices to manage and control physical resources proliferate, attacks on public and private infrastructure, either for profit or as an aspect of covert or overt warfare, are anticipated to grow (Cisco Systems Inc. 2018; Greenberg 2017). While each nation faces different threats and has varying responses, cyber threats themselves are increasingly borderless and transnational (Subrahmanian et al. 2015).

Transparent markets for anonymously contracting freelance cybersecurity actors to perpetrate malicious acts via the dark web are burgeoning (Radware 2019). The dark web denotes web content and resources that utilize Internet carriage but require specialized software, configurations, encryption protocols, and authorization to access. In aggregate, dark web facilities constitute overlay networks that sit atop yet are inaccessible to public users of the World Wide Web (Egan 2019).

Even as risks multiply, many organizations continue to pursue a strategy of digital expansion. The sheer pace of digital development to pursue economic incentives often bypasses security fundamentals. Motivations include increasing speed to market, convenience, and connectivity (Schneier 2018). Supporting remote workers from home is a rapidly growing demand which further expands vulnerabilities. Security professionals face pressure to mediate between increasing demands for connectivity and the resulting increase of exposure to a range of opportunistic threats (Grahn et al. 2017; Shostack 2014).

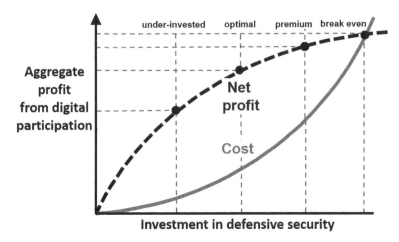

Fig. 2.1 Theoretical optimality between profit from digital participation and security costs

In managing this risk from an economic vantage, the marginal costs of defense should ideally be allocated to address the likelihood weighted impact of incidents as a measure of exposure (Fisk 2019). This theoretical trade-off can be summarized in an economic model that recognizes the rising costs of securing digital infrastructure as greater profits aggregate from higher levels of digital participation, as implied in a World Economic Forum report on cyber resilience (2015) (Fig. 2.1).

2.2.3 Adversarial Advantages

A central managerial challenge in defending networked infrastructure is the increasing imbalance between defense and attack cost economics (Hubbard and Seiersen 2016; Poppensieker and Riemenschnitter 2018). While the proliferation of digital access, data, and devices is a driver of beneficial economic opportunities for firms, new risks and threats also emerge due to perverse incentives for adversaries.

Cybercrime can be a lucrative endeavor, and the relative costs associated with developing and maintaining capabilities are attractively low (Booz Allen Hamilton 2019; Subrahmanian et al. 2015). The return on investment of cybercrime acts as a powerful motivator, with one source citing a figure of 1425% return on investment (Trustwave 2015). In 2019, it was estimated that ransomware damages alone exceeded $11.5 billion (Morgan 2017c). By 2021, it is estimated that the aggregate cost of cybercrime damages will rise to $6 trillion (Morgan 2017a).

A central challenge to stewards charged with defending networked infrastructure is the asymmetric advantage of threat agents versus defenders (Fisk 2019; Forrester 2020). The analyst firm ESG reports that 94% of cybersecurity professionals feel that "cyber adversaries have a significant or marginal advantage over cyber defenders in general" (Oltsik 2019a). Whereas defenders must vigilantly protect their infrastructure's expanding breadth with a slim threshold for failure, threat agents have the luxury of persistence, surveillance, powerful tools, scale, and porous targets.

Often only a single weakness needs to be opened to get a foothold. Research conducted jointly by IBM and the Ponemon Institute suggests it takes the average company 206 days to discover a breach and 73 to fix it, causing damage on the average of $3.92 million (IBM Security and Ponemon Institute 2019). Attackers have an inherent advantage as they can attack from anywhere using readily available advanced tools against prolific vulnerabilities (Schneier 2018).

It is increasingly infeasible to protect a particular device or user from compromise by erecting a hardened barrier. Digital devices are increasingly less singular and discrete nexus points and more vertices of prolifically connected services, applications, and data. Attacks can occur from multiple points on or across networks and may involve complex combinations of devices and applications. New applications and devices access diverse resources, spanning cloud-based facilities, virtual systems, containers, and composite microservice-based applications.

Cyber adversaries rapidly adapt, permutate, and evolve around defenses. Attacks are growing in volume, size, and sophistication (Poppensieker and Riemenschnitter 2018; Schwartz 2019a). Research claims that data breach incidents increased by 54% in 2019, with over 3800 incidents reported (Sanders 2019). Social vectors are by far the preeminent mechanism to gain unauthorized access to systems, with research suggesting that 99% of cyberattacks rely upon human interaction to initiate (Help Net Security 2019).

Industrial and utility infrastructures are an increasing adversarial target, particularly by state-sponsored actors as part of covert nation-state subversion campaigns (Osbourne and Banjanovic 2016; Schwartz 2019b). There is growing evidence that state-sponsored actors are mechanizing spear-phishing, targeted attempts to compromise credentials through social vectors such as personalized emails, to gain entry to and compromise industrial and utility targets (Spadafora 2019). A burgeoning third-party market offers increasingly sophisticated tools and services to facilitate such activities (The Economist 2019b, c).

Another observed trend, which is anticipated to expand, is the advent of semi-automated and self-propagating threats at scale (IDC 2014). Botnets, large groups of commandeered devices, can be assembled at scale as a platform for mass attacks (Puri 2003). Such malicious platforms are actively virulent—botnets during inception actively spider out to search for, compromise, and incorporate additional devices (Schwartz 2019b; Vasilomanolakis et al. 2015).

Mass attacks often seek to exploit zero-day vulnerabilities – hidden weaknesses exploited and lying dormant until attackers launch at scale (Greengard 2016). These threats utilize automated scripts, or bots, to identify, infect, and exploit hidden vulnerabilities, leading to incursions which may not be apparent in human monitoring until a sudden and dramatic attack. The Mirai botnet and associated permutations, which often target and exploit vulnerabilities in Internet of Things (IoT) devices, is one example of destructive new forms of staged and scaled threats targeting zero-day vulnerabilities (Antonakakis et al. 2018).

As digital infrastructure proliferates, cyber vectors are rapidly becoming vehicles for both traditional and novel forms of crime (Schwartz 2019a). Cybercrime proliferates and advances as new economic incentives emerge surrounding focused vulnerabilities. Following the 2018 cryptocurrency bubble, a criminal adaptation was cryptojacking, commandeering computing power to covertly mine cryptocurrencies (Schwartz 2018).

In these examples, malware and bots are conjoined with phishing attacks to identify, infect, and expand an imprint across vulnerable candidate systems at scale (Booz Allen Hamilton 2019; Schwartz 2019b). Utilizing such ensemble methods and responding to shifting incentives, hybrid attacks continually combine technologies and techniques in unexpected, novel forms (Schwartz 2019a).

The adaptive, incentive-driven nature of cyber threat trends is highlighted by the relative decline of cryptojacking incidents, mirroring the cryptocurrency bubble's deflation in 2018. Underscoring both the responsiveness and lack of compunction of threat actors, the 2020 Coronavirus (COVID-19) crisis saw a rapid increase in ransomware attacks targeting hospitals and research facilities (Gallagher and Bloomberg 2020; Jowitt 2020). Phishing emails adopted subject matter related to the coronavirus in an attempt to deliver malware and other malicious payloads (Reiner 2020). Also, state-level actors quickly adapted to the 2020 Coronavirus crisis as an opportunity to sow disinformation (Beaumont et al. 2020) and to stage attacks to raid medical research (Cohen and Marquardt 2020).

2.2.4 Managerial Responses to Growing Challenges

To counter expanding threats, institutions and firms expend significant sums to assure and protect. Global expenditures on information security reached $114 billion in 2018 and are expected to reach $170 billion by 2022 (Deshpande et al. 2018). In aggregate, it is anticipated firms will expend $1 trillion globally on cybersecurity defense between 2017 and 2021 (Morgan 2018a). This has also influenced opinion concerning government policy and intervention. A January 2019 survey of 775 C-suite executives and policy influencers revealed that a majority supported more significant governmental investments in national security infrastructure and software (Violino 2019a).

However, such investments fall short of automation: human stewards remain necessary for effective monitoring, detection, and remediation. While attack surfaces and threats are expanding exponentially, human monitoring of cybersecurity threats cannot scale at the same rate. Cybersecurity professionals are increasingly challenged to protect prolifically networked systems, devices, and data between the expansion of attack surfaces and the asymmetric advantage of evolving threats.

Even as the burden of cybersecurity defense increases, a shortage in trained personnel grows (Conklin et al. 2014; Garande 2019; Giles 2018b; Oltsik 2019b; Ponemon Institute 2018; Zinatullin 2019). Many cybersecurity professionals operate under the onus of fear, uncertainty, and doubt (FUD), with the implication that information technology infrastructure may already have been compromised (Connors and Endsley 2014; Shostack 2012).

Shortages in trained personnel exacerbate pressures on security professionals. Increasing pressures on existing cybersecurity staff have led to an epidemic of burn-out (Shostack 2012). A recently cited study reported that 64% of present cybersecurity professionals have considered quitting their current position and 63% have considered leaving the profession entirely (Palmer 2019).

By 2021, it is estimated that there will be 3.5 million unfilled cybersecurity jobs globally (Morgan 2017b) and that a growing proportion will involve requirements for data-focused approaches (Greengard 2016). In the US market alone, as of December 2019, there were over 500,000 cybersecurity job openings, with 16% of the openings entailing analysis-related job duties, per the NICE workforce framework (CyberSeek 2019).

There is a growing human resources imbalance resulting from both the expansion of digital attack surfaces and the increasingly asymmetric power of attackers (Oltsik 2019b). This has led directly to an interest in applying machine-driven methods to semi-automate aspects of the growing protection and assurance gap (Greengard 2016; Verma et al. 2015).

To supplement the limitations of traditional castle-and-moat approaches, managers have increasingly looked to analytics methods to extend the reach and effectiveness of prevention, detection, and remediation efforts (Greengard 2016). In particular, many large organizations have turned to machine learning techniques to automate both detection and prevention (Cisco Systems Inc. 2018; Spring et al. 2019; Stamp 2018).

Data-focused approaches, namely, the application of statistical, machine learning, and advanced analytics techniques under the banner of data science, are of particular interest (Heard et al. 2018). Such methods address the need for more focused detection mechanisms and align with emerging paradigms for security, which evoke resilience, situational awareness, and readiness.

In keeping with the growing interest in data-focused approaches, "cybersecurity data scientist" is represented explicitly as a specialized security professional role. In October 2018, online employment and information portal Glassdoor listed over 430 "cybersecurity data scientist" job openings (Glassdoor 2018). Further, a 2018 LinkedIn search for professionals with a job title containing "cybersecurity" and "data scientist" or "analytics" revealed 350 people globally operating in this capacity. Cybersecurity professionals and managers hold that data-focused and analytics skills will continue to grow in importance as demands for efficiency and automation increase (Heard et al. 2018).

2.2.5 Market Responses and Distortions

Evidence of a growing global interest in applying data science methods to cybersecurity can be anecdotally seen in an examination of Google Search trends. Over the 3 years between April 2016 and April 2019, there was a 93% correlation between

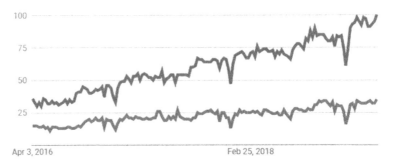

Fig. 2.2 Google web searches for "data science" and "cybersecurity" (4/2016–4/2019)

the fluctuations in the relative volume of searches for the keywords "data science" and "cybersecurity" (Google Trends 2018). Figure 2.2 charts the proportional search volumes with "data science" charted above and "cybersecurity" below.

Commercial vendors have responded to the interest in conjoining data science and cybersecurity (Oltsik 2019b). Beginning in earnest in 2017, many commercial software and systems vendors put forward a growing array of offerings to address the demand for data-driven cybersecurity solutions (Morgan 2018a). The terms AI, machine learning, and analytics are frequently advertised as core functions in commercial cybersecurity offerings (Giles 2018a).

From the perspective of methodological rigor, this popularization of data science in commercial technology markets is a double-edged sword. While it encourages interest and investment, the promotional nature of associated information surrounding capabilities risks misleading distortions, misunderstandings, and unrealistic expectations. The use of the term analytics and data science in commercial advertising typically promotes unclear marketing concepts rather than rigorously validated methods or guidance on organizational best practices (Oltsik 2019b).

There is also the danger of rushing commercial solutions to market without having validated operational efficacy—sacrificing methodological integrity for time to market (Giles 2018a). As an example, starting in 2017, deep learning became heavily promoted as a machine learning method in the technology industry. However, many data science practitioners have cautioned that the swell of interest in this method has led to exaggerations concerning the universal efficacy of the technique and misunderstandings concerning its proper operational application (Allaire 2018).

Recent articles in the professional press have highlighted the growing misalignment between the commercial marketing of machine learning and the lack of accompanying evidence of cybersecurity efficacy. Practitioners have raised a critique that case studies concerning the effective implementation of machine learning in cybersecurity operational environments are lacking (Ross 2019; Violino 2019b). This mirrors a more extensive critique emerging concerning the dangers of investing blind, enthusiastic trust in algorithmic and machine learning decision-making systems (O'Neil 2016; Smith 2018).

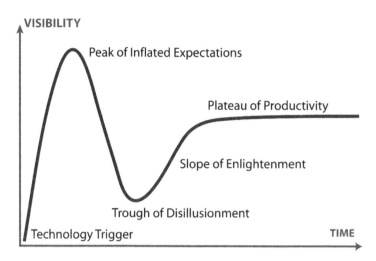

Fig. 2.3 The Gartner technology hype cycle. (Jeremy Kemp at Wikipedia/CC BY-SA)

Caution is advised concerning the present versus aspirational capabilities of data science methods in the cybersecurity domain, especially when the advocacy has commercial provenance. As with many innovations in the technology industry, a classical "hype cycle" is observed, whereby initial inflated exuberance leads to disappointment before realistic, practical bearings can be regained.

The Gartner hype cycle model (Kemp 2007), per Fig. 2.3, is a well-recognized IT industry measure from the eponymous IT research firm. The model observes informational distortions associated with the commercial promotion of new technologies and IT innovations. The hype cycle reflects that emerging technologies initially are often promoted beyond their base capabilities until the development of best practices leads to a more mature understanding of practical uses and implementations.

A caveat to the growing rise of interest in employing analytics in the cybersecurity domain is the implication that a peak may be evolving. The Gartner hype cycle principle is that focused efforts to undertake both holistic and systematic analyses of innovations are required to ground and balance over-inflated expectations. From a managerial perspective, this caveat implies and advocates examining CSDS outside purely market-driven commercial frames. The specific over-focus on machine learning methods in commercial cybersecurity technology offerings (e.g., as packaged software features) with little regard for the organizational implementation and validation of these solutions represents a growing lacuna between commercial advocacy and organizational context.

In terms of pragmatic concerns, a growing realization is emerging that adversarial actors themselves are increasingly examining and employing machine learning techniques in attack vectors. Such approaches include automating traditional mechanisms such as password attacks, phishing, vulnerability scanning, and malware command and control (Chio and Freeman 2018; Drinkwater 2018; Forrester 2020; Frenkel 2017; Gopalakrishnan 2020; Open Data Science 2018; Strout 2019; Trayan 2019; Yin et al. 2019).

Increasingly there is also the emerging specter of meta-threats: adversaries using machine learning to attack systems and services driven by machine learning, for instance, to confuse or poison machine learning algorithms and training data (Barreno et al. 2006; Brundage et al. 2018; Chebbi 2019; Samuel 2019; Strout 2019). The practical implication is a growing sense of an inevitable technical cold-war brewing, whereby both attackers and defenders are impelled to utilize advanced machine learning mechanisms to gain marginal advantages (Ashford 2019; Zrahia 2019).

The emerging adoption and application of machine learning methods by attackers can be viewed as opportunistic behavior motivated by economic incentives. The drivers incentivizing attackers to adopt advanced analytics mirror drivers impelling cybersecurity managers: automation, efficiency, addressing shortages in trained personnel, operating at great speed and scale, optimizing the value from large sets of data, and facilitating productive decision-making in environments of fundamental complexity. To the degree that market distortions perpetuate confusion, it is advocated that defending managers maintain a pragmatic focus on discrete cost and efficiency-based outcomes.

2.2.6 Management of Information Systems (MIS) Context

Cybersecurity data science (CSDS) is framed as the hybridization of data science methods with cybersecurity goals. The aggregation of the two domains, however, creates a challenge of interdisciplinary breadth. The cybersecurity domain's expanding scope encompasses technical, physical, economic, and sociopolitical systems on a global scale (Lehto 2015).

Data science itself encompasses a broad and continually evolving set of interdisciplinary methods for deriving insights from data (Voulgaris 2017), spanning from statistical to machine learning techniques (Davenport 2013; Kaisler et al. 2014). Cybersecurity and data science, united under the CSDS umbrella, opens an expansive purview, merging the diversity of data science methods with an expanding set of cybersecurity needs. An immediate challenge concerns the risk of combinatorial diffusion and confusion.

Given the hybridized CSDS domain's conceptual breadth, an immediate challenge concerns defining disciplinary focus (Bechor and Jung 2019). Framing context helps ground systematic inquiry. As a high-level focus concerns deploying information systems and solutions in organizational settings, the management of information systems (MIS) discipline provides a pragmatic and logical disciplinary locus to examine CSDS.

MIS is an interdisciplinary field focused on the orchestration of processes, technology, and people to realize organizational value from information systems (ITIL

2019; Keen 1981; Leavitt 1964; Looijen and Delen 1992). Both technical and behavioral approaches are utilized to study information systems (Laudon and Laudon 2017). Core sub-domains of the MIS field align with CSDS goals, in particular decision support systems (DSS), expert systems, management support systems, executive information systems, and artificial intelligence (AI) (Haag and Cummings 2012; Hsu 2013; Pearlson et al. 2016; Sousa and Oz 2014).

Data science from an MIS perspective embodies the application of statistical, analytical, and algorithmic methods to drive scientific decision support processes (Goes 2014). As a practitioner-led discipline, the field focuses on the implementation of analytics methods to drive value-creating decision insights from data (Davenport et al. 2010; Kelleher and Tierney 2018). Data science offers efficiencies to improve managerial and expert decision-making. As hybridized in CSDS, data science aims to drive efficacy and automation in cybersecurity decision processes.

The cybersecurity field aligns closely with the MIS tripartite model (Fig. 2.4) as it spans organizational, process, and technical contexts. A central principle is that assuring cybersecurity creates a set of organizational demands which are fulfilled through security information systems and organizational processes. This invokes M. Looijen's information systems management model, which frames a supply-demand framework aligning business information management goals with IT solutions (Looijen and Delen 1992) (Fig. 2.5).

Focusing this inquiry within the MIS discipline addresses a research and practitioner lacuna: outlining a set of best practice prescriptions for implementing and managing CSDS. Conflating MIS with CSDS addresses a lacuna in cybersecurity research literature, which has otherwise focused on advanced analytics almost exclusively as an engineering challenge (Greengard 2016; Verma et al. 2015; Zinatullin 2019). As the implementation of decision support and expert systems necessitates organizational decision methods and processes, MIS addresses the broader management context.

Research addressing applied machine learning for cybersecurity is a growing and popular area. Broader organizational management perspectives on cybersecurity analytics are otherwise underrepresented. There is a research gap concerning the organizational management of cybersecurity analytics, for instance, to address focused topics such as human-in-the-loop processes, model management and validation, and data management. Academic research has focused on addressing the demand for ever more advanced engineering methods, while practitioners struggle with fundamental issues of organizational coordination, implementation, model management, and data management best practices (Oltsik 2019b; Ross 2019).

As substantiated in the literature and qualitative review conducted herein, there is a research gap related to the framing, management, and application of CSDS methods. This lacuna is observable more generally in the lack of scholarship on the topic of machine learning in business and management journals (Salmon 2019). That this gap requires accommodation is supported by the observation that organizational factors are preeminent to the successful implementation of analytics solutions (Kiron et al. 2014; Kiron and Shockley 2011; LaValle et al. 2010; Roy and Seitz 2018). Offering CSDS methods and solutions without an organizational

Fig. 2.4 Classical MIS
tripartite model

Fig. 2.5 Looijen
information systems
management model

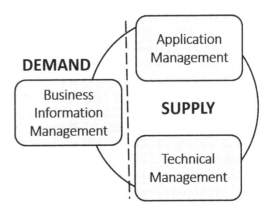

decision-making and process context falls into an increasingly critiqued trap associated with data science: emphasizing technology and methods while ignoring organizational context (Kelleher and Tierney 2018; O'Neil 2016; Voulgaris 2017).

2.2.7 *Professional Maturity Model*

CSDS has put forward as an *emerging professional field*. As cited previously, CSDS is a profession in that there are named practitioners operating in a professional capacity as *cybersecurity data scientists* and active job openings exist for this role. As hybridized from two independent domains, *cybersecurity* and *data science*,

CSDS derives from two fields undergoing rapid professional transformations. As an emerging domain, a key research question regards the relative state of CSDS maturity on a path to professionalization.

It has been proposed that research and managerial evidence substantiates that cybersecurity domain pressures impel the growth and expansion of CSDS. Increasing threats, staff shortages, and an environment of growing complexity impel the need for more sophisticated, data-driven decision approaches in the cybersecurity field. However, to position the emerging field of CSDS on a path to professionalization, a greater context is necessary concerning the concept of a profession and professional maturity.

Research on professionalization, the process of why and how professions evolve, holds divergent views concerning the definition of a profession (Van der Krogt 2015). This is due to the interdisciplinary nature of the topic, as well as shifting historical, political, economic, and cultural factors that have changed the perceived nature of professions (Freidson 2001; Muzio and Kirkpatrick 2011). For this inquiry, a clear and practical definition was judged desirable in keeping with the pragmatic focus of the MIS domain, which generally eschews complex sociocultural theory (Laudon and Laudon 2017; Lim et al. 2013; Wu et al. 2006).

A straightforward generally accepted definition of a profession is that of *a group of people with specialized training that gain income from the sale of that expertise* (Cannon 1978). CSDS meets this simple criterion. A broader definition offers that "professions are somewhat exclusive groups of individuals applying somewhat abstract knowledge to particular cases" (Abbott 1988). While both descriptions are suitable, the second, more flexible definition aligns with the malleable and shifting nature of information age labor (Freidson 2001; Muzio and Kirkpatrick 2011).

Professionalization research views the process of professionalization as a distinct topic of inquiry. Maturity in professional domains is a well-researched topic (Freidson 2001; Greenwood 1957; Muzio and Kirkpatrick 2011; Vollmer and Mills 1966). At a high level, mature professions are typically denoted by the existence of recognized professional bodies, formal certifications and degrees, training and educational programs, and best practices codified in the form of shared standards (Beer and Lewis 1963). In mature professions, these aspects intertwine, facilitating interactions between theoretical research and operationalized practice. When robust, such professions span across public, private, and academic institutions. The medical profession, a vast, storied, and highly formalized field, can be considered an exemplar.

Shifting historical frames have led to disparate views concerning the professionalization process. Clear distinctions are made between the evolution of professions in pre-industrial and postindustrial, information age venues (Freidson 2001; Muzio and Kirkpatrick 2011; Vollmer and Mills 1966). In pre-industrial terms, the process of professionalization has been attributed to structural emergence in the context of institutional facilitation.

For instance, it has been noted that natural philosophers, the predecessors of professional scientists, were initially a loose hodgepodge of aristocratic courtiers, savants, skilled tradespeople, and itinerant academics (Shortt 1983). Key to the historical emergence of professional scientists was an increasingly formal institutional demarcation between engineering-focused craftspeople and formal recognitions bestowed upon natural philosophers by chartered professional societies (Beer and Lewis 1963).

As the majority of the first scientific societies historically relied upon aristocratic charters and patronage, there is a valid observation that segmentation based upon social class and class hierarchies plays a role in professionalization. The distinction between engineering and craftspeople from elites who utilize the mechanisms and byproducts of craft, yet focus on applied theory and administrative politics, is otherwise central to the pre-industrial notion of a profession.

In the modern frame of postindustrial capitalism, it has been noted that professions emerge and shift more quickly in reaction to rapid economic and technological changes (Muzio and Kirkpatrick 2011). Using the example of science, modern scientists have diversified into academic, governmental, and industrial research typologies, and various grades and levels distinguishing each have emerged on top of these variants (Beer and Lewis 1963). Information age economic and institutional pressures continue to transmute the scientific profession alongside other knowledge-focused professions.

In postindustrial economies, workers no longer remain in the same professional designation for the entirety of their careers. This circumstance challenges the ability to offer strict definitions bounding a profession and highlights the growing importance of continual *professional emergence* as a central concept. A distinction is made in the literature between the *sociological view* of professions, which concentrates on longer-term structural-societal factors, and the *organizational view*, which focuses on shorter-term, rapidly evolving professional service and topical specializations (Muzio and Kirkpatrick 2011). The latter context is relevant to emergent CSDS professionalization as the field is highly influenced by rapid technological change. CSDS boundaries are being shaped by the shifting demand for professional services, which derive from rapidly evolving and emerging technology-focused forces.

Professionalization scholar T. P. W. M. van der Krogt distinguishes four major theoretical strands underpinning professionalization research: *trait approach, functionalist theory, power theory,* and *Freidson's theory* (Freidson 2001; Van der Krogt 1981, 2015). To varying degrees, all four frameworks reflect *sociological view* perspectives, particularly the importance of societal, economic, political, and cultural factors in the context of power dynamics. Social forces and conflict are emphasized more heavily in the last three frameworks. However, the first, the *trait approach*, offers a straightforward practical framework that also aligns with the *organizational view* of professions. In the trait approach, E. Greenwood holds that:

> the skills that characterize a profession flow from and are supported by a fund of knowledge that has been organized into an internally consistent system, called a *body of theory*. A profession's underlying body of theory is a system of abstract propositions that describe in general terms the classes of phenomena comprising the profession's focus of interest. Theory serves as a base in terms of which the professional rationalizes his operations in concrete situations. Acquisition of the professional skill requires a prior or simultaneous mastery of the theory underlying that skill. Preparation for a profession, therefore, involves considerable preoccupation with systematic theory, a feature virtually absent in the training of the nonprofessional. (1957)

From the perspective of management theory, the *trait approach* can be considered to align with the *knowledge-based view* (KBV) of the firm (Grant 1996, 1997; Jensen and Meckling 1992), which is a refined specification of the *resource-based view* (RBV) (Barney 1991, 1999; Conner 1991; Penrose 1959; Wernerfelt 1984; Williamson 1975). The RBV aligns with the notion that professions, especially in modern frames, are highly influenced by institutional competition for resources (Beer and Lewis 1963). The KBV, as a refinement, recognizes the value-enhancing capabilities of information systems in expediting firm knowledge to make informed decisions.

The KBV core assertion emphasizes the importance of incentives, acting in concert with information and decision rights, to understanding the capabilities and behavior of organizations. The importance of decision rights, information access, and incentives in orienting organizational actors connects KBV to considerations of *organizational design* (Burton et al. 2011), *management control systems (MCS)* (Merchant and van der Stede 2003), *organizational culture* (Cameron and Quinn 2011), and *change management* (Hiatt 2006; Kotter and Cohen 2002). Each of these topics has more than a passing relevance to the emergence of the CSDS field.

Adopting an *organizational view* of professions informed by the *trait approach* (as framed by Greenwood) aligns with the KBV paradigm. The *trait approach* of professions is well-recognized in research and aligns well with management perspectives, raising issues of conflict and power in the context of firm management dynamics. Systematic theory in professions is emphasized, which ties closely to the necessity and centrality of theory in the organizational application of data science to practical challenges in the cybersecurity domain. The concept of a *body of theory* highlights the crucial distinction between the classical concept of *technê*, knowing how to do, versus *logos*, Aristotelian structured reason, and *epistêmê*, knowing what and why.

The KBV's emphasis on decision-making aligns well with CSDS as a decision-facilitating service whose value proposition rests upon the ability to improve decisions in the increasingly complex and uncertain cybersecurity domain. For the purposes of this inquiry, treating CSDS as an emergent set of organizational traits and behaviors with central theoretical and decision-making aspects provides a mechanism for assessing its emerging maturity.

To drive the comparative examination of CSDS, the *trait approach*, as aligned with the *organizational view* of professions and the KBV, fits well with pragmatic MIS research interests and needs. Table 2.3 frames Greenwood's five distinguishing attributes for a profession (1957) as partially refined by Van der Krogt (2015).

Table 2.3 Model for assessment of professional maturity		
	1	Systematic body of theory
	2	Authority and judgment recognized by client
	3	Community sanctions authority
	4	Ethical code of stewardship
	5	Professional culture supported by associations

Table 2.4 Model for assessment of professional emergence

1	Active, focused interest from diverse participants
2	Active professionals with associated job titles and roles
3	Emerging and informal training
4	Informal professional groups
5	Professional and industry literature
6	Research literature
7	Formalized training
8	Formal professional groups
9	Professional certifications
10	Standards bodies
11	Independent academic research disciplinary focus

Although the *trait approach* has been critiqued from the sociological perspective, it is generally regarded to identify key elements of the professionalization process. As the goal of this inquiry is a compact model to support MIS research, this straightforward framework is asserted to be adequate and appropriate. The practical goal of this inquiry is to examine criteria for status as a profession; deeper explorations of issues of social power and control are considered out of scope.

Further, concerning *professional emergence* in the context of the *trait approach*, to introduce rigor in the assessment of the current versus future status of emerging and developing professions, a rough, hierarchical continuum is proposed (Table 2.4).

This practical, staged model to chart *professional emergence* is extrapolated from *trait theory* as espoused by professionalization scholar E. Greenwood (1957). Representation of key categories and staging is validated in interpretations of the seminal works of other noted scholars of professionalization, in particular, Beer and Lewis (1963), Vollmer and Mills (1966), and Freidson (2001). Also, Muzio and Kirkpatrick's (2011) more recent interpretations of the dynamic interplay between institutions, professional organizations, and professionals are observed in this scaled framework. The concept of professional emergence has particular relevance in a modern, information age context, wherein professional services-framed roles are noted to evolve and emerge more rapidly and fluidly.

Notable in the proposed *professional emergence* model is the iterative, multi-domain, and staged interplay of professional evolution, namely, the interaction between self-organizing professional initiatives, formal professional societies, institutions, and research academics. The complex interplay between these various forms of social and institutional organization has long been noted in historical research on professionalization, reaching back to works investigating the rise of the scientific and medical professions (Beer and Lewis 1963; Freidson 2001; Van der Krogt 2015).

Compact and straightforward, these two models combined are submitted as a practical professional maturity assessment litmus test. With these simplified practical models derived, the fields of cybersecurity, data science, and CSDS will subsequently be assessed and compared according to their relative level of professional maturity and development. The focused, practical goal is to analyze the current *as-is* state of the CSDS field, based on comparative research literature, and to extrapolate *to-be* gaps.

2.3 Cybersecurity as a Profession

2.3.1 Origins and Definitions

To properly frame CSDS as an emerging hybrid field, a focused overview of the cybersecurity domain is essential. Cybersecurity has a rich history, tracing its modern manifestation back to the early 1970s (Hafner 1998; Leiner et al. 1999) and further back to the nineteenth century when encompassing computationally intensive cryptography (Simon Singh 1999).

A logical focus on technology harkens to the cybersecurity domain's roots, which spring from engineering-driven innovation. Networked communications and computational devices, particularly their rapid conjoint evolution in the late twentieth century, otherwise are the dominant focus. A robust industry definition of *cybersecurity* promulgated by the International Telecommunication Union (ITU-T) reflects the technological engineering foundations of the field:

> Cybersecurity is the collection of tools, policies, security concepts, security safeguards, guidelines, risk management approaches, actions, training, best practices, assurance and technologies that can be used to protect the cyber environment and organization and user's assets. Organization and user's assets include connected computing devices, personnel, infrastructure, applications, services, telecommunications systems, and the totality of transmitted and/or stored information in the cyber environment. Cybersecurity strives to ensure the attainment and maintenance of the security properties of the organization and user's assets against relevant security risks in the cyber environment. The general security objectives comprise the following: availability; integrity, which may include authenticity and non-repudiation; and confidentiality. (2008)

In a more compact definition, the following is a composite extrapolated from content analysis conducted on recent cybersecurity domain research:

> The approach and actions associated with security risk management processes followed by organizations and states to protect confidentiality, integrity and availability of data and assets used in cyber space. The concept includes guidelines, policies and collections of safeguards, technologies, tools and training to provide the best protection for the state of the cyber environment and its users. (Schatz et al. 2017)

Both definitions, reflective of the engineering roots of cybersecurity, are focused on technical aspects. Content analysis research efforts to extrapolate cybersecurity definitional commonalities resident in literature have validated this observed technical focus (Craigen et al. 2014; Schatz et al. 2017). Issues of risk and assurance are raised in some definitions yet are not typically extended beyond mechanical questions of data integrity and access control. Implications concerning managerial economics, aspects of information (rather than data) access, and the imperfect trade-offs embedded therein are frequently sidestepped in domain research literature.

However, reflecting the increasing reach and scope of networked digital technologies, cybersecurity now impacts a broader range of factors and systems, many of which are non-mechanistic. The domain's purview has expanded from traditional, highly technical standards and specifications to aspects variably concerned with organizational, legal, regulatory, economic, sociocultural, and political factors. Recently, ethical issues related to the right to privacy and information hygiene, i.e., addressing the proliferation of so-called fake news, have sprung to the fore (The Economist 2019a). Such public debate has flourished in lockstep with the advent of new regulatory strictures, such as the EU's landmark General Data Protection Regulation (GDPR) framework (The Economist 2018).

As the cybersecurity domain continues to evolve and expand through proliferation and advancement, broader definitions are increasingly necessary (Craigen et al. 2014). There is a need to address emerging areas such as information assurance, organizational processes, human resources, ethical factors, and media theory. There is an awareness that compliance and assurance, beyond benchmarking and audit, increasingly occupy the attention of cybersecurity stakeholders and managers.

In a managerial context, mirroring the growing relevance of cybersecurity to enterprise risk, the need for an expanded domain model to amplify such definitions is advocated, one which specifically recognizes the increasing interplay between technical aspects and organizational factors. Beyond linear and deterministic notions of control and exposure, this implies formal attempts to capture probabilistic aspects of risk. Also, the relevance and timeliness of actionable operational information must be broached, recognizing that the context of cybersecurity events spans from the instantaneous (tactical) to the broadly historical (strategic). This can be encapsulated in the notion of framing cybersecurity as an explicit challenge of *security assurance*, rather than as a simple mechanical audit.

2.3.2   Managing and Assuring

The traditional technical and engineering bias in the cybersecurity domain can be framed as both a help and hindrance. The overtly mechanistic mindset has inculcated a pragmatic, systematic focus in efforts to identify and overcome cybersecurity technical vulnerabilities. In this context, security administration is viewed as an unambiguous technical task, focused on issues of access and control, but is not otherwise concerned with economic or ethical trade-offs.

However, the mechanistic paradigm, focused on pragmatic efficacy and determinism, has self-limiting characteristics. Management of resource access, which must satisfice economic quandaries between the dueling poles of the *value of practical expediency* and *controlling the risks of exposure*, is given short shrift in classical definitions of cybersecurity (Craigen et al. 2014). Audit is straightforward when focused on binary options: largely a mechanical box-checking exercise.

However, as networked computational technology has proliferated exponentially, organizational, legal, regulatory, economic, social, and political impacts comingle increasingly with technical infrastructure. Also, growing inherent technical complexity, increasingly multidimensional rather than linear, has the potential to create mechanical paradoxes and conflicts of access and control. Access provided or denied to source systems or data can improperly share or interrupt services in a downstream system when increasingly distributed, composite service architectures are the growing norm.

Enterprise risk transforms the stakes of access and control from binary decisions to judgments that must be measured economically and in a programmatic context. In the enterprise risk paradigm, access control administration is a set of trade-offs and considerations that increasingly touch on issues of risk distribution, organizational trust, competitive equity, and information integrity (as distinguished beyond data integrity).

Cybersecurity management, by necessity, must scale from the highly technical to the organizational, which raises social and ethical aspects. Management-focused perspectives mediate between these poles, establishing measurable standards that address sound prudential goals underpinning assurance and stewardship. The ability to not only audit but benchmark organizational capabilities according to both ideal and practical peer standards is otherwise an essential requirement (Bishop 2003). This serves the purpose of addressing broader macro-social and governmental as well as owner and shareholder interests, in the case of commercial concerns.

Specifically, as the MIS domain is promoted as a disciplinary locus in this inquiry, a broader definition of cybersecurity that embraces organizational management interests, such as economic trade-offs, ethics, legal compliance, and human resource management, is advocated. This highlights the importance of processes such as risk, assurance, decision, and information management. In particular, the goal of assurance captures the combined technical and organizational scope of cybersecurity in both tactical and strategic settings. Such framing leads to a forward-thinking representation of cybersecurity in the classical MIS tripartite model, per Fig. 2.6.

As audit and assurance both rely upon operational accountability via role-based ownership and stewardship, understanding key cybersecurity roles and skills helps categorize focused areas where analytics can be applied. The NICE Cybersecurity Workforce Framework (NIST 2017) offers a categorical perspective on key cybersecurity work role functions:

- Secure and provision
- Operate and maintain
- Oversee and govern
- Protect and defend
- Analyze
- Collect and operate
- Investigate

Fig. 2.6 Cybersecurity in
the MIS tripartite model

Data science methods ostensibly can provide value across all the functions cited. However, there is a corresponding lack of clarity as to which particular data science methods best align to corresponding work roles.

A more focused conceptual model is useful to focus on the security context for the implementation of data science treatments. Aggregating and extrapolating from a broad set of standards and frameworks, security assurance functional areas are framed in a proposed quadrant model per Fig. 2.7. The derived quadrant specification for security assurance distinguishes temporal concerns (more tactical versus more strategic) from planning and implementation targets (more technical versus more organizational). Four general areas of security assurance activity are specified as aligning with the quadrant positions (clockwise from bottom to left): security monitoring, operational security design, management and stewardship, and readiness and response. The quadrants roughly observe the four cyclical stages of the enterprise security cycle as specified in *The Practice of Network Security* (Bejtlich 2013).

Within the quadrant designations, a catalog of common enterprise security-related goals and tasks have been positioned. These tasks are extrapolated from a range of accepted industry views, in particular the (ICS)[2] CISSP domains (ICS2 2019), SANS Institute training themes (SANS Institute 2019a), and workforce categories specified in the NICE Cybersecurity Workforce Framework (NIST 2017). The resulting aggregation in the quadrant model offers a novel summary positioning of security assurance tasks according to varying time horizons and goals.

The derived quadrant model counterposes and overlays two central tensions in cybersecurity assurance which are useful to orienting and mapping data science prescriptions:

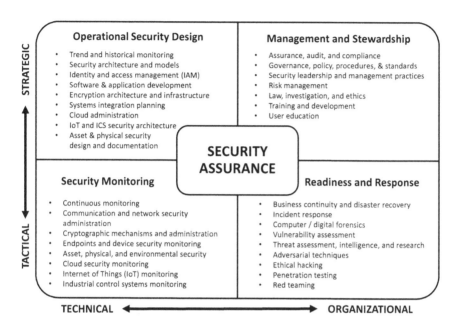

Fig. 2.7 Security assurance quadrant framework

1. *Assurance as a service*: timeliness (*tactical*) versus continuity and credibility (*strategic*) locus
2. *Cybersecurity assurance*: mechanical (*technical*) versus contextual (*organizational*) aspects

The quadrant model is useful in efforts to position data science prescriptions to address contextual cybersecurity requirements. This otherwise allows for a more refined MIS supply-and-demand context, previously raised by the Looijen model (Fig. 2.5), for framing CSDS prescriptions, and will be invoked later in this inquiry.

2.3.3 Academic Programs and Research

Universities globally have responded to a growing demand for cybersecurity professionals and concomitant interest from students. An array of programs from the undergraduate to PhD levels are available worldwide. Master of Science (MSc) programs are particularly popular, providing postgraduate qualifications for continuing and returning students seeking to specialize. Nearly 150 universities in the United States are noted as offering MSc degree programs in cybersecurity (Morgan 2018b) with a similar number listed in the United Kingdom (Find a Masters 2019). Online delivery is an increasing trend, with 42 of the 148 US-based MSc degree offerings (~28%) listed as online programs. Outside degree programs, universities increasingly offer classroom and online professional continuing education options such as workshops and professional certificate programs.

Cybersecurity is an interdisciplinary field with connections to the disciplines of MIS, information technology, computer science, and computer information systems. This, combined with connections to wider fields such as law, public policy, psychology, and audit, has led to long-standing claims that the discipline suffers a disciplinary identity crisis (Benbasat and Zmud 2003). More concerning have been charges that universities are not preparing students properly for operational cybersecurity work in industry (Conklin et al. 2014). This has led to government initiatives to improve academic alignment with industry, such as the US NIST National Initiative for Cybersecurity Education (NICE).

There are many cybersecurity-focused academic research journals, notable examples edited and published through the ACM, Elsevier, IEEE, IEE/IET, Oxford Academic, and Springer (Kuhn 2016). In addition to research journals, well-regarded cybersecurity research is also published as papers collected in conference proceedings, such as Springer Lecture Notes on Computer Science and USENIX conference proceedings. The format and nature of conference papers provide a vehicle for fast-emerging research, which often has an engineering focus. This is as opposed to peer-reviewed journals, which are often more demanding concerning theoretical content and whose timeline for publication is more extensive.

While the cybersecurity academic research domain's scope is interdisciplinary, voluminous, and growing, two central themes are dominant: designing reliant systems and protecting those systems from persistent risks of compromise. Encompassing combinations of networks and computer devices, research spans practical guidance on defense to highly technical research studies on advanced intrusion and exploitation techniques. Earlier comments regarding the engineering bias in the cybersecurity field are thus also observed in the venue of published research, although broader topics are being inevitably admitted to the field due to the pressures of pragmatic necessity.

2.3.4   Professional Organizations and Certifications

While not exhaustive, what follows represents well-recognized cybersecurity professional institutions, organizations, and societies. Many have strict membership or participation standards and are active in promulgating best practices and standards surrounding professional skills, knowledge, and behavior. In their capacity to establish and promote professional standards, these organizations act as gatekeepers and stewards of professional status, systematizing and standardizing disciplinary boundaries. The listing is a sample of the breadth of cybersecurity organizations and bodies, with no claims of exhaustive representativeness implied.

The formalization of professional expertise standards spans a continuum from *best practices* codified in professional training curricula to certification exams attesting a professional's capture of skills and theoretical knowledge, which extends to *body of theory*. Many organizations frame training programs, host-recognized professional certificate offerings, and/or act as certification authorities that set standards and benchmark professionals. The cybersecurity domain, being diverse and broad, hosts a range of certifications from the high level to the highly specialized across a range of subjects. While not exhaustive, examples of well-recognized certificates are included for those bodies hosting such offerings.

Governmental
- **ENISA**: The European Union Agency for Network and Information Security (www.enisa.europa.eu) is a center of expertise delivering advice and solutions to EU member states. The organization offers a range of cybersecurity publications and documents.
- **NCSC (UK)**: The National Cyber Security Center (www.ncsc.gov.uk) is a UK governmental body fostering and advocating best practices and assisting with national security incident response.
- **NCSC (NL)**: The Nationaal Cyber Security Centrum (www.ncsc.nl) is an organization within the Dutch Ministry of Justice and Safety (Ministerie van Justitie en Veiligheid) promoting the safety and stability of the digital domain in the Netherlands.
- **NIST**: The National Institute of Standards and Technology (www.nist.gov) is a US government-affiliated body supporting standards and research into technology and research. A range of resources are shared related to cybersecurity via the NIST Computer Security Resource Center (csrc.nist.gov).

Government Affiliated
- **NICE**: The National Initiative for Cybersecurity Education (www.nist.gov/nice) is a US public-private-academic partnership initiative led by NIST and focused on promoting cybersecurity education, training, and workforce development.
- **MITRE**: A set of US federally funded nonprofit R&D centers and public-private partnerships focused on technology and research in the areas of security and safety (mitre.org).
- **SEI CERT Division**: The Carnegie Mellon Software Engineering Institute (SEI) Computer Emergency Readiness Team (www.sei.cmu.edu/about/divisions/cert) is a US government federally funded nonprofit R&D center. The organization supports and circulates research, offers training, and engages in public-private partnerships to improve software and Internet security measures.

Trade and Standards Bodies
- **CompTIA**: The Computer Technology Industry Association (www.comptia.org) is a trade association offering research, tools, best practices, training, and certifications. *Examples of topical certifications offered include network infrastructure configuration and management (Network+), IT security and risk management (Security+), and advanced security practitioner (CASP+).*
- **ISO/IEC JTC 1**: A joint committee of the International Organization for Standardization and the International Electrotechnical Commission (www.iso.org) dedicated to developing and publishing international standards related to information technology (IT) and communications technology (ICT).

- **ITU-T**: The International Telecommunication Union (ITU) Telecommunication Standardization Sector (www.itu.int) develops and publishes international standards surrounding infrastructure related to information and communication technologies (ICTs).
- **PCI Security Standards Council**: The Payment Cards Industry Security Standards Council (www.pcisecuritystandards.org) is a global body facilitating the development and dissemination of tools and best practices promoting payment card account security.

Nonprofit Industry and Professional Bodies

- **CIS**: The Center for Internet Security (www.cisecurity.org) offers best practices, tools, and information sharing on security threats.
- **CSA**: The Cloud Security Alliance (cloudsecurityalliance.org) is dedicated to defining and raising awareness of best practices to help ensure secure cloud computing environments. *Certifications offered include the Certificate of Cloud Security Knowledge (CCSK) and the Certified Cloud Security Professional (CCSP).*
- **IAPP**: The International Association of Privacy Professionals (iapp.org) is an organization hosting resources, tools, conferences, training, and certifications for information privacy professionals. *Example certifications include the Certified Information Privacy Professional (CIPP).*
- **IACRB**: The Information Assurance Certification Review Board (iacertification. org) is an industry-standard organization for information security professionals. *IACRB offers a range of certifications, including the Certified Application Security Specialist (CASS), Certified Penetration Tester (CPT), and Certified Computer Forensics Examiner (CCFE).*
- **ISACA**: An information systems audit and control association (isaca.org) offering guidance, benchmarks, frameworks, and tools related to information security, assurance, risk management, and governance. *Certifications include the Certified in the Governance of Enterprise IT (CGEIT), Certified Information Systems Auditor (CISA), Certified Information Security Manager (CISM), and Certified in Risk and Information Systems Control (CRISC).*
- **ISF**: The Information Security Forum (www.securityforum.org) is focused on information security and risk management best practices. The ISF offers research and consulting services related to security advising, support, and training.
- **(ISC)²**: The International System Security Certification Consortium (www.isc2. org) is an international professional association for information security leaders. The organization provides networking and collaboration opportunities, along with development and leadership tools. *A range of professional certifications are offered, including the Certified Information Systems Security Professional (CISSP), Information Security System Management Professional (ISSMP), and Systems Security Certified Practitioner (SSCP).*
- **OASIS**: A consortium (oasis-open.org) promoting open standards initiatives associated with cybersecurity, IoT, cloud computing, and other areas.
- **OWASP**: The Open Web Application Security Project (www.owasp.org) is a global organization focused on improving web application security.

Commercial Industry and Professional Organizations

- **EC-Council**: The International Council of E-Commerce Consultants (www. eccouncil.org) is a commercial cybersecurity professional organization and certification body. *Example certifications offered include the Certified Ethical Hacker CEH), Licensed Penetration Tester (LPT), and Computer Hacking Forensic Investigator (CHFI).*
- **Offensive Security**: This company (offensive-security.com) provides penetration testing training, tools, and certifications, particularly as associated with the Kali Linux distribution (kali.org). *Example certifications include the Offensive Security Certified Professional (OSCP) and Expert (OSCE).*
- **SANS Institute**: The SysAdmin, Audit, Network and Security Institute (sans. org) is a commercial body specializing in information and cybersecurity training, research, and certification. *The associated Global Information Assurance Certification (GIAC) program offers a range of certifications, such as the GIAC Security Expert (GSE) and GIACC Information Security Fundamentals (GISF).*
- **Security Onion Solutions**: A company (securityonionsolutions.com) offering intrusion detection, security monitoring, and log management tools, training, and consulting, particularly as associated with the Security Onion Linux distribution (securityonion.net).

Commercial Technology Organizations

- **Cisco**: Cisco Systems, Inc. (cisco.com) is a commercial technology conglomerate producing networking, telecommunications, and associated services and products. *Examples of IT professional certifications related to products and services include Certified Entry Networking Technician (CCENT), Network Associate Cybersecurity Operations (CCNA Cyber Ops), and Certified Design Associate (CCDA).*
- **Microsoft**: A global commercial software and technology vendor, Microsoft (microsoft.com) promotes security best practices and offers security-related solutions and products. *A range of security-related certifications are offered, including the Microsoft Certified Solutions Expert (MCSE) and Microsoft Certified Technology Specialist (MCTS).*
- **Palo Alto Networks**: A commercial technology vendor and service provider (paloaltonetworks.com), the firm also offers best practices, education, and certifications in the domain. *The Palo Alto Network Certified Cybersecurity Associate (PCCSA) is an example of a certification offered.*

As selectively and partially raised here, there are a broad set of professional certifications on offer in the cybersecurity field, befitting an increasingly diverse domain. As certification programs also involve financial incentives for the hosting bodies, spanning training, certification and testing fees, conferences, and technological solutions and applications, the relative robustness, quality, and industry acceptance of various certifications and certification bodies is a topic of debate. A broader listing of recognized cybersecurity certifications is hosted on Wikipedia (Wikipedia 2019), and a listing of training providers and certification training programs is provided by Cybersecurity Ventures (Morgan 2019b). A comprehensive listing and review of the various certifications, associated training offerings, and comparative subject matter covered is otherwise outside the scope of this inquiry.

2.3.5 Frameworks

Frameworks for capturing systemic representations of complex domain phenomena are both a byproduct of and a facilitator of professional knowledge formalization. Revisiting the medical profession as an exemplar, a plethora of frameworks exist to represent the different interacting systems of the body (respiratory, circulatory, digestive, organs); the incidence, progression, distribution, and control of disease (epidemiological models); the processes for intervention (e.g., surgical and emergency procedures); and the governance, management, and administration of healthcare (hospital management, standards of clinical care, emergency response, ethical frameworks, etc.).

The cybersecurity domain similarly strives to capture and address an expanding and increasingly complex taxonomy of technologies, array of systems, and distribution of interacting domains. Codifications of the various associated systems and phenomena are central mechanisms to store, standardize, and categorize knowledge in this respect. Standardizing representational models supports professional and institutional knowledge capture and exchange. Frameworks are particularly relevant in a management context, as the organizational implementation of operational security programs typically derives from such structured representations, both for guidance and measurement (Stallings 2019).

The interplay between engineering standards and protocols, institutional stewardship and guidance, management and governance principles, and practitioner standards and skills is apparent in the range of frameworks available. Each artifact originates from one or more of the public or private institutions mentioned previously:

Technical and Engineering Standards
- **ISO/IEC 27000 Information Security Family of Standards**

  - **27001**: Information security management control specifications suitable for audit and certification
  - **27002**: Best practice information security controls for implementers
  - **27032**: Guidelines for improving the state of cybersecurity

- **ISO/IEC Business Continuity and Risk Management Standards**

  - **22301**: Business continuity management system specifications
  - **31000**: Risk management guidelines, principles, and framework

- **ITU-T Security Standards and Guidance Documents**: A set of documents from the ITU-T related to security for various communications systems and technologies, particularly documents from Study Group 17 (SG17) related to Internet cybersecurity standards and recommendations.

- **OSI**: The Open Systems Interconnection model characterizes and generalizes the functions of communication systems according to a seven-layer representation. Abstraction promotes design interoperability of communications systems with industry-adopted standards.
- **TCP/IP**: The transmission control protocol/Internet protocol, also known as the Internet protocol suite, is a four-layer conceptual model for the set of communication protocols used on the Internet and computer networks. The abstraction layers classify protocols according to the scope of the traffic involved (i.e., link, Internet, transport, and application).

Standards and Guidelines for Controls and Management
- **CIS Critical Security Controls for Effective Cyber Defense**: Also known as the CIS Top 20 or SANS Top 20, best practices from SANS and CIS related to organizational controls and infrastructure security benchmarks.
- **CompTIA Channel Standard**: A set of best practices and guidance offered by CompTIA and categorized within the NIST Framework.
- **CIA Triad**: A conceptual framework for information security policy denoting the goals of confidentiality, integrity, and availability.
- **COBIT**: The Control Objectives for Information and Related Technologies is a good-practice framework for IT management and governance created and maintained by the ISACA.
- **Cybersecurity Culture Guidelines—Behavioral Aspects of Cybersecurity**: ENISA report covering the organizational and social-behavioral aspects of cybersecurity.
- **Cybersecurity Maturity Model Certification (CMMC)**: A set of best practice-based standards for the assessment of cybersecurity maturity to support risk awareness and audits (Carnegie Mellon University and John Hopkins University 2019).
- **GDPR**: The EU General Data Protection Regulation (Regulation 2016/679) concerns rules for handling and processing data concerning EU individuals by companies and organizations.
- **HIPAA Privacy Rule**: US national standards to protect medical records and personal health information.
- **IACD**: The Integrated Adaptive Cyber Defense is a strategy and framework for operationalizing cybersecurity interoperability, automation, and response (Johns Hopkins University 2014).
- **ISF Benchmark**: A tool available to ISF members to assess and analyze comparative security capabilities against peers and best practices.
- **ISF Standard for Good Practice (SGP)**: ISF guidance on information security controls and information risk management best practices.
- **MITRE Cyber Resiliency Engineering Framework**: A guide to support the discussion and analysis of cyber resiliency goals, objectives, metrics, practices, and costs in organizations (Bodeau and Graubart 2011).
- **NICE Cybersecurity Workforce Framework**: Standards and best practices related to cybersecurity education and workforce challenges. A categorization of

Fig. 2.8 NIST Framework

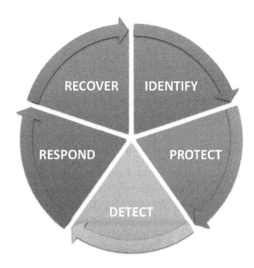

cybersecurity work functions, tasks, and roles described in NIST Special Publication 800-181.

- **NIST Framework for Improving Critical Infrastructure Cybersecurity**: Security framework and documents supported by NIST, offering a set of linked goals surrounding organizational defense and response efforts and cited as a guide for other benchmarks (NIST 2018) (Fig. 2.8).
- **Payment Card Industry Data Security Standard (PCI DSS)**: An information security standard to protect data and reduce card fraud administered by the PCISSC and applied in compliance audits.
- **Security Guidance for Critical Areas of Focus in Cloud Computing**: A set of guidelines for secure cloud operations offered by the CSA.
- **SOAPA**: An enterprise architectural framework from ESG representing Security Operations Analytics Platform Architecture (Oltsik 2018).
- **SOAR**: An enterprise framework from Gartner representing Security Orchestration, Automation, and Response (Barros and Chuvakin 2018).

Threat Models
- **Attacker Personas**: This is a model framed by Microsoft employees to assess and categorize the types of cybersecurity threats and their motivations (Aucsmith et al. 2003).
- **Cyber Kill Chain**: A framework created by Lockheed Martin for identifying and preventing intrusions through a process-based perspective on attackers' objectives (Lockheed Martin 2012).
- **Diamond Model of Intrusion Analysis**: A model for analyzing intrusion incidents and activities (Caltagirone et al. 2013).
- **MITRE ATT&CK**: The Adversarial Tactics, Techniques & Common Knowledge from MITRE frames methods for cyber incursion and compromise in terms of a process followed by adversaries (MITRE 2018).

- **ODNI Cyber Threat Framework**: A framework for characterizing and categorizing cyber threat events and trends (Office of the Director of National Intelligence (ODNI) 2018).
- **STRIDE**: A model originally framed by Microsoft categorizing security threat mechanisms according to the acronym: Spoofing of user identity, Tampering, Repudiation, Information disclosure (data leak), Denial of service, Elevation of privilege (Kohnfelder and Garg 1999).

Semantic Frameworks and Ontologies
- **Common Attack Pattern Enumeration and Classification (CAPEC)**: Structured framework for describing known tactics, techniques, and procedures (TPP) applied by adversaries.
- **Cyber-Investigation Analysis Standard Expression (CASE)**: Open standard ontology/specification language for sharing cybersecurity case investigation information.
- **Cyber Observable eXpression (CybOX)**: A semantic framework for describing objects and properties in the cybersecurity domain (merged into STIX).
- **Incident Object Description Format (IODEF)**: This is a standard data format for the exchange of incident information between security teams.
- **Integrated Cyber Analysis System (ICAS)**: A US DARPA initiative for documenting infrastructure to aid attack forensics and tactical cyber defense incident response.
- **Malware Attribute Enumeration and Classification (MAEC)**: A semantic framework for describing structured malware behavior.
- **OASIS Customer Information Quality (CIQ)**: A language for representing information about individuals and organizations.
- **Structured Threat Information Expression (STIX)**: A structured language specification for describing cyber threat information so it can be shared, stored, and analyzed in a consistent manner. This initiative embeds or ties to several of the other initiatives listed and is overseen by the OASIS Cyber Threat Intelligence Technical Committee (OASIS 2019).
- **Unified Cyber Ontology (UCO)**: A common ontology to unify and represent disparate cybersecurity domain knowledge.
- **Vocabulary for Event Recording and Incident Sharing (VERIS)**: A formal metrics framework for describing security incidents and their effects in a structured manner.

To provide context regarding the final category, semantic frameworks are structured, computer-readable specifications for ontologies. Ontologies are formal representations of complex semantic relationships in a specified domain. Ontologies can be considered taxonomies in which the relationships between multiple objects are formally specified. Ontologies support the encoding of complex conceptual models, specifying subject-verb-object relationships (Allemang and Hendler 2011; Antoniou et al. 2019; Gorelik 2019). Taxonomies specify tree-based, two-dimensional subject-object relationships. Ontologies can be the outcome of integrating multiple taxonomies where the relational overlaps between objects are formally defined.

Encoded ontologies can be used for various practical purposes such as data integration, search and retrieval, and automated decisioning and inference. In IT engineering and data science efforts, ontologies promote data quality, specify data integration points, and facilitate data presentation in support of analytics (Rogova 2019; Ruvinsky et al. 2019). In efforts to integrate and improve cybersecurity operations, ontologies provide a foundation for knowledge engineering-derived AI (symbolic AI), which facilitates interoperability and hybrid human-machine reasoning (Riley 2019a).

The creation and application of ontologies utilizing semantic frameworks in the cybersecurity domain is a specialized topic of growing interest (Bechor and Jung 2019; Dipert 2013; D'Amico et al. 2009; Georgescu and Smeureanu 2017; Kolini and Janczewski 2017; Nakamura et al. 2015; Obrsta et al. 2012; Park et al. 2009; Riley 2019a, b; Sikos 2018a, b; Sikos et al. 2018). Several of the cybersecurity specifications documented here are recognized by, embedded within, or coordinate with the OASIS STIX initiative (OASIS 2019).

2.3.6 Professional Status

Against the standards and measures of professional maturity, despite being relatively new compared to medical and scientific fields, the cybersecurity domain can be considered to have a relatively high level of maturity in the breadth of its professional trappings. The domain evidences central aspects of formalization and spans all types of institutional facilitation. A host of institutions and organizations frame best practices through an array of standards and frameworks. A broad range of professional training options and certificates are on offer. There is a mature and extensive *body of theory*.

Tables 2.5, 2.6, and 2.7 summarize a view of cybersecurity professional maturity and professional emergence according to the previously derived models (based upon Tables 2.3 and 2.4, and referring to the pie chart heuristic represented in Table 2.7).

Based on this high-level litmus test, the cybersecurity profession can be considered to have a high degree of maturity and a high level of professional emergence. However, several caveats are also raised when assessing according to the derived models. Three major factors appear as impediments to continuing professional maturity in the domain:

1. *Responsibility without power:* The authority and ethical stewardship vested in cybersecurity professionals can be regarded as weak and subject to challenges, particularly in commercial environments where conflicting incentives overrule prudence. While great responsibilities are thrust upon cybersecurity professionals, the reality is that the cybersphere is largely unregulated globally and locally. As such, commercial organizations, particularly, pursue self-interest often in advance of understanding the vulnerabilities and impacts of threats being assumed. Stewards are left to chase behind, attempting to plug vulnerabilities

Table 2.5 Assessment of professional maturity in the cybersecurity domain

1	Systematic body of theory	◗	Highly developed, diverse theoretical frameworks systematized for a range of functional purposes by a range of institutions. However, there is an increasing need for new theory to accommodate the rapid expansion of the domain and challenges associated with the pressure to automate
2	Authority and judgment recognized by client	◐	The profession is vested with de facto authority to protect against persistent, demonstrated threats. Professionals frequently hold formally recognized certifications, and major organizations increasingly recognize the authority of the domain at the executive level. However, mixed practical authority in the respect that guidance and prescriptions are taken by clients at will. In the context of guidance on hypothetical prevention with uncertain impact, many organizations choose to accept vulnerabilities in the context of pursuing commercial opportunities. This results in variable deference to the authority of cybersecurity stewards
3	Community sanctions authority	◐	Fairly strong in that the credentialing system is prolific. However, wavering to the degree, there are pronounced labor shortages that restrict the degree to which credentialism can be expected or mandated. Lack of resources results in situations where the requirement for credentials is often waived in deference to practicality, and there are many anecdotal indications that organizations balk at funding the often expensive and time-consuming credentialing process
4	Ethical code of stewardship	●	Organizations influence standards and culture through certifications, membership requirements, and codes of behavior. Legal, regulatory, compliance, and strong ethical strictures bound activities. Misbehavior leads not only to professional censure but often harsh legal repercussions
5	Professional culture supported by associations	●	Many formal institutions across academic, public, and private spheres. A broad range of formal and informal professional organizations supporting continuing education and networking through conferences

assumed by often unaccountable stakeholders. Recently there has been increasing agitation for greater regulatory oversight nationally and internationally, but, with some exceptions (i.e., EU's GDPR), this is yet mainly aspirational and evolving. Within organizations, the ascent of the Chief Information Security Officer (CISO) role has offered some hope of balancing perverse commercial incentives with enterprise-level accountability. As well, there are intimations that corporate boards may be held increasingly responsible for cybersecurity breaches.

2. *Rapidity of technical proliferation and change:* The largest pragmatic challenge facing the profession is the rapidity of change concerning the scale and scope of the underlying technologies and systems being safeguarded by the profession. As has been substantiated, the profession has been challenged in its mission to safeguard infrastructure given the exponential expansion of the cybersphere, prolific interconnection, and the accompanying growth of sophisticated and prolific threats. The professional maturity of the domain, being enmeshed and formalized amongst a complex of formal international institutions and organizations,

Table 2.6 Assessment of professional emergence in the cybersecurity domain

1	Active, focused interest from diverse participants	●	Well substantiated—an active professional community
2	Active professionals with associated job titles and roles	●	Entrenched and global
3	Emerging and informal training	●	Many informal training options
4	Informal professional groups	●	Many informal groups
5	Professional and industry literature	●	Many sources of professional and industry literature
6	Research literature	◑	Several dedicated academic research journals, although primarily focused on technical and engineering topics. Emerging topics are less represented in formal journal research
7	Formalized training	●	Many options cited for formal training
8	Formal professional groups	●	A broad number of professional organizations
9	Professional certifications	◑	There are a broad number of well-recognized certifications. One weakness is that the credentialing system is heavily privatized and disagreements on primacy exist
10	Standards bodies	●	Many standards bodies across a range of topics
11	Independent academic research disciplinary focus	◑	There are named degrees from undergraduate to doctoral levels. There are some critiques of disciplinary independence and focus given the dependence on multiple allied domains. Critiques include overemphasizing technical factors, the relevance of body of theory in the context of rapid change, and the degree to which programs prepare students for practical operational settings (Conklin et al. 2014)

Table 2.7 Key for interpreting categorical professional maturity status

Symbol	Interpretation of categorical professional maturity[a]
○	Little or no evidence of meeting the criteria denoted (~0%)
◔	Relatively low or immature level regarding the criteria (~25%)
◑	A mid-point of maturity, though still lacking in many regards (~50%)
◕	Fairly robust and mature regarding the criteria, with some gaps (~75%)
●	Strong and robust maturity regarding the criteria denoted (100%)

aThe pie-chart guide is provided as a visual heuristic to summarize to what degree the professionalization category has been met on a rough scale based on evidence gathered and summarized in the related table cannot scale and evolve at the speeds necessary to keep pace with the manifestation of new threats. The proliferation of networked digital technology has led to a situation where traditional paradigms and cybersecurity approaches are failing (Lehto and Neittaanmaki 2015; Leuprecht et al. 2016). In the wings is a rising practical pressure to rethink definitions and methods in the domain (Conklin et al. 2014), most certainly to focus more heavily on formal decision management and automation, which otherwise are central topics of concern to CSDS.

3. *Expanding breadth and disciplinary diffusion:* Debates concerning the academic disciplinary status and relevance of information security have been cited (Benbasat and Zmud 2003; Conklin et al. 2014). As has been remarked at length, the field is experiencing dramatic and rapid expansion due to the general proliferation of digital technology in modern society (Craigen et al. 2014). This leads to a risk of diffusion of focus, where cybersecurity concerns are everywhere, and thus nowhere, for lack of a systematic locus. The seeming omnipresence of digital interconnection risks provoking an identity crisis in the profession. Many of the profession's principles were defined during a now outmoded period in which digital proliferation was not yet prolific.

The caveats observed otherwise serve to underscore the drive towards an emerging data-driven paradigm in the domain. This impetus addresses internal and external pressures, namely, the need to refresh skills and methods to keep pace with changing dynamics and the need for greater efficiency to operate in environments of increasing scale and complexity. In the context of the *trait approach* view of professionalization, the *body of theory* associated with the cybersecurity domain is being pressured to evolve out of stark necessity.

2.4 Data Science as a Profession

2.4.1 Origins and Definitions

Data science is a popular, prolific, and fast-expanding field contemporary to this inquiry. Since being hyperbolically dubbed "the sexiest job of the twenty-first century" in an eponymous 2012 Harvard Business Review article (Davenport and Patil 2012), the domain has exploded in commercial and academic mindshare and influence. Data science has become a highly visible professional phenomenon across a range of industries. It is popularly, and controversially, perceived as a panacea to improve decision-making in increasingly complex and diverse environments, from manufacturing, public transport, and finance to advertising, sports, and politics. Here we grapple with assertions of its suitability in the cybersecurity domain.

The specific origins and definition of the term data science are contingent upon the context invoked. Debates and confusion in practitioner settings are common, as the central definitions are various and conflicting, yet also overlap (Bloor 2013; Mongeau 2015; Press 2013b). This has led to industry (Press 2013a) and academic

(Acito and Khatri 2014; Bayrak 2015; Keller 2013; Liberatore and Luo 2010) attempts to trace the broader historical, disciplinary, and ontological foundations. However, as a phenomenon still heavily propelled by commercial markets and interests, attempts at a focused definition resist tight academic circumscription.

As to the current definition of data science, clarity depends upon whether one considers the term to imply (1) a distinct *academic field and research discipline*, (2) a new *profession with an accompanying set of practitioner skills*, (3) a programmatic *organizational management paradigm*, or (4) a *commercial marketing umbrella* for a set of technologies and software. The challenging status quo is that the invocation of the term often entails overlaps between multiple theatres.

Tracing provenance is instructive to improving understanding. The emergence of data science is a relatively recent outgrowth and manifestation of what has been termed the *analytics movement* (Liberatore and Luo 2010). This movement manifested as a business practitioner hybridization of several specialized, overlapping academic domains, including computer science, business intelligence, statistics, econometrics, financial analysis, machine learning, and operations research (Rose 2016).

The analytics movement gained visibility through a series of influential popular management press books and articles beginning in 2007. The works of T. Davenport (Davenport and Harris 2007; Davenport et al. 2010), along with a series of articles in the Harvard Business Review (Blenko et al. 2010; Boyd and Crawford 2014; Davenport 2009, 2013; Davenport and Patil 2012), and the MIT Sloan Management Review (Kiron et al. 2014; Kiron and Shockley 2011; Kiron et al. 2011; LaValle et al. 2010; LaValle et al. 2011) served to establish analytics as a major topic in popular applied management.

Highlighting practical organizational decision-making aspects, Davenport and Harris defined analytics as "the extensive use of data, statistical and quantitative analysis, explanatory and predictive models, and fact-based management to drive decisions and action" (Davenport and Harris 2007). Another definition framed analytics as "the use of data and related insights… to drive fact-based planning, decisions, execution, management, measurement and learning," noting that analytics may be "descriptive, predictive or prescriptive" (Kiron et al. 2011).

Turning to technical and engineering aspects, the terms *analytics* and *big data* became increasingly intertwined in allied publications starting in 2011. This denoted a growing focus on performing analytics on particularly large, robust, and/or extensive datasets (Baesens 2014; Brynjolfsson and McAfee 2012; Davenport 2014; Gerhardt et al. 2012; Henke et al. 2016; Hey 2010; Hey et al. 2012; Manyika et al. 2011; Singh and Singh 2012). A *three Vs* model was invoked, namely, that big datasets can be "big" in terms of *volume* (amount), *variety* (range of data types), and *velocity* (speed) (Gandomi and Haider 2014; Kelleher and Tierney 2018). This was expanded subsequently in some versions to include up to *seven Vs*, adding *veracity* (quality of and trust in data) (Chen and Zhang 2014; Jin et al. 2015; Morabito 2015; Voulgaris 2017), *variability* (shifting context), *visualization* (ability to summarize visually), and *value* (ability to extract value) (McNulty 2014). The focused framing of big data in multiple contexts attested to the growing importance of data generally in enterprise settings and a desire to qualify the concept of big data in more specific terms.

The big data phenomenon emerged from new technical approaches to storing, retrieving, and searching large datasets, particularly Map Reduce, Apache Hadoop, the Apache Hadoop ecosystem of technologies and tools, and the subsequent proliferation of derivative cloud-based services (Boyd and Crawford 2014; Chen and Zhang 2014; Kambatla et al. 2014; Loshin 2013; Loukides 2011; Morabito 2015; Sivarajah et al. 2017; Strawn 2016; Zicari 2014; Zicari et al. 2016).

The terms data science, analytics, and big data are frequently invoked interchangeably in practice (Baesens 2014), and there are unresolved debates concerning the distinctions between the three (Press 2013a, b; Rose 2016), resulting in critiques and confusions (Ekbia et al. 2015). Here the term *data science* will be invoked as the professional application of *analytics* methods with the assumption that big data analysis and engineering are often a central component.

Data science, analytics, and big data have all been the focus of intense commercial interest and are associated with a massive industry purveying hardware and software solutions, open-source projects with commercial partners, and consulting services. This has also propelled and encouraged big data as an umbrella term for programmatic enterprise efforts to manage and analyze very large sets of data (Barlow 2013; Chalvatis 2017). This phenomenon of market-propelled exuberance and its effects are not dissimilar from the discussion previously raised with reference to the cybersecurity industry. Namely, informational distortions are perpetrated through the marketing of solutions and services, which tend to proliferate confusions while attempting to simplify, excite, and sell (Schutt and O'Neil 2014). Here, we can refer once again to the Gartner technology hype cycle raised earlier (Fig. 2.3).

Gartner removed big data as a focused trend in its 2015 hype cycles for advanced analytics and data science (Linden 2015) and emerging technologies (Walker and Burton 2015). According to Gartner, this was not because the topic diminished in importance. Rather, it was observed that big data was seen as being increasingly subsumed in most all substantial data science and analytics efforts (Willis and Burton 2015). Similarly, data science and analytics are themselves not tracked by Gartner as granular categories but represent an umbrella term for a proliferation of associated innovations, efforts, and technologies. The specific topic of big data engineering and storage is now more commonly associated with the term *data lake*, which has emerged as a catch-all for large enterprise big data storage and analysis infrastructures and architectures (Gorelik 2019; Heudecker and Ronthal 2018). Meanwhile, the term "AI" has emerged as a popular marketing umbrella to encompass all these trends.

The sheer diversity of domains to which data science is now applied has and continues to diversify and expand. In Gartner's representation, some applications of data science peak and decline on the respective hype cycle. New applications continue to evolve, focused security-related applications related to CSDS being an exemplar. Despite commercial distortions, substantial academic, professional, and methodological underpinnings reinforce the stance that the data science field, agglomerative and conceptually permissive as it is, marks a lasting paradigm shift in management and organizational decision-making (Bayrak 2015; Liberatore and Luo 2010). The assertion here is that data science, analytics, and big data, by any other names (e.g., under the umbrella of "AI"), embody lasting and impactful sociotechnical advancements, as opposed to transient fads.

2.4.2   Organizational Job Role

Taking a conciliatory stance by embracing multiple definitions, data science can be variously said to refer to an evolving interdisciplinary academic umbrella, a set of practitioner algorithms and methodologies for applied computational statistics, a programmatic organizational approach for computationally intensive decision making, and/or a professional nom de guerre for all or some of the previous frames, with implied practitioner job tasks and roles associated.

It is in this latter sense, that of a professional phenomenon and job role, where we will focus here, although the full scope of definitions will be touched upon and interrogated in the broader course of this inquiry. As the focus here concerns practitioners in an MIS context specifically, the term data science will be used pragmatically to denote a de facto umbrella for a growing profession that inherently involves applying (big data) analytics.

For all the popular attention extended to data science, fundamental debates concerning the discrete nature of work performed (i.e., core job tasks) persist. Where debates tentatively agree, key themes and boundaries can be framed. Identifying common ground is facilitated by distinguishing teleological (goal-oriented) versus ontological aspects (nature in terms of origins). This bifurcation mirrors the classical etymological distinction between *technê* (skill and craft in practice) and *logos* (structured reason) supporting *epistêmê* (knowledge) (Tulley 2008).

In terms of ontological aspects as related to epistemological underpinnings (*logos* and its extension to *epistêmê*), extending the previous historical overview, data science as a profession began to be popularized in 2012. Data science was then used to describe what data analytics practitioners did, a job or role, which was at that point seen as a multidisciplinary agglomeration of tools and techniques (Davenport and Patil 2012). The practitioner domain data science was a byword for conducting data analytics, which itself encompassed a grab-bag of statistics, machine learning, etc.

As the term data science is frequently used in reference to a job role, it lexically implies a process of discovery by invoking the term *science*. Analytics has more traditionally been focused on implementing practitioner methodologies. In contrast, data science ostensibly projects more broadly to embrace the full scope of data management and the utilization of analytics within the scope of a scientific process.

Concerning teleological aspects as related to practice (*technê*), whether framed under the conceptual umbrella of analytics, big data, or data science, the unifying goal is a critical interest in applying computational technologies and methods to the analysis of robust sets of data in order to surface value-creating insights to guide decision-making (Blenko et al. 2010). As data science has proliferated as a profession, allied domains and thus goals have been added to the discipline. The field is considered to include aspects of data engineering, visualization, text and semantic analytics, programming, systems development, and, now increasingly, expert systems and AI, including deep learning methods.

In the context of enterprise computing and MIS, in aggregate, the unified goal is a desire to not only strengthen organizational decision-making quality in the face of growing business complexity and uncertainty but to realize core competitive value creation (Kiron and Shockley 2011; Kiron et al. 2011). Management research advocates analytics as a competitive differentiator and thus a driver of economic value (Davenport 2009; Davenport et al. 2010; Kiron and Shockley 2011; Kiron et al. 2011; LaValle et al. 2011). Comparative organizational research has asserted that higher levels of organizational analytical maturity are correlated with value creation and competitive outperformance (Henke et al. 2016; Kiron et al. 2011; LaValle et al. 2011; Shanks et al. 2010).

A key driver that has propelled the analytics movement is improved decision-making leading to increased profits and reduced costs for organizations (Albright et al. 2011; Blenko et al. 2010). This encompasses a claim that data science, the umbrella for the application of advanced analytics, contains the means and promise to improve a broad range of core firm functions such as strategy, financial planning and analysis (FP&A), marketing and advertising, manufacturing and distribution, supply chain management, and personnel management (HR functions). At root is the claim that structured, evidence-based, quantitative decision-making with a scientific basis outperforms individual judgment and intuition, which is subject to biases. This latter claim has been tied in analytics management research to a growing body of behavioral economic research (Bonabeau 2003; Kahneman 2011; Kahneman and Klein 2009; Lovallo and Sibony 2010a).

To ground the analysis of data science to goal-focused tangibles, attempting to navigate the turbulence between commercial and academic frames, it is useful to extrapolate a set of discrete job tasks and skills associated with data scientists. A definition offered by the US Bureau of Labor Statistics frames key tasks associated with the data science professional role:

> Develop and implement a set of techniques or analytics applications to transform raw data into meaningful information using data-oriented programming languages and visualization software. Apply data mining, data modeling, natural language processing, and machine learning to extract and analyze information from large structured and unstructured datasets. Visualize, interpret, and report data findings. May create dynamic data reports. (Bureau of Labor Statistics 2018)

Invoking the refinement of information as a set of process-driven techniques frames data science itself as a process. This functional view allies data science with MIS, which hosts the subfields of data mining, decision support systems (DSS), and business intelligence (BI) (Davenport and Harris 2007; Diaz 2018; Laudon and Laudon 2016; Sharda et al. 2014; Skyrius et al. 2013). Given this association, it is useful to apply the classical MIS tripartite framework of organization, process, and technology in order to understand how data science as a job role fits functionally within an organization. A non-exhaustive outline of functions aligned to the MIS framework (Fig. 2.9) suggests the following breakdown of associated tasks:

Fig. 2.9 Data science in
the MIS tripartite model

Technical
- Extracting, transforming, and loading data (ETL)
- Programming related to data exploration, cleaning, and validation
- Operationalizing the streamlined processing of large volumes of data
- Designing and deploying statistical and machine learning mechanisms
- Developing data visualization methods and tools such as dashboards

Process (Methods)
- Applying an iterative life cycle approach to acquiring, selecting, exploring, assessing, treating, validating, modeling, and interpreting data
- Statistical analysis and diagnostics
- Extracting and select features/feature engineering
- Selecting, testing, and applying a range of specialized analytics methods
- Analyzing, comparing, using, and validating various algorithms and modeling techniques
- Evaluating and formally validating model performance
- Developing and testing prototypes for the systematized analysis of data
- Implementing solutions into production

Organizational
- Identifying, framing, and scoping business problems suitable for analytics
- Collaborating cross-functionally to frame, develop, validate, operationalize, interpret, and communicate analytical methods and models
- Framing analytical outcomes desired (describe, explain, predict, optimize, and/or influence)
- Determining the scope of analytics models (technical, economic, and/or behavioral)
- Understand and apply domain expertise in the context of data analysis
- Scoping and designing data-focused systems and solutions

- Facilitating the design and implementation of big data management, real-time processing, and analytics tools and infrastructure
- Collaborating to monitor and maintain production models and to address model- and data-related issues
- Collaborating and communicating with experts and colleagues outside the organization
- Organizing and guiding team-based data analysis projects
- Teaching, training, and socializing analytics approaches

This breakdown of data science functional tasks was extrapolated from a range of perspectives framed in academic and professional literature (Barlow 2013; Harris et al. 2013; Kelleher and Tierney 2018; Loukides 2011; Magoulas and King 2016; Provost and Fawcett 2013; Schutt and O'Neil 2014; Skiena 2017; Svolba 2017; Voulgaris 2017).

Observations derived from this overview emphasize four aspects of data science as a job role:

1. *Collaborative dependencies:* There is an emphasis on organizational collaboration to drive various processes during the analytical inquiry. These processes are assumed to be backed by a vested trust in the data scientist to interpret and maintain standards for scientific inquiry. The nature of dependencies positions the data scientist both as a service provider and as a steward of scientific rigor, framing the potential for stakeholder incentive conflicts.
2. *Rigor without power:* It is notable that although there appears a great deal of titular theoretical, methodological, and technical deference to the data scientist as expert, decision-making rights and control of firm resources are not explicitly granted. This would not be the case if the data scientist were vested with a management role, such as a Director of Data Science, but there are no clear guarantees that there is a domain-specific management umbrella for data science in any particular organization. A data scientist more typically either reports up to a Center of Excellence (CoE) as an internal consultant, is situated in the IT department, or operates from a functional business group such as marketing or finance (Kiron et al. 2014). There are many aspects of data science tasks that imply a responsibility for rigor (quality), but not necessarily power or control within the firm to mandate the allocation of resources or conformity to analytical guidance.
3. *Disciplinary breadth*: The sheer conceptual breadth of the domain in terms of expectations concerning responsibilities and duties is startling. This otherwise highlights observations that the singular data scientist is tantamount to a "unicorn" (and is thus mythical). It has been observed that the data scientist role is not, in fact, singular in most mid- to large-sized organizations, but is more commonly a set of specialized roles with different expertise taken on by a team in a greater effort (Harris et al. 2013; Magoulas and King 2016).
4. *Computational emphasis:* Although there are references to rigor and integrity in analysis, the central scope of deliverable work emphasizes engineering and technical aspects, including potentially deploying systematized IT solutions.

This profile frames several potentially conflicting incentives, both within the role of data scientist and as a role in an organization. A theme that will return and is otherwise central is the danger of vesting responsibility for scientific rigor in such a role, yet not necessarily linking the role to decision rights and control over firm resources. Namely, scientific rigor imposes an inherent overhead, and when conflicts occur concerning time to delivery, resource management, and quality, a key danger is that the vested data scientist, as a steward, is potentially over-ruled concerning efforts to maintain and safeguard standards surrounding scientific rigor. Otherwise, this centrally calls into question to what degree rigorous scientific inquiry can be conducted or expected in non-research institutions, particularly commercial organizations. There are inherent firm pressures to reduce rigor and quality in the name of time to delivery.

2.4.3 Academic Programs and Research

The rapid popular emergence of data science from commercial practice has left rigorous academic qualification lagging (Gandomi and Haider 2014; Jin et al. 2015; Sivarajah et al. 2017). Academic viewpoints lack strong concordance regarding the disciplinary boundaries surrounding the field and debates are rife. As a practice which claims lineage variously from the disciplines of statistics, computer science, business intelligence (BI), econometrics, machine learning, and operations research (Liberatore and Luo 2010; Press 2013a; Schutt and O'Neil 2014), there are disagreements as to whether any particular progenitor deserves privileged status.

Computer science has driven innovative data science engineering and methodological underpinnings, and so can claim a key role in proliferating praxis. Big data technologies and systems, analytics tools and solutions, and several key machine learning innovations are largely the results of computer engineering efforts. Operations research has also asserted a stake (Liberatore and Luo 2010). The field of statistics has additionally advocated itself as a central disciplinary locus point for data science (Donoho 2015).

Ongoing debates between computational and statistical positions on data science are insightful in understanding struggles to frame data science as an emerging academic discipline. In the broadest purview, data science implies the application of computational technology and methods to the analysis of data in the service of scientific inquiry. From this broader vantage, data science and the analytics movement extend back to the origins of modern computing and the emergence of computationally intensive methods in the field of statistics. The statistics perspective on data science positions computational tools and algorithms as facilitators, rather than independent agents. Conflicts appear to the degree that machine learning advocates assert the ability of algorithms to produce valid predictions absent explanatory theory.

The disciplinary affiliations and locus of data science are still a matter of active debate. However, this has not forestalled a rush to frame degree offerings. The intense popular and commercial interest in data science has led to many new academic certificate and degree offerings. A 2019 listing hosted by the Institute for Advanced Analytics (IAA) contains nearly 250 university programs offering master degrees in analytics and data science in the United States alone (Institute for Advanced Analytics 2019). Another resource, DataScience.Community, lists over 600 global university programs associated with analytics and data science, including online, professional certificate, and executive programs (DataScience. Community 2019). Although there are several emerging data science PhD programs, academia still lacks a hard consensus concerning the disciplinary boundaries circumscribing such a research degree (Donoho 2018).

There are a growing set of academic research journals and proceedings focused on data science and analytics topics. Focused areas of interest include data science, analytics, data mining, machine learning, and big data. In the main, the most recognized journals focus on methodological and/or technical topics. Organizational management views on data science are primarily addressed in management practitioner journals such as the *Harvard Business Review* and the *MIT Sloan Management Review*, as cited earlier. Research focused on data science in business research journals on strategy, accounting, finance, marketing, general management, or HR is relatively muted, suggesting a lacuna (Salmon 2019).

Complicating the ability to refine data science into a coherent academic discipline is the tendency to co-opt new concepts while leaving strict definitions unclear. For example, artificial intelligence (AI) has recently become a popular byword for machine learning (ML). ML is historically considered a branch of the academic discipline of AI, itself a focused field within computer science. However, in a commercial context, AI is used increasingly to imply a broad movement to package multiple technical innovations into platforms that evidence focused human expertise, often simulating complex decision-making and judgment. Confusions occur as the use of the term AI might variously be intended to refer to expert systems, natural language processing (NLP), machine learning, or a data science initiative.

2.4.4   Professional Organizations and Certifications

In comparison to the cybersecurity profession, the relatively young field of data science offers fewer professional associations and certificates, although offerings are fast-growing. Apart from a few select allied professional organizations that have re-framed themselves in the analytics space (i.e., INFORMS), training and certifications range widely in terms of quality and industry acceptance. Offerings range from intensive data science "boot camps" to formally sponsored university certificate programs in data science.

Areas of focus associated with certificate offerings and professional organizations mirror the breadth of the data science umbrella, with specialties in data engineering, big data, cloud solutions, analytics, and data science explicitly, along with a broad array of specialty programs in focused areas such as marketing analytics, financial analysis, operations management, and fraud analytics.

Public and Private Professional Certificates (Non-exhaustive Examples)

- **Harvard University**: Offers a Professional Certificate in Data Science and a Data Science Certificate (Harvard Extension School).
- **John Hopkins University**: Offers a Data Science Specialization through online learning provider Coursera.
- **Simplilearn**: A certification training provider offering a Data Scientist Master Certificate program.
- **University of Chicago**: Offers the Machine Learning for Analytics MasterTrack Certificate and others.
- **University of California Berkeley**: A range of data science and analytics offerings for executives and practitioners, including a Certificate Program in Data Science via UC Berkeley Extension.

Industry and Professional Bodies

- **ACM**: The Association of Computing Machinery (acm.org), the world's largest professional and scientific computing society, offers a range of articles, research, lectures, and conferences related to data science and allied topics.
- **DASCA**: The Data Science Council of America (dasca.org) is an independent third-party standards and credentialing body for the data science profession offering Big Data Engineering, Big Data Analyst, and Data Science certifications.
- **Data Science Association**: A nonprofit professional association (datascience-assn.org) promoting best practices, accreditation programs, and a data scientist ethical code.
- **IAPA**: The Institute of Analytics Professionals of Australia (iapa.org.au) offers an Analytics Certificate course.
- **iCAS**: The Casualty Actuarial Society Institute (thecasinstitute.org) is a professional organization offering the Certified Specialist in Predictive Analytics (CSPA) credential.
- **IEEE**: The Institute of Electrical and Electronics Engineers (ieee.org) is a well-respected global organization that publishes professional and academic articles on engineering and technology topics, including data science, analytics, and big data.
- **INFORMS**: The Institute for Operations Research and the Management Sciences (informs.org) is a global professional association for operations research and analytics professionals. INFORMS offers the Certified Analytics Professional (CAP) certification, the first widely recognized credential in the analytics field.
- **International Statistical Institute**: This institute is a professional organization (isi-web.org) for the field of statistics. The ISI hosts a declaration on professional ethics (International Statistical Institute 2010).
- **OR Society**: The Operational Research Society (theorsociety.com) is an international professional society for the field of operations research.

2.4.5   Frameworks

Frameworks and models in the data science domain abound. Analytics models themselves are techniques for processing data into insights and are an inherent aspect of the field. Analytics models can be seen as procedural, data-driven vehicles for deriving insights and can be framed as active decision agents in some cases (Latour 2005; Morrison and Morgan 1999). The notion of models having independent agency vis à vis theory creation is debated in the field of statistics (Breiman 2001; Donoho 2015, 2018). However, it is possible, and increasingly common, to build and operationalize machine learning decision models with semi- or fully autonomous agency.

The range of analytics models (techniques applied by practitioners) is a substantial topic, and a comprehensive overview is not currently in scope. For background reference, some examples of well-known techniques cited in other works include regression analysis, multivariate statistics, anomaly detection, time-series analysis, forecasting, longitudinal analysis, association analysis, process analysis, graph analytics, geospatial analysis, survival analysis, A-B testing, optimization, simulation, probability analysis, Bayesian analysis, neural networks, natural language processing (NLP), text analytics, and machine learning methods (Baesens 2014; Baesens et al. 2015; Bartlett 2017; Hastie et al. 2009; Leodolter 2013; Maheshwari 2015; Provost and Fawcett 2013; Schutt and O'Neil 2014; Skiena 2017; Svolba 2017; van der Aalst 2016; Voulgaris 2017).

Machine learning models specifically occupy a great deal of focused attention in the data science domain. Machine learning at times requires less primary data preparation effort as many methods do not make assumptions concerning the parametric nature of generated data—data generation is treated as an unknown process. These methods are popular in commercial environments due to their flexibility and efficacy in extracting results from large and varied datasets. The popularized notion that such models are cost-effective to prepare, deploy, and maintain is an attractive aspect, although bypassing basic calibration efforts leads to questionable validity.

Machine learning models are generally complex and computationally intensive, albeit most are at base sophisticated algorithmic composites of primary statistical, multivariate, and linear techniques. Machine learning models are considered to fall into four major types, listed here with representative example implementations (Table 2.8).

Concerning classification (supervised) approaches, labeled data captures a prior outcome or observation as associated with a multivariate vector, such as a case of observed fraud or confirmation that a particular customer purchased an item. The labels, along with supporting data, are used to train an algorithm to recognize similar cases and to make predictions when the outcome is not known.

Distinct from analytical models, meta-models, and frameworks are high-level representations of methods and processes intended to categorize, socialize, organize, structure, or manage analytical efforts. Meta-models can be applied to guide organizational efforts to build, test, and deploy models. These can be considered professional best practices, or body of theory, for the management of analytics in

Table 2.8 Examples of machine learning algorithms categorized by type

Unsupervised	Supervised	Semi-supervised	Reinforcement
Extracts patterns from datasets when little is known, resulting in suggested segments	*Using "labeled" examples, the algorithm is trained to classify new data*	*Hybrid unsupervised and supervised using a small set of "labeled" data*	*A method for active training based on the maximization of a reward*
• Clustering (hierarchical, k-means, mixture models) • Anomaly detection—LOF • Neural networks (autoencoders, deep belief nets, GAN, SOM) • Association rules • Latent variable models (expectation-maximization, PCA, SVD)	• Decision trees • Linear regression • Support vector machines (SVM) • Neural networks • Naïve Bayes • Deep learning • K-Nearest neighbor • Linear discriminant analysis • Similarity learning	• Generative models • Low-density separation • Graph-based methods • Heuristic approaches	• Q-Learning • Monte Carlo based • SARSA • Temporal difference (TD) • Deep adversarial networks • Deep Q network

organizational settings. As well, meta-models assist in promoting understanding concerning the types of analytical techniques available and their proper use.

Four central types of analytics, along with the central questions they address, have been frequently cited in practitioner research: descriptive—what happened?, diagnostic—why is it happening?, predictive—what are trends?, and prescriptive—how to optimize? (Arindam Banerjee et al. 2013; Chandler et al. 2011; Hostmann 2012; Sallam and Cearley 2012; Sivarajah et al. 2017). T. Davenport has broadened this overview more recently to include cognitive analytics, the combination of methods to automate expert tasks (Sarkar 2016). Beyond this, the proliferation of analytics has increasingly raised the importance of data quality and the need for visualizations in order to summarize complex results. A composite representation of the range of analytics methods is offered in Fig. 2.10 (Mongeau 2014).

In this representation, semantic analytics is positioned as an additional area dealing with context and meaning. This domain is associated with addressing questions of complex interpretation and representation. In contradistinction, cognitive analytics as a concept implies bundled aggregations of multiple methods represented here and is thus not shown.

Beyond the types of analytical models (techniques) and methodological categories, another meta-aspect of analytical models concerns the scope of analytical inquiry in terms of the phenomenon being analyzed. Analytic methods can be applied across *techno-economic-behavioral* phenomena: in technical settings, e.g., factories, transport infrastructure, or computer networks; in economic frames, e.g., financial risks, or profits/costs; and/or in behavioral contexts, e.g., as associated

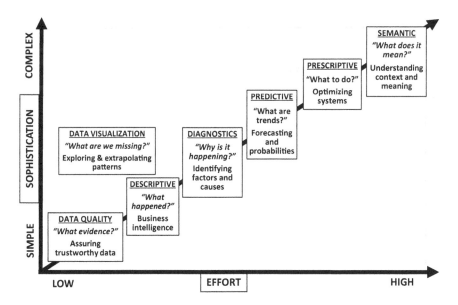

Fig. 2.10 High-level overview of analytics methods

with how agents or people make decisions in aggregate under different conditions such as political, consumer, or risk-reactive behavior.

As data science has proliferated into broader domains, the scope of examined phenomena often spans across *techno-economic-behavioral* frames. This allows for a broader purview in an inquiry but also creates more complex models which encompass wider spans of variability and uncertainty. A sufficiently complex set of linked analytical models covering a broad domain becomes, in a sense, a type of simulation of complex phenomena, in that the composite analytical model is an active, modeled representation of real-world phenomena (Law 2007; Robinson 2004; Shannon 1975; Weisberg 2013).

Composite techno-economic-behavioral analytical models of phenomena have become more prevalent in areas involving complex real-world systems where the stakes for success or failure are high. Examples include military applications (McDermott et al. 2013), medical management (Phillips et al. 1998), and energy forecasting (Offer et al. 2011).

Techno-economic-behavioral model scope has been raised in the cybersecurity domain as well, observing the increasing embedding of technical dependencies in "real-world" processes (Robinson 2004; Rosenzweig 2013; van den Berg et al. 2014). More generally, model scope in terms of phenomenological breadth has been raised as an inherent aspect of big data (Chen and Zhang 2014; Chen et al. 2012; Ekbia et al. 2015; Gandomi and Haider 2014; Jin et al. 2015; Liu et al. 2016) and as a growing aspect of IT research in the context of the general proliferation of and importance of networked computer technology (Banker and JKauffman 2004; Lim et al. 2013).

Specifying model scope helps focus analytical framing and development. The model scope (area of use) implies particular data sources, techniques, and outcomes to ground analytic problem specification. By cross-referencing analytics methods, one is able to specify the target analytical use case. As an example, Fig. 2.11 positions different cybersecurity analytical use cases specified by cross-referencing analytical methods (from Fig. 2.10) and model scope (usage domain). This represents a novel conceptual tool for facilitating analytical model development.

2.4.6 Professional Status

The professional status of data science is more contentious than that of cybersecurity. While many of the formal aspects of a profession have or are developing, the sheer rapidity of data science's emergence and growth has attracted detractors and critiques. Despite its commercial popularity, an ongoing debate exists concerning the degree to which data science can be considered an independent, mature profession. While there is a practical reality, namely, many named professionals working with the job title *data scientist*, under deeper examination, many of the other trappings surrounding professional fields are less well developed.

Critiques of data science have asserted that it is less a profession and more a collection of commercialized methods associated with data broadly, which is to

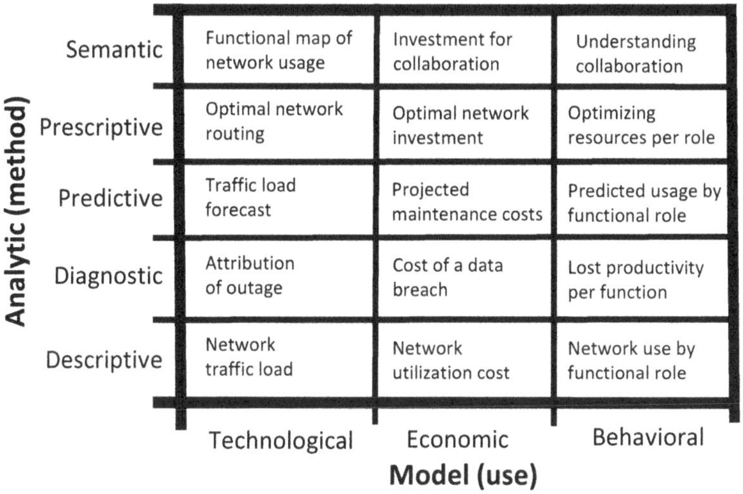

	Technological	Economic	Behavioral
Semantic	Functional map of network usage	Investment for collaboration	Understanding collaboration
Prescriptive	Optimal network routing	Optimal network investment	Optimizing resources per role
Predictive	Traffic load forecast	Projected maintenance costs	Predicted usage by functional role
Diagnostic	Attribution of outage	Cost of a data breach	Lost productivity per function
Descriptive	Network traffic load	Network utilization cost	Network use by functional role

Analytic (method)

Model (use)

Fig. 2.11 Analytical methods cross-referenced with model (usage) for IT domain

say a good breadth of general IT professional work. Another challenge is that of confusion and diffusion related to the tendency to chase buzzwords and trends framed as marketing terms. As an example, AI and deep learning have recently been heavily promoted as important techniques centrally associated with data science. However, their relevance in a professional context is much less well defined, and the long-term significance of these trends has yet to be fully understood. Data science, being led heavily by commercial marketing trends, tends to have a magpie aspect, whereby it quickly latches onto and envelops shiny new trends in IT generally.

Tables 2.9 and 2.10 provide a summary maturity assessment of the data science profession (Mongeau 2019b). The assessment framework is based upon the professionalization model derived earlier in Tables 2.3 and 2.4, and utilizes the pie-chart assessment criteria specified in Table 2.7.

Whereas cybersecurity achieves many of the hallmarks of a profession and has marked itself as a focused area of academic study, data science remains quite a broad designation. A central question is naturally posed as to what degree cybersecurity data science can progress to professional status while the data science field remains broad and disputed concerning its level of formalism.

On the side of advocacy, individuation and maturation in specialized data science domains has been robust. As exemplars, the specialized areas of marketing analytics, financial analysis, fraud analytics, and operations management have demonstrated strong professional individuation. These examples demonstrate a professional maturation path through specialized fields. In these cases, specialty fields individuate through the application of analytics to focused domain challenges.

This principle has been observed in research on the organizational implementation of analytics. As opposed to a path whereby general analytics experts identify focused applications, Kiron et al. identify a *specialized path*, whereby analytics

Table 2.9 Assessment of professional maturity in the data science domain

1	Systematic body of theory	◑	In his seminal article, Cleveland advocated the central importance of theory to data science (2001). The data science umbrella hosts a highly sophisticated and diverse set of interdisciplinary methods spanning a range of academic fields. However, critiques linger concerning the ability to rigorously validate methods (e.g., challenges related to primary data quality, correlation not causation, or the inherent opacity of more complex methods). A central challenge is the sheer breadth of the domain. Whereas breadth might otherwise be considered a strength, it also poses challenges related to complexity and the ability to validate models in straightforward and replicable ways. Whether due to lack of overhead or inherent model opacity, difficulty substantiating rigor creates doubts concerning efficacy. These doubts can lapse into questions of ethical integrity. Theoretical integrity in the domain is methodologically diverse yet practically is often shaky due to issues surrounding rigor and ad hoc or at-will standards for organizational compliance
2	Authority and judgment recognized by client	◐	Data science currently evidences a great deal of titular respect and authority due to popularity. However, this may very well prove to be transient and superficial (for instance, an aspect of the technology hype curve generally). Management research attests to the practical difficulties of promulgating data science-derived guidance into organizational decision making (Chalvatis 2017; Lazer et al. 2014; Sivarajah et al. 2017). Technical decision guidance can be gamed or perverted by vested interests, with a lack of consensus concerning formal organizational governance principles. The degree to which decision guidance resulting from applied data science practice is incorporated in organizations is anecdotally variable: management can decide to ignore, reframe, or modify results to meet its own needs or interpretations. A substantial critique is that time and resource pressures do not easily facilitate rigorous scientific validation in commercial settings. Detaching data science from scientific rigor undercuts the authority of recommendations, relegating assertions to informed opinions from technocratic experts rather than objective outcomes resulting from rigorous, peer-reviewed scientific inquiry
3	Community sanctions authority	◐	Data science, being primarily a free-market movement and typically detached from formal strictures, lacks traditional bodies for enforcing ethical and methodological rigor. The exceptions are highly specific to regulated industries, such as pharmaceuticals, financial services, and public services, and in these cases, strictures typically pre-date the data science movement. Despite many options, there are a lack of broadly accepted, universally agreed professional certificates. Diversity and breadth outstrip agreements and consensus on quality. Deeper acceptance of particular professional certifications as standards would promote the notion of accepted practice licenses, which would help advance assertions of professional authority. Although a rough consensus exists that STEM-associated advanced academic degrees are well-regarded, there is a lack of standards regarding the application of the diverse fields composing data science to applied practice. Formal, as opposed to collegial, authority is lacking (i.e., independent certification bodies maintaining methodological and ethical standards). Universities worldwide have responded to the great interest in the domain with a raft of new advanced degree offerings as well as sponsored certificate programs. However, standards for such degrees and certifications lack universal consensus. Debates persist concerning the discrete nature and disciplinary focus of the domain

| 4 | Ethical code of stewardship | ◑ | There have been attempts to define ethical guidelines, for instance, the professional code of conduct by the Data Science Association (2019). There have also been recent attempts to profile the dangers and gaps of absent ethical standards, for instance, in O'Neil's book *Weapons of Math Destruction* (2016). Further, ethical standards are clear and well-socialized in the statistics discipline (International Statistical Institute 2010). However, the resident challenge is immaturity in disciplinary rigor and acceptance/enforcement through professional organizations. Policing of standards are poorly supported to non-existent. Ethical stewardship seems mainly to be bound to a loose agreement to apply scientific methods in practitioner efforts. However, given the conflicting pressures for timely results versus the inherent costs of academic-level validation, such claims are dubious, especially in commercial settings. This is a common critique concerning the invocation of the term "science" as associated with the profession—there are loose and dubious grounds for maintaining scientific rigor in the profession, apart from self-policing. Self-policing is threatened when commercial pressures supersede attempts at rigor. Greater institutional support and acceptance for a clear set of standards is needed, and this is forestalled to the degree there are continuing debates concerning the boundaries of the discipline |
| 5 | Professional culture supported by associations | ◑ | Several professional societies have responded to the swelling interest in this domain by accommodating and adopting analytics and data science into their purview. Examples include INFORMS and The OR Society (operations research), IEEE (engineering), and ACM (computing/IT). However, there is room for improvement and expansion, especially the degree to which these organizations can promote the authority of their certificate offerings, as well as frame and enforce ethical standards |

develops as an outgrowth of focused functional domain needs in an organization (Kiron et al. 2014). This suggests a promising avenue for the emergence of CSDS: CSDS emerging from cybersecurity as a set of challenges that frame focused data science solutions, rather than from data science proper as a shotgun approach.

Advocacy for the growing importance of organizational integration of data science into the cybersecurity profession is a key topic in state-of-the-art assessments of the field. Data science and machine learning are typically positioned as key strategic goals for improving resilience and responding to emerging challenges (Stamp 2018). As expressed in the 2019 *Cyber Intelligence Tradecraft Report* from the Carnegie Mellon Software Engineering Institute:

> A practice of high-performing organizations is to ensure there is proactive collaboration between strategic analysts and data scientists. The data scientists build tools for both strategic analysts and threat analysts. They also help with identifying trends and correlations. Indeed, one high-performing organization explained that "you need data scientists to win wars." (Ettinger 2019)

Table 2.10 Assessment of professional emergence in the data science domain

1	Active, focused interest from diverse participants	●	A massive, potentially historically unprecedented, amount of commercial and practitioner interest has been invested in this domain
2	Active professionals with associated job titles and roles	◕	Entrenched and global, albeit there are growing debates regarding the diversity of qualifications and cogency of job design specifications associated with titled practitioners
3	Emerging and informal training	●	A massive variety of academic and commercial training options and opportunities
4	Informal professional groups	●	Many informal groups—very active self-organized groups and clubs in almost all global industrialized urban areas (i.e., meetups, professional society chapters, and commercially sponsored presentations)
5	Professional and industry literature	●	Many sources of professional and industry literature from a broad range of institutions. Many best practices and whitepapers sponsored by providers of commercial services (i.e., consulting) and products (i.e., software and hardware)
6	Research literature	◕	A range of academic research journals and conference proceedings, although there are critiques regarding disciplinary focus. Some indication that management research has struggled to connect methods and technologies to effective mechanisms for organizational implementation and oversight
7	Formalized training	◕	Many options are cited for formal training, both commercial and academic. However, there are disputes concerning how body of theory is standardized in curricula, leading to shaky claims concerning training integrity and quality
8	Formal professional groups	◐	There are numerous professional organizations, albeit having in some cases co-opted analytics/data science into a primary domain. Examples include INFORMS and The OR Society (operations research), IEEE (engineering), and ACM (computing/IT)
9	Professional certifications	◔	Many options, but a lack of consensus on quality and acceptance. Uneven standards and weak consensus between fast-emerging academic options (formal degree and certificate programs) and commercial offerings (e.g., data science boot camps)
10	Standards bodies	◔	A bête noire—seemingly lacking in a formal sense, except for a few initial attempts (e.g., Data Science Association ethical guidelines). Some attempts to assert standards through certifications (i.e., PhD research programs and professional certifications, such as the INFORMS CAP), but otherwise lacking in firm, enforceable, recognized standards
11	Independent academic research disciplinary focus	◔	Despite a spate of new PhD programs and research journals, data science has struggled to assert and shake the critique of being an interdisciplinary grab-bag, begging and borrowing at will from a range of domains. Currently, there are weak grounds for asserting a cogent research locus to unite disparate methods under a coherent disciplinary umbrella

2.5 CSDS as an Emerging Hybrid Profession

2.5.1 Origins and Definitions

CSDS has been framed as an emerging hybrid profession. Simple status as a profession is de facto in that there are named professionals employed with the explicit job title *cybersecurity data scientist*, as can be attested by routine searches on professional networking websites such as LinkedIn (LinkedIn 2018) and Glassdoor (Glassdoor 2018). CSDS is also cited as a key emerging specialty in the cybersecurity profession, per the NIST NICE cybersecurity framework (NIST 2017) and the professional cybersecurity press (Morgan 2019a). In the US market, as of December 2019, 16% of cybersecurity job openings ($n \cong 82,000/500,000$) were aligned with the "analyze" functional category per the NICE workforce framework (CyberSeek 2019).

However, being an emerging profession, CSDS lacks many characteristics associated with mature professions. Uncertainty remains concerning core job skills and practitioner methods. Efforts to refine a definition of the emerging domain are an important step to iterating *professional emergence*. A key step in professional emergence entails knowledge individuation to frame a distinct practitioner *body of theory*, per Greenwood's professionalization model (1957).

Current CSDS professional practice is defined through the functional hybridization of the cybersecurity and data science professional fields (Bechor and Jung 2019). Emerging CSDS practitioners have access to best practices and professional credentials by operating at the intersection of the two parent domains. Pragmatic legitimacy is asserted to the degree data science methods evidence solutions to growing gaps in cybersecurity efforts.

In the current state, CSDS is a set of practical prescriptions for cybersecurity challenges. To utilize these prescriptions to advance a working definition, it is appropriate to review and reinforce key challenges observed facing the cybersecurity domain. As validated in survey research of security practitioners, cybersecurity efforts are challenged by a complex of factors including growing threats (Kirchhoff et al. 2015; Security Brief Magazine 2016), expanding vulnerabilities (SANS Institute 2016), shrinking human resources (SANS Institute 2015), data overload (Ponemon Institute 2017), and challenges orchestrating automated decisioning (SANS Institute 2015; SAS Institute Inc. 2016; Security Brief Magazine 2016; UBM 2016).

As observed in research and professional literature, traditional cybersecurity approaches oriented towards erecting two-dimensional *castle-and-moat* barriers miss subtle and sophisticated, multidimensional, and staged incursions, especially those involving social engineering (Leuprecht et al. 2016). Raising alerts via signature and rule-based detection methods creates a surfeit of warnings (Mahmood and Afzal 2013; SANS Institute 2015, 2016; Security Brief Magazine 2016). Data quality, disconnected data sources, poor or absent diagnostics, and a lack of recorded incidents challenge those seeking to refine and target data-driven alerting (Ponemon Institute 2017).

The expansion of Internet carriage, data volumes, and devices interacting with and dependent upon networked infrastructure has exponentially expanded the attack surfaces presented to agents seeking to damage, exploit, or otherwise misuse digital infrastructures (Poppensieker and Riemenschnitter 2018; Schneier 2018). The broad expansion of digitization as measured by Internet usage, data volumes, and the proliferation of surrounding networked infrastructure and devices has seen risks multiply (Cisco Systems Inc. 2017; Hubbard and Seiersen 2016; Lehto 2015).

Growing vulnerabilities parallel technical development and innovation (Cisco Systems Inc. 2018). For example, mobile, smart device, and Internet of Things (IoT) technologies create new targets, avenues to linked resources, and methods of attack (Schneier 2018). While threats and vulnerabilities accumulate, there is a rising lack of skilled practitioners to accommodate and monitor the expanding risks manifested by prolific digital saturation and the impact of allied technologies (Conklin et al. 2014) (Fig. 2.12).

The pressures afflicting cybersecurity professionals frame use cases for data science solutions. Data science offers a range of practical methods to address focused cybersecurity challenges. Big data analytics, data engineering, pattern analysis, anomaly detection, and predictive algorithms address noted cybersecurity gaps.

Industry research has reinforced assertions concerning the efficacy of security analytics. Per Fig. 2.13, a 2017 global survey of 621 IT security practitioners conducted by the Ponemon Institute and sponsored by SAS Institute reflected a dramatic reduction in false-positive security alerts following the implementation of security analytics programs (2017).

The same survey highlighted the relative perceived importance of various cybersecurity analytics goals, with *detecting events in progress* and *determining the cause of past events (forensic attribution)* being cited as the most pressing.

Framing CSDS as a pragmatic response to cybersecurity operational gaps, several objectives are suggested as key value propositions (Fig. 2.14):

1. **Data engineering**: Orienting data engineering objectives (e.g., building ETL pipelines)
2. **Reduced data volumes**: Refining fast and big data into "smart data" enriched with context (e.g., defining meaningful data schemas and structures)
3. **Discovery and detection**: Orchestrating a cyclical process of discovery and detection
4. **Automated models**: Facilitating the development, validation, and implementation of analytical models for pattern extraction, anomaly generation, and event detection
5. **Targeted alerts**: Leveraging data analytics tools and methods to produce targeted evidence-based alerts and visualizations to support investigation and remediation activities
6. **Resource optimization**: Routing focused incidents to the right resources at the right time for rapid review and remediation

Addressing and integrating cybersecurity challenges with focused data science value propositions, a proposed working definition of CSDS can be framed: *the*

Fig. 2.12 Key cybersecurity challenges

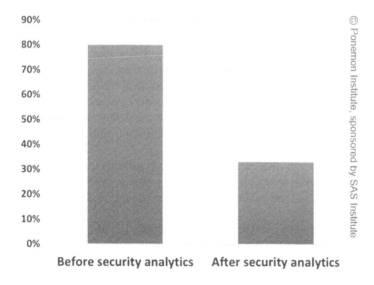

Fig. 2.13 Security analytics reduces false alerts

practice of data science to assure the continuity of digital devices, systems, services, software, and agents in pursuit of the stewardship of systemic cybersphere stability, spanning technical, operational, organizational, economic, social, and political contexts.

Fig. 2.14 Key CSDS value propositions

2.5.2   The Data-Driven Paradigm in Cybersecurity

As has been reinforced, the interest in applying data science to cybersecurity has been a steadily rising trend and can be considered a paradigm shift in the cybersecurity profession. Data science offers a set of practical methods to address a range of growing cybersecurity challenges. The traditional cybersecurity castle-and-moat paradigm has been rendered outmoded given the expanding distributed and promiscuously interlinked nature of contemporary digital infrastructures. The emerging paradigm frames data analytics facilitated continual surveillance as a mechanism for vigilant situational awareness. Inverting the castle and moat, this implies an inside-out monitoring approach, metaphorically akin to Jeremy Bentham's notion of the panopticon—surveillance as a centralized and continual activity (McMullan 2015).

As a key aspect of security operations, data collection and analysis are rooted in traditional network monitoring. In *The Practice of Network Security Monitoring*, enterprise security is framed as a circular, continuous process in which the collection and analysis of data are centrally placed, per Fig. 2.15 (Bejtlich 2013). Analysis of data is positioned as the central mechanism for detecting vulnerabilities and intrusions, which are then remediated.

Bejtlich comments concerning the enterprise security cycle that "although it depicts a smooth progression from one phase to the next, in the real world, all four activities occur simultaneously because organizations often experience different intrusion states at once. IT and security teams plan new defenses while existing

Fig. 2.15 Network security monitoring— enterprise security cycle

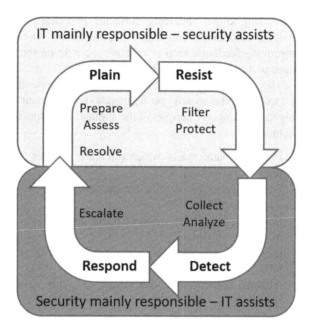

IT mainly responsible – security assists

Plain

Resist

Prepare
Assess

Filter
Protect

Resolve

Escalate

Collect
Analyze

Respond

Detect

Security mainly responsible – IT assists

countermeasures repel some intruders. While working to detect one set of intruders, CIRTs are responding to other intruders already in the organization" (Bejtlich 2013).

Both data-driven continuous monitoring and the circular nature of scientific discovery align with and address this notion of operational simultaneity. Situational awareness is both a constant monitoring activity, seeking to detect, and a continual exploratory effort, conscious of ever-shifting conditions and seeking to discover new phenomena. When data science is promoted as a central facilitator, the main concerns of enterprise security become collecting and processing data in support of analytical model development, testing, and validation.

Analytical models become active participants or agents in decision-making aimed at vulnerability analysis, anomaly detection, incident detection, and remediation. The great promise of advanced algorithms and machine learning is that discernment and implicit decision-making expertise, traditionally the highly manual burden of the cybersecurity professional, becomes semi-automated and embedded in the systemic monitoring process.

As new incidents are recorded and the results of treatments noted, machine learning models improve, as per the operant *learning* moniker. Complex decision-making is facilitated as a mechanism to accommodate the challenges presented by speed and volume. Machine learning is applied and utilized in the cybersecurity realm by supporting the identification of complex patterns indicative of known cyber-attacks (supervised machine learning), as well as identifying unusual new patterns and anomalies potentially indicating novel exploits (unsupervised machine learning).

Together, combinations of these approaches can be linked in processes that directly support key aspects of cybersecurity: alerting cybersecurity operators concerning known incidents, surfacing potential new and emerging threats, and making automated, self-improving decisions. Orchestrated cyclically, a self-improving feedback loop is enabled, a condition that results in operational resilience under changing conditions.

The great benefit of machine learning to cybersecurity is to orient and facilitate the work of cybersecurity professionals such that their efforts are more efficient and targeted. Common examples of the applied use of machine learning in cybersecurity include:

- Filtering spam emails, which often are vehicles for fraud, virus, and/or phishing attacks
- Detecting suspicious online financial transactions based on behavioral and technical factors
- Assessing suspicious network traffic and device behavior indicating potential malware infections
- Identifying and blocking dangerous websites or web traffic, for instance, DDoS attacks
- Alerting unusual user and device behavior based on a comparison to past behaviors

The notion of CSDS orchestrating organizational, process, and technical elements to address cybersecurity assurance via data science invokes the classical MIS tripartite model (Fig. 2.16). As opposed to the popular implication from marketeers, machine learning is not ipso facto a path to replacing human professionals in cybersecurity. In most cases, identification of an incident requires a final human determination of malicious intent and an appropriate organizational response. At most, machine learning models, as active decision facilitators, repositions cybersecurity professionals as stewards and active caretakers of analytical models. As models degrade and new circumstances emerge, there is a need for human oversight to monitor and proctor the continual maintenance and training of models (Oltsik 2019b).

The practical promise of machine learning for cybersecurity is increasingly efficient *human-in-the-loop* operant learning. This implies a cooperative process whereby machine learning provides operators with swift context and guidance to triage and react to security events, per the feedback cycle offered in Fig. 2.17. In turn, operators provide feedback on effectiveness. Through this mechanism, operators train and mentor the algorithms to become increasingly effective in a cyclical feedback process (Winn 2020). There will always be opportunities for increasing automation, but as long as human agents are the primary perpetrators of novel malicious acts, it will require human stewards to make the determination of complex adversarial intent and to improve models as new adversarial behaviors emerge.

In a linear representation, these efforts aim to streamline and semi-automate triage-focused decision-making associated with vulnerability detection, anomaly investigation, and incident remediation. In this context, cybersecurity can be framed as a decision process whereby operators, whether human and/or machine, are presented with evidence of potential misuse or compromise of networks or networked

Fig. 2.16 CSDS in the
MIS tripartite model

Fig. 2.17 Cyclical
data-focused cybersecurity
detection and prevention

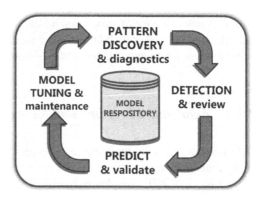

Fig. 2.18 Simplified linear process for operationalizing CSDS insights

devices. Aligning data science within a process in this linear context, a highly simplified model is framed in Fig. 2.18.

As a programmatic vision, this model is instructive to orient managers and practitioners. However, in implementation, in survey research conducted by the Ponemon Institute and sponsored by SAS institute, many organizations report that challenges remain. Per Fig. 2.19, data management is a frequently cited shortcoming impeding the implementation of security analytics, particularly as related to issues of data quality, integration, and volume (Ponemon Institute 2017).

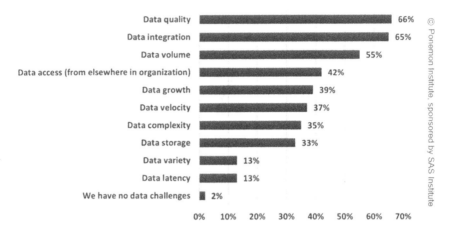

Fig. 2.19 Most frequent challenges with data for security analytics

The data-driven paradigm for cybersecurity in this sense is still aspirational and can be characterized as an assemblage of promising methods that, in implementation, still lack evidence of strong organizational and technical management fundamentals. It is implied that CSDS, in order to advance professionally, must address gaps associated with organization orchestration.

2.5.3 Academic Programs and Research

Data analysis increasingly appears as a focused subject in the curricula for cybersecurity academic degree and certificate programs (Bicak et al. 2015; Conklin et al. 2014; Kim et al. 2005; van den Berg et al. 2014; Verma et al. 2015). Forensics and monitoring are traditionally taught in cybersecurity programs, topics that emphasize collecting and analyzing data.

The scope of data analysis in the security domain is broadening as mechanisms for storing and analyzing complex sets of data become increasingly available and practically desirable (Mahmood and Afzal 2013). This has seen expanded coverage of data collection, storage, and analysis topics, notably fundamentals surrounding big data and data science (R. Verma et al. 2015). External drivers also encourage this trend, as has been expounded at length.

Cybersecurity data analysis is increasingly formally recognized as a cybersecurity job specialty, for instance, in the US NIST NICE Cybersecurity Workforce Framework (Bicak et al. 2015; NIST 2017). Data science and analytics appear explicitly in the NICE framework as a specialized cybersecurity workforce area associated with the aggregate functional goal of "Protect and Defend; Analyze, Collect and Operate, and Investigate." The associated core task identified by the NICE framework is "utilize technical documentation or resources to implement a new mathematical, data science, or computer science method."

Such formal recognition has led to greater attention to content and standards for associated training programs. Concerning recommending standards for cybersecurity academic programs, the NSA in the United States recently released criteria for 4-year degree programs in cybersecurity that include probability and statistics as core requirements (NSA 2019). This encapsulates a growing sentiment that quantitative risk and probabilistic analysis are becoming essential for analysts in the cybersecurity domain and that traditional deterministic, rules-based methods are no longer enough.

Research interest in CSDS rapidly grows, mirroring the growing commercial attention dedicated to cybersecurity data analytics. However, as CSDS is a specialized interdisciplinary domain, the relative coverage of focused research topics varies widely. An imbalance in journal literature exists between research focused on engineering implementations and applied methodological advances versus work addressing organizational management perspectives, the latter being a lacuna. This mirrors a gap observed in general cybersecurity scholarship (see Sect. 2.3.3).

There is no lack of engineering-focused research on data science for cybersecurity. Academic literature searches reveal interest in "cybersecurity data science" has rapidly spiked, with 20,800 topical research articles appearing since 2015, more than 75% of these having been published between 2018 and mid-2019 (Google Scholar 2019a). Similarly, "cybersecurity machine learning" research interest has risen rapidly, with more than 50% of the 15,400 topical articles appearing on Google Scholar having been published between 2018 and mid-2019 (Google Scholar 2019b). A number of meta-assessments concerning dominant themes arising in cybersecurity machine learning research have been conducted (Berman et al. 2019; Buczak and Guven 2016; Sarker et al. 2020; Spring et al. 2019; Stamp 2018).

As this effort is focused on CSDS from the MIS domain perspective, CSDS-related engineering research, being otherwise well-covered and profiled, is less of interest than organizational implementation and orchestration perspectives. It is asserted that MIS-oriented organizational perspectives are a lacuna in CSDS-related literature. In order to substantiate this lacuna, a semi-structured thematic gap analysis of 33 manuscript-length works in the CSDS domain was undertaken. Methods and results are summarized in Sect. 2.6.4.

2.5.4 Professional Organizations and Certifications

Widely accepted specialized professional CSDS certifications are yet forthcoming. This can be attributed to the relatively recent emergence of CSDS as a specialization. Established cybersecurity certificate authorities are likely best positioned to frame such offerings in the future. The continuous monitoring cybersecurity function has the closest historical connection to CSDS. This function requires utilizing a variety of tools to monitor and analyze telemetry and log data in order to detect vulnerabilities, anomalies, and potential incidents.

Cybersecurity incident forensic analysis also entails the analysis of data in the form of evidence. In operation, when incident root causes are determined through forensic means, detection evidence ideally is captured and utilized to improve continual monitoring focused on future prevention. This principle is manifested in machine learning by providing predictive models with training datasets in the form of labeled incidents.

In the cybersecurity realm, the SANS Institute GIAC Continuous Monitoring (GMON) and Detection Analyst (GCDA) certifications cover a number of topics associated with user, device, and network monitoring, including tools and methods for analyzing log and telemetry data (SANS Institute 2019b). The SANS Continuous Monitoring and Security Operations training class offering (SEC511) is allied with the GMON certification and covers topics such as log analysis, monitoring methods, and analysis tools (SANS Institute 2019a). Connecting with industry standards, the NIST SP 800-137 guidelines for continuous monitoring are referenced. The SANS SIEM with Tactical Analytics course (SEC555) is allied with the GCDA certification and covers topics related to data analytics such as endpoint analytics, log analysis, baselining, and user behavior monitoring.

Online CSDS training offerings are emerging, with O'Reilly (Givre 2017) and Udemy (Tsukerman 2019a) having hosted focused classes developed by cybersecurity data scientists. SAS Institute has also hosted focused training on CSDS (2019). An advantage of training offerings by independent facilities and educators is that the CSDS domain is rapidly evolving. Independent providers and online facilities can better adapt and update curricula as challenges, methods, and technologies swiftly evolve.

A challenge of formal professional certification programs and associated training is that they typically have a lifespan measured in years rather than months. Certification-associated training programs are premised upon certifications designed by larger, bureaucratic institutions and thus are highly structured and less agile to change. When an emerging body of knowledge shifts rapidly, attempting to capture, codify, and freeze a set of best practices programmatically is inherently challenging.

Outside training offerings, professional organizations allied with both cybersecurity, and data science increasingly produce webinars, articles, and research on the topic of CSDS. IT and analytics professional organizations such as IEEE, ACM, and INFORMS identify CSDS as a growing area. Likewise, the major cybersecurity professional organizations generally recognize the growing importance of CSDS and machine learning to the profession's future. As has been cited, this interest is mainly pragmatic, observing the efficacy of data science methods to address a range of growing cybersecurity challenges.

Increasingly, to the degree cybersecurity challenges are overwhelming private and commercial spheres, governmental agencies are stepping forward to offer proactive guidance concerning best practices for the application of CSDS to cyber defense. These efforts variously are led by regulatory and supervisory bodies, such as NIST (United States), ENISA (EU), and NCSC (United Kingdom); military organizations, such NATO's Cooperative Cyber Defence Centre and US Department of Defense-sponsored projects and organizations such as MITRE; and intelligence authorities, such as sponsorship and guidance from the NSA in the United States.

2.5.5 Frameworks

Cybersecurity operational functions associated with continuous monitoring and forensics have traditionally emphasized data analysis as a key component. However, data analysis approaches applied have typically been simplistic, focusing on raising alerts from deterministic rules and signatures. An observed shortcoming is that rules-based detection mechanisms spawn too many alerts as data volumes and environmental complexity expands (Ashford 2019; Forrester 2020; Oltsik 2019b; Ponemon Institute 2018; Riley 2019b). Security operations in many organizations cannot keep pace with the volume of alerts generated by deterministic, rules-based mechanisms.

Data science-based detection methods support the focused refinement and targeting of alerts. Probabilistic and multivariate statistical techniques allow for complex, scenario-driven approaches. Such approaches support risk-based assessments and triage mechanisms. Machine learning-based approaches have particularly evidenced an ability to encapsulate some aspects of expert decision-making in environments where data volume, complexity, and limited resources demand automation (Khan and Parkinson 2018; Stamp 2017, 2018).

Advocacy for an increasingly central role for analytics and machine learning in cybersecurity is growing in research and professional press literature (Ashford 2019; Bechor and Jung 2019; Ferguson 2019; Mahmood and Afzal 2013; Spring et al. 2019; Stamp 2018; Verma et al. 2015). Firms are mirroring this interest, with a 2019 survey of security stakeholders indicating that 38% of C-suite executives were planning to invest in AI and ML to improve security in the next 24 months (Violino 2019a).

In response, cybersecurity frameworks which position data analysis centrally are emerging from traditional continuous monitoring models. An example is the Carnegie Mellon Software Engineering Institute (SEI) representation of the cyber intelligence framework as derived from the US government model for the cyber intelligence cycle. In this model, depicted in Fig. 2.20, human and machine teaming supported by data-driven analysis is central to both tactical and strategic functions (Ettinger 2019).

This model positions machine-driven data analysis as servicing all four quadrants of the security assurance framework proposed earlier (see Fig. 2.7). Additionally, this model captures the emerging paradigm in which machine learning algorithms as decision agents facilitate and assist human stewards. In turn, data-oriented professionals are assumed to take on an expanded role in the active training, monitoring, assessment, and modification of these agent models.

While numerous operationally focused representations for the interaction of humans and machine-driven agents are evolving, the majority are ensconced more generally in the data science domain. As such models evidence practical efficacy within the cybersecurity domain, they will be refined and customized in time to impute domain focused context. This can be considered a working out of CSDS *body of theory*, stipulating and constituting macro-practitioner context. For instance, general machine learning and data mining process models socialized in the

Fig. 2.20 SEI Cyber Intelligence Framework

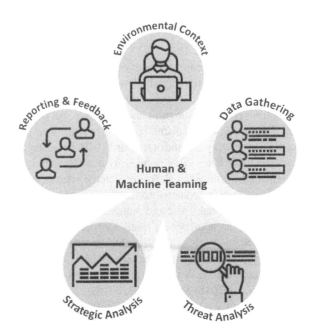

analytics domain, such as CRISP-DM, SAS SEMMA, and KDD (Azevedo and Santos 2008), have led to examinations specific to the cybersecurity domain (Buczak and Guven 2016).

One framework which is practically useful to emphasize and extend in the CSDS domain is that of the analytic model framework represented earlier in Fig. 2.11. Data analytics in the cyber domain, as represented in the Cyber Intelligence Framework (Fig. 2.20), is evidencing expansion from being a tactical mechanism to addressing broader investigative and strategic goals. This represents a dramatic expansion in the assertion of the breadth of problems data science can address in the cybersecurity domain.

The broadening of cyber analytics from monitoring and forensics to strategic aspects signifies an expansion to broader timeframes, more complex models, and increasingly advanced methods. The assertion of a broader purview for CSDS thus necessitates a better understanding of which analytics (methods), models (uses or use cases), and temporal scopes (timeframes) are being addressed with each methodological implementation. Expanding the analytic model framework, CSDS models can be charted in a three-dimensional context as having methodological, utilitarian, and temporal context, per Fig. 2.21.

While this assertion of expanded CSDS utility is encouraging, issues of model complexity are created. A critique raised in the earlier overview of the data science profession relates to the sheer conceptual and methodological breadth of the domain. Data scientists are often at risk of losing their way amongst an array of complex models, particularly where assemblages of linked models are built to guide decision-making.

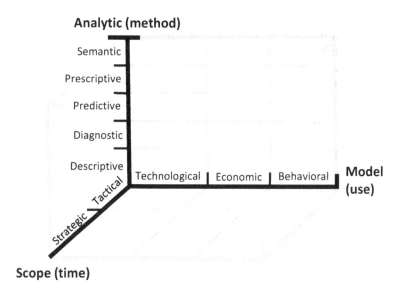

Fig. 2.21 CSDS analytical, utilitarian, and temporal model context

Model complexity takes root in particular where multi-staged models encompass a chain of technological, economic, and behavioral phenomena. Complex multi-stage models invoke inherent variability and begin to take on the aspect of a rough systems dynamic simulation, a toy model that must admit the limitations of inherent abstraction. By carefully specifying and segmenting method, use, and temporality, per the framework in Fig. 2.21, the CSDS analyst can ensure they are not overextending model complexity.

Whereas CSDS practitioners may be tempted to construct master-composite models for monitoring and detection, amounting to grand theory, a best practice observed in data science, is that composite models are only as reliable as their weakest link. This observation has deeper connections to scholarship on model management and validation in computer simulation (Balci 1998; Kleindorfer and Ganeshan 1993; Kleindorfer et al. 1998; Law and Kelton 2000; Robinson 2004; Sargent 2007; Weisberg 2013). Compartmentalizing complex composite models into focused segments which are fit for purpose prevents attempting to build a false oracle or tower of babble: aggregate uncertainty in model scope compromises reliability and accuracy. The advocacy and utility of the three-dimensional framework offered in Fig. 2.21 is to focus and specify, not expand, model breadth.

2.5.6   Professionalization Status

Summarizing the analysis conducted in Sect. 2.5, the drivers propelling CSDS professionalization are multifaceted and include adversarial, economic, operational, commercial, academic, and technical pressures. CSDS can be framed as being *hybrid-emergent* as it results from the hybridization of cybersecurity and data science root parent domains. From a practical standpoint, the emerging field can be considered a specialized subfield gaining increasing visibility within the more mature cybersecurity profession.

The CSDS domain reveals many facets. It can be characterized depending on the standpoint of various perspectives:

• *Technical:* CSDS is a toolkit of methods and technologies addressing focused challenges.
• *Process:* CSDS is a set of methodological procedures applied in combinations.
• *Organizational:* CSDS is a structured program for achieving operational goals.
• *Research:* CSDS is an emerging academic sub-domain embodied by topical focus areas.
• *Institutional:* CSDS is a programmatic effort framed by institutional interests.
• *Paradigm:* CSDS is a transition signifying the shift towards data-driven decision facilitation.
• *Profession:* CSDS is an aggregation of above factors as an emerging professional role.

Extrapolating from the overview in Sect. 2.5, CSDS professional maturity is assessed in Tables 2.11 and 2.12 according to the professionalization assessment models derived previously (based upon Tables 2.3 and 2.4, and referring to pie-chart key in Table 2.7).

In assessing CSDS professionalization, a question is posed as to when professional status is achieved in emerging fields such as CSDS. In professionalization literature, this point is conceptualized as *occupational closure*: the demarcation point when specialized skills and professional knowledge are codified, particularly in focused professional training curricula and linked certification programs and licensure. Codification implies adherence to standards, formal credentials, and experiential prerequisites. Allied is an assertion of bureaucratic authority projected by professional organizations, training programs, and credential or certificate authorities. Being in the early stages of emergence and lacking a dedicated professional standards body or bodies, CSDS currently has weak claims to bureaucratic authority, lacking what Greenwood terms *sanction of the community* (1957).

Occupational closure is, in this sense, the result of attaining institutional sponsorship and formalization on the path to professional emergence. In pursuing professional emergence, a chicken-and-egg paradox is resident whereby a systematic body of theory supported by institutions is necessary to achieve knowledge individuation (i.e., to advance authority, individuate professional communities, enforce ethical codes, and foster professional culture). However, lacking professional institutional facilitation, knowledge individuation is forestalled as body of theory development and codification is inhibited.

In this context, body of theory represents codified theoretical knowledge—knowing what to do and why, *logos* with a purview over *technê*, as represented in Fig. 2.22. In professionalization scholarship, body of theory maturation is crucial to the early stages of professionalization. As Greenwood notes:

Table 2.11 Assessment of professional maturity in the CSDS domain

1	Systematic body of theory	◐	A rich body of theory can be derived from the combination of the CSDS root parent domains. However, greater specification is required to standardize the prescription of data science methods to focused cybersecurity use cases. The data science body of theory breadth threatens pragmatic focus in CSDS and hence the ability to frame and adopt standard practices. The critique that the data science domain lacks rigor due to breadth can be addressed within CSDS by focusing on specific use cases and by treating each challenge as phenomena susceptible to scientific methods and processes
2	Authority and judgment recognized by client	◐	Two challenges concerning the ability to assert authority issue from the CSDS parent domains. As the cybersecurity profession is increasingly overwhelmed by rising practical challenges, the integrity of the traditional profession is challenged. In the data science domain, critiques concerning professional permissiveness and lack of rigor challenge the authority of associated methods. However, parent challenges are also CSDS opportunities. Conjoined, data science methods suggest prescriptions for growing cybersecurity challenges. The CSDS claim to authority is initially contingent upon evidencing pragmatic treatments that lead to demonstrable, efficacious solutions. As there is prior evidence of the efficacy of machine learning to address focused cybersecurity challenges (i.e., detection of spam email, viruses, and malware), a foundation for asserting pragmatic authority is building. However, to command domain authority, CSDS must transition from spot solutions to a more formalized body of theory vis à vis stronger links between standardized theory and practice. Demonstrating standard practices in the form of applied scientific methods to address focused categories of cybersecurity phenomena will be key to strengthening future appeals to CSDS domain authority
3	Community sanctions authority	◑	Pragmatic cybersecurity challenges have increasingly led both academic and institutional authorities to position and advocate a central role for data science in the domain. What is lacking are standardized training and certification bodies to promote a shared body of theory to establish a unified professional community. In turn, focused academic and professional efforts to strengthen CSDS body of theory standards are a prerequisite to attracting institutional support
4	Ethical code of stewardship	○	No explicit codes for CSDS ethical standards have been framed. Of the two parent domains, cybersecurity is the logical locus for grounding ethical stewardship given precedence. A CSDS certification program could conceivably be a vehicle for framing and enforcing such principles. Beyond good cybersecurity stewardship principles, CSDS requires conjoined ethical strictures related to the conduct of data science within the domain. Tying ethical stewardship to a commitment to scientific rigor would be a key step. This implies strengthening CSDS body of theory to explicate standards for the application of scientific rigor
5	Professional culture supported by associations	○	Currently, there are no widely known dedicated CSDS professional associations. CSDS practitioners must synthesize professional membership by attaching simultaneously to both cybersecurity and data science professional bodies. This results in professional culture being an ersatz fusion of the two domains. The gap suggests the need for focused special interest groups hosted within existing professional bodies

Table 2.12 Assessment of professional emergence in the CSDS domain

1	Active, focused interest from diverse participants	●	Interest in CSDS is growing rapidly in commercial, academic, and institutional domains
2	Active professionals with associated job titles and roles	◕	Professionals with the job title "cybersecurity data scientist" are working globally in a range of industries across many organizations. Evidence of this job role can be seen directly by accessing online professional networking and job posting web resources (Glassdoor 2018; LinkedIn 2018). The professional cybersecurity press recognizes CSDS as a key growing job specialty (Morgan 2019a). Government institutions have validated data science as a job skill of growing importance (Ettinger 2019; NIST 2017). However, there are a lack of recognized standard skills and job duties for the job role. This highlights the importance of developing a stronger CSDS body of theory and ideally to frame training and certification offering
3	Emerging and informal training	◑	Several informal commercial training options and opportunities have emerged, particularly via online channels
4	Informal professional groups	◑	There are indications of ad hoc meetings in industrialized urban areas (i.e., meetups, conferences, vendor-sponsored events). However, there seem to be few groups meeting on a regular basis
5	Professional and industry literature	◕	There is a growing body of literature issuing from the professional cybersecurity press (Ferguson 2019; Morgan 2019a). IT industry analyst firms such as Gartner, Forrester, and IDC regularly address CSDS tangentially under various topical guises such as *security analytics*. Much professional press and analyst coverage is focused on commercial technological solutions, with organizational and methodological topics being relatively less represented
6	Research literature	◕	As covered previously, journal research on CSDS and machine learning methods is fast growing. A body of CSDS monograph literature has emerged in the past decade, namely, the corpus of 33 works reviewed immediately hereafter. CSDS organizational, risk, scientific methodology, and data management factors were identified as research lacunas
7	Formalized training	◔	Academic and certification bodies have introduced the topic of cybersecurity data analysis as an area of growing importance. However, CSDS is typically introduced tangentially as an aspect of monitoring or incident forensics. Formalized training from recognized bodies will likely be a key next step in CSDS professionalization as this will facilitate body of theory codification
8	Formal professional groups	○	There are no known formal CSDS professional groups. There are computer security special interest working groups attached to computer science or analytics associated professional associations which at times raise CSDS as a focused topic of interest (i.e., ACM, IEEE, INFORMS)
9	Professional certifications	○	There are no known focused CSDS professional certifications. SANS Institute GIAC Continuous Monitoring (GMON) and Detection Analyst (GCDA) certifications address some aspects of data analytics for cybersecurity
10	Standards bodies	○	No known dedicated CSDS standards bodies exist. However, some governmental institutional bodies and allied sponsored organizations are increasing their focus on CSDS (e.g., in the United States: NIST NICE, MITRE, Carnegie Mellon SEI)
11	Independent academic research disciplinary focus	○	While there is evidence of latent interest, there are no known focused CSDS research journals or academic research degree programs

Fig. 2.22   CSDS occupational closure model

In the evolution of every profession there emerges the researcher-theoretician whose role is that of scientific investigation and theoretical systematization. In technological professions a division of labor thereby evolves, that between the theory-oriented and the practice-oriented person. Witness the physician who prefers to attach himself to a medical research center rather than to enter private practice. This division may also yield to cleavages with repercussions upon intraprofessional relationships. However, if properly integrated, the division of labor produces an accelerated expansion of the body of theory and a sprouting of theoretical branches around which specialties nucleate. (1957)

Greenwood's account observes key factors now affecting the status of CSDS. For one, there are resident indications of intraprofessional tensions between traditional cybersecurity professionals and cybersecurity data scientists that signify a growing paradigm shift. This is manifested in resistance and conflict reported within the cybersecurity profession itself, whereby established cybersecurity professionals and the professional press increasingly evidence skepticism concerning the efficacy and relevance of data science-associated approaches (Ross 2019; Violino 2019b). Secondarily, CSDS theoretical body of knowledge emerges from the hybridization of cybersecurity and data science practice. As will be discussed in greater detail, methodological issues still under debate in the realm of data science thus impact the status and emerging viability of CSDS.

2.6   Literature Analysis Outcomes

2.6.1   Result 1: CSDS Comparative Professional Maturity Gaps

Aggregating the professional maturity assessments undertaken previously is useful to understanding the relative maturity of the CSDS in relation to its two parent domains. Understanding discrete gaps in the comparative professionalization maturity of CSDS serves several purposes. For one, a discrete understanding of gaps in professionalization helps orient and motivate attention towards areas where improvement is needed. In a more general sense, specifying the particular areas of professional immaturity substantiates the research effort.

To the degree body of theory is an identified gap, this particular area can be highlighted as an area of attention for broader research efforts, whether grounded in the MIS, organizational research, or cybersecurity domains.

Several immediate research goals are furthered by understanding the comparative maturity of the CSDS field. The professional maturity model, being conscious of the need to codify professional knowledge, also motivates an effort to identify a central body of CSDS literature and to understand themes raised across such literature. These needs are addressed in *Results 3* and *4* following. Understanding CSDS *as-is* maturity and associated gaps also provides a foundation for *Phase II* interview research, a goal addressed in *Result 5*.

To frame a comparative assessment of CSDS professional maturity gaps, Sects. 2.3.6 and 2.4.6 undertook a maturity assessment of the cybersecurity and data science professions, utilizing a model based on Greenwood's notion of professional maturity (as explained in Sect. 2.2.7). As the CSDS parent domains, results supported a structured overview of the CSDS field in Sect. 2.5. This culminated in a maturity assessment of CSDS profession in Sect. 2.5.6.

The aggregate maturity assessment results provide for a final, high-level comparative assessment of the CSDS profession positioned in relation to the two parent domains. The summarized maturity levels utilize the heuristic pie charts used previously, as framed in Table 2.7, and reproduced attached to Table 2.14 for convenience. Aggregating the professionalization pie-chart assessments across the three domains produces results summarized in Tables 2.13 and 2.14, revealing CSDS as relatively less mature and emerged as a profession.

The aggregated summary tables below utilize the professionalization maturity and emergence models framed originally in Sect. 2.2.7 through Tables 2.3 and 2.4. Professional maturity-level pie chart summaries are aggregated from the related domain maturity assessments captured previously: Sect. 2.3.6 cybersecurity (Tables 2.5 and 2.6), Sect. 2.4.6 data science (Tables 2.9 and 2.10), and Sect. 2.5.6 CSDS (Tables 2.12 and 2.13).

It is apparent in the aggregate summary that the CSDS domain is in a lower state of relative development as compared to the parent domains. Of interest, despite the great amount of attention paid to the data science domain, it is framed as being less mature than the cybersecurity domain. This raises the question as to whether the data science domain may be comparatively more culpable in forestalling advancement of the CSDS domain.

Table 2.13 Summary assessment of professional maturity across three domains

#	Criteria	Cyber Sec	Data Sci	CSDS
1	Systematic body of theory	◕	◕	◐
2	Authority and judgment recognized by client	◐	◐	◐
3	Community sanctions authority	◐	◐	◔
4	Ethical code of stewardship	●	◐	○
5	Professional culture supported by associations	●	◐	○

Table 2.14 Summary assessment of professional emergence across three domains

#	Criteria	Cyber Sec	Data Sci	CSDS
1	Active, focused interest from diverse participants	●	●	●
2	Active professionals with associated job titles/roles	●	◕	◕
3	Emerging and informal training	●	●	◕
4	Informal professional groups	●	●	◕
5	Professional and industry literature	●	●	◕
6	Research literature	◕	◕	◕
7	Formalized training	●	◕	◔
8	Formal professional groups	●	◕	○
9	Professional certifications	◕	◔	○
10	Standards bodies	●	◔	○
11	Independent academic research disciplinary focus	◕	◔	○

Pie symbol interpretation key

Symbol	Interpretation of categorical professional maturity
○	Little or no evidence of meeting the criteria denoted (~0%)
◔	Relatively low or immature level regarding the criteria (~25%)
◐	A mid-point of maturity, though still lacking in many regards (~50%)
◕	Fairly robust and mature regarding the criteria, with some gaps (~75%)
●	Strong and robust maturity regarding the criteria denoted (~100%)

2.6.2 Result 2: CSDS Functional Demand Model

CSDS is a de facto profession in active practice. As the field is rapidly expanding, it is useful to understand the discrete factors driving interest and growth. Understanding the factors impelling growth is useful for developing a refined understanding of the goals and actions representative of the field. This improves an understanding of how CSDS, at least in the ideal, positions data science solutions to address cybersecurity needs.

This literature analysis has highlighted an interest across commercial, governmental, and nonprofit bodies to advance cybersecurity capabilities to address growing challenges. Growing demand for improved security approaches, in turn, propels interest in CSDS. CSDS addresses key gaps facing the cybersecurity field—too many alerts, resource constraints, increasingly sophisticated threats, expanding digital infrastructure, etc. This organic matching of needs and capabilities is promising in terms of promoting the development of the CSDS domain.

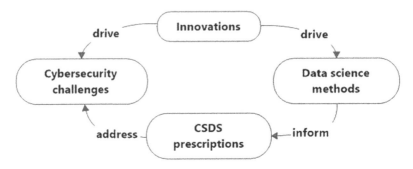

Fig. 2.23 Integrated CSDS motivational symbiosis

A structured understanding of how CSDS prescribes solutions to cybersecurity challenges improves insight into discrete use cases and unique functions on offer. Viewing cybersecurity as a set of challenges (demand) and data science as a set of prescriptions (supply) serves to position a model for CSDS services. Figure 2.23 frames the concept of an integrated symbiotic relationship between cybersecurity challenges and data science prescriptions vis à vis CSDS.

In this conception, forces of technical innovation are interposed as a common element which both advances analytical approaches to defense (data science methods) and propels increasing exposure, vulnerability, and threats (cybersecurity challenges).

This notion of CSDS as a fulfillment of supply and demand forces can be expanded to refine an understanding of the discrete actions and elements implied. Observing Looijen and Delen's principle of information systems management as fulfilling functional demands through the supply of IT solutions (Looijen and Delen 1992), CSDS can be positioned as the fulfillment of security assurance demands through the supply of applied IT functions and processes (implicitly mediated by applications and technology).

Figure 2.24 populates a representation of the Looijen and Deelen model with a representation of various threats, goals, roles, methods, and objectives (deriving from concepts raised in literature and as annotated in the figure). Implicit in this model of CSDS as a fulfillment of supply-and-demand forces is that prescriptions must evidence demonstrable efficacy.

2.6.3   Result 3: CSDS Literature Corpus

A central goal of this research is to advance CSDS professionalization by developing body of theory. It was determined that a useful contribution in this regard would be to assemble and validate a representative corpus of CSDS literature. Composing, refining, evidencing, and communicating this corpus would be a useful contribution to promoting CSDS body of theory development. Beyond the value inherent in

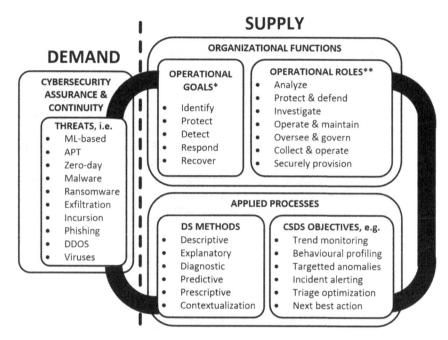

* Operational goals adapted from NIST Cybersecurity Framework (NIST, 2018)
** Operational roles from NIST NICE Workforce Framework (NIST, 2017)

Fig. 2.24 CSDS MIS supply-demand model

framing the corpus, reviewing and summarizing subject matter contained therein would serve to inform the larger research effort. As a central component of this literature analysis, such an effort was undertaken. The process and results are noted here.

An immediate challenge in composing a CSDS corpus was the stark absence of theory-laden CSDS perspectives in journal research. In particular, there was little journal research which covered CSDS from MIS perspectives, namely, combining technical, process, and organizational factors. As has been asserted, engineering-focused and applied implementation research are well-addressed in CSDS research, mirroring the parent cybersecurity domain (de Bruijn and Janssen 2017; Foroughi and Luksch 2018).

In particular, there is a large and growing body of technical and engineering research on applied machine learning for security (see Sect. 2.5.3). Also, a number of literature-based meta-studies on themes surfaced across CSDS-related technical research have been produced (Berman et al. 2019; Buczak and Guven 2016; Sarker et al. 2020; Spring et al. 2019; Stamp 2018). CSDS technical case study research and meta-research is thus well-covered.

CSDS technical research, being well-addressed and focused on engineering to the exclusion of organizational and theoretical content, was less of interest. Beyond demonstrating new methods or technologies to improve security assurance, these works contain few methodological best practices, theoretical assertions, or organizational and management perspectives. More worrying, few generalizable findings or theoretical models are framed in such research, indicating a lack of falsifiability and scientific rigor. In short, CSDS technical journal research, on the whole, had little to contribute to CSDS body of theory development, a necessary step for iteration of the field.

However, more substantial, manuscript-length, CSDS-related works did contain coverage on organizational, theoretical, and methodological topics. By nature of scope and length, even those CSDS-related books with a technical orientation delved into broader topics related to organizational implementation, theory, best practices, and conceptual models. The book-length format, as opposed to journal publications, supports and encourages digressions from engineering-focused content. Based on an initial survey, it was determined there were enough book-length works with a broad subject-matter purview to suggest a CSDS corpus of interest to body of theory development efforts.

The identification and validation of the CSDS corpus was a two-stage process. The first stage, initial identification, involved identifying and procuring a set of 20 initial book-length, CSDS-related works. The initial works were identified through keyword-based literature searches. The search logic applied sought for book-length works matching subject-matter keywords containing "cybersecurity" or "security" or "network monitoring" or "intrusion detection" and "data science" or "science" or "analytics" or "analysis" or "machine learning" or "statistics" or "statistical" or "mathematics" or "AI" or "big data" or "data-driven" or "deep learning."

Searches were conducted on several platforms, including Google Scholar, EBSCO Connect, and Amazon.com. Some works were also identified by referral from practicing CSDS professionals. This was the result of a feedback process in which the initial list of works was distributed via online CSDS special interest communities. CSDS practitioners replied with recommendations to add to the list or commented on the suitability of the list. This helped to refine the initial set in a stepwise fashion.

Works were previewed through summary overviews as well as a review of the table of contents, when possible, to confirm they contained substantial CSDS-related content. The books were selected if they aligned significantly with CSDS topics as framed in Sects. 2.2–2.5 of this literature analysis. An initial corpus of 20 books was thus identified, validated, selected, and reviewed, with the benefit of ad hoc feedback from a community of CSDS domain experts.

As the composition and review of the corpus occurred over time, a second, concluding phase to assess and add new works was undertaken. A reality of the fast-developing CSDS domain was that during the initial 2-year process, a non-trivial number of additional new works were published. Also, new recommendations were received from CSDS practitioners concerning older works that were worthy to add to the corpus. This resulted in the identification and inclusion of 13 additional works.

As there was a validation process by several respected members of the CSDS practitioner community and no other comparable corpus has been put forward, the collected and summarized set of 33 works is proposed as a contemporary representative corpus of the CSDS field. The relative recency and growing popularity of cybersecurity data science is evidenced in observing that 66% of the titles profiled, 22 out of 33, were published in the 5-year period between 2016 and 2020. Nearly 40% of the works, 13 out of 33, were published in the last 2 years between 2018 and 2020.

To extend the exercise, a subject-matter survey of the works was conducted (see *Result 4*, following). As a byproduct of that effort, a short summary of each work was produced. These summaries help frame and communicate subject-matter spanning the corpus. To promote CSDS body of theory development, the resulting corpus has been circulated via online communities, via the researcher's professional blog (Mongeau 2020a) and via presentations at professional conferences (Mongeau 2020b, c).

For the benefit of grounding this research effort and profiling the range of practices in the CSDS domain, summaries of the CSDS corpus works follow (in chronological order from 2001 through 2020):

1. *Computer Intrusion Detection and Network Monitoring: A Statistical Viewpoint* (Marchette 2001)
   The first widely circulated monograph-length work that addresses the application of data-centered methods to computer and network security. Provides an introduction to networking alongside statistical and machine learning approaches to network traffic monitoring. Addresses network and host monitoring for intrusion detection, as well as virus, worm, and Trojan detection.

2. *Statistical Methods in Computer Security* (Chen 2005)
   A compellation of contributed works covering a range of topics related to data-driven and statistical approaches to network security. This work is notable for combining statistical perspectives with organizational risk management views. Topics covered include security policies, risk management, wireless security, Internet security, firewalls, secure architectures, viruses and worms, password security, and infrastructure monitoring.

3. *Machine Learning and Data Mining for Computer Security: Methods and Applications* (Maloof 2006)
   The first widely circulated monograph-length work encountered that explicitly addresses the application of machine learning to problems of computer security. This edited work collects a series of research papers addressing methods for detecting behavioral anomalies in complex network and computer datasets. An introduction to the domain as related to information assurance is provided followed by a series of focused approaches for detecting various malicious events and intrusions. Behavioral factors are profiled, as are the topics of data management and decision modeling. The range of topics span practical and theoretical guidance and maintain relevance to contemporary challenges.

4. *Intrusion Detection: A Machine Learning Approach* (Yu and Tsai 2011)

This work explicitly addresses the application of machine learning to intrusion detection. Covers the application of predictive and learning algorithms in the context of intrusion detection systems (IDS). Examines the application of a range of specific machine learning algorithms to detecting intrusion on both wired and wireless networks.

5. *Data Mining and Machine Learning in Cybersecurity* (Dua and Du 2011)
   This work addresses the application of machine learning methods to a range of security challenges. Covers the application of both unsupervised and supervised machine learning methods to computer misuse, anomaly detection, scanning detection, network traffic profiling, and hybrid detection. Presciently, the topic of privacy preservation in the conduct of data analytics is raised. The conclusion presages challenges that have since become central to the domain: malware, botnets, cyber warfare, mobile technology, privacy preservation, and hybrid cybercrime.

6. *Network Anomaly Detection: A Machine Learning Perspective* (Bhattacharyya and Kalita 2013)
   A practitioner-focused work covering network fundamentals, vulnerabilities, and methods for detecting anomalies. Machine learning methods and statistical methods for network anomaly detection are covered, as are the topics of feature selection and model performance evaluation. Feature engineering and model management, being treated explicitly, observes the practical importance of these topics. An explanation of attacker methods and tools leads to a discussion of emerging challenges.

7. *Applied Network Security Monitoring: Collection, Detection, and Analysis* (C. Sanders and Smith 2013)
   This book observes the network security monitoring process, covering the topics of data collection, detection, and analysis. Detection methods covered are largely signature-based and statistical. The work reviews a range of open-source tools and packages for collecting and analyzing network data. Links are provided to practitioner examples and exercises. While not covering data science or machine learning methods explicitly, the focus on primary objectives in network monitoring is valuable to practitioners and managers seeking to establish a foundation for undertaking CSDS efforts.

8. *Network Security Through Data Analysis* (Collins 2014)
   Recognizing security analytics as an emerging discipline, this work covers the fundamentals and mechanics of data, tools, and analytics for network security. Practicalities in staging data-focused analysis and conducting basic statistical and analytical methods are the focus. Covers data collection via sensors, data aggregation, and data analysis. Machine learning is not raised explicitly, although a range of foundational topics and methods are explored in depth.

9. *Data Analysis for Network Cybersecurity* (Adams and Heard 2014)
   A collection of workshop research papers, the focus of this work is monitoring and analyzing network traffic data at the intersection of statistics, data mining, and computer science. A range of techniques are explored to support the prevention of or rapid identification of malicious network activity. Focused topics

covered include graph analytics, attack detection, scan statistics, anomaly detection, and device monitoring.

10. *Data-Driven Security* (Jacobs and Rudis 2014)
    A well-rounded guide on a range of topics for undertaking a security analytics program with a core focus on data analysis and visualization. Subjects covered include analytics tools, data management, statistics, machine learning, visualization, and implementing solutions. Gives a valuable perspective concerning the distinctions and overlaps between exploratory and predictive methods in operational environments.

11. *Fraud Analytics Using Descriptive, Predictive, and Social Network Techniques* (Baesens et al. 2015)
    Multiple practical commonalities bind the fraud and cybersecurity domains. Committing acts of financial impropriety is often a key motivation for cyber adversaries. Also, the statistical and analytical techniques for detecting and remediating fraud are quite suitable for cross-application to cybersecurity use cases. Fraud analytics is a mature domain with a set of well-tested techniques and best practices. This work is thus an invaluable technical guide for a range of methods suitable for CSDS consideration and adoption, spanning insights from data gathering and processing to model building and implementation.

12. *Essential Cybersecurity Science* (Dykstra 2016)
    The first widely circulated practitioner-focused work which frames CSDS explicitly. The work is significant to CSDS body of theory in explicitly introducing the scientific method as a process and framing guidelines for evidence-based inquiry. Focused topics covered include visualization, forensics, cryptography, malware analysis, and systems engineering. Several hands-on examples and cases are explored with references to additional resources.

13. *Dynamic Networks and Cybersecurity* (Adams and Heard 2016)
    An edited collection of research papers, this work focuses on the analysis of computer network traffic data, with a primary focus on temporal and graph statistical methods. Although machine learning is mentioned, there is a core focus on primary statistical methods. Insightful coverage is offered on emerging methods for monitoring and anomaly detection in network data.

14. *Cybersecurity and Applied Mathematics* (Metcalf and Casey 2016)
    This work comprehensively explores a range of primary mathematical and statistical analysis techniques in the context of cybersecurity. The methods covered include metrics, probability, data analysis, graph theory, game theory, visualization, string analysis, and homologies. While not covering machine learning explicitly, this work provides a primary foundation for diagnostic and feature selection techniques which can be applied to bolster machine learning efforts.

15. *How to Measure Anything in Cybersecurity Risk* (Hubbard and Seiersen 2016)
    Addressing a literature gap, this work focuses on quantitative methods for characterizing cybersecurity risk. A strong case is made that monitoring and remediation efforts are destined to flounder without a corresponding understanding

of how to characterize and prioritize probabilistic risk. A demarcation is made between decision science and data science, with this work espousing the two as being complementary and mutually beneficial. As the focus is on framing metrics, the work addresses approaches to better manage organizational efforts.

16. *Data Analytics and Decision Support for Cybersecurity* (Carrascosa et al. 2017)
An edited compilation of research papers, this work melds the topics of data analytics and decision support for cybersecurity. Papers range from theoretical to practice-focused and cover specialized topics including intrusion detection, insider threats, malware, visualization, monitoring, data quality, and decision support.

17. *Introduction to Machine Learning with Applications in Information Security* (Stamp 2017)
Focusing on applied machine learning for cybersecurity, a broad range of techniques for preparing, building, testing, and validating models are addressed. Particular methods covered in hands-on examples include Hidden Markov models (HMM), principal component analysis (PCA), support vector machines (SVM), clustering, Naive Bayes, and regression analysis.

18. *Information Fusion for Cybersecurity Analytics* (Alsmadi et al. 2017)
Intended for cybersecurity researchers, this work is an edited collection of papers on the topics of attack prediction and information fusion. Focused topics covered include network anomaly detection, graph analytics, trust management, wireless sensor networks, digital forensics, and intrusion detection.

19. *Research Methods for Cybersecurity* (Edgar and Manz 2017)
This comprehensive and valuable reference is notable for providing in-depth coverage of the application of scientific research methods in the security domain. A broad-based introduction to scientific methods and cybersecurity leads to coverage on a range of methodological approaches. Observational, mathematical, experimental, and applied research methods are discussed in detail. Goals, methods, approaches, diagnostics, and examples are covered. This work is significant in addressing a gap in security research concerning focused attention to methodological rigor.

20. *Big Data Analytics in Cybersecurity* (Savas and Deng 2017)
This edited collection of chapters addresses a well-rounded range of topics associated with the application of analytics to large cybersecurity datasets. Aspects covered include network forensics, vulnerability assessments, root cause analysis, data visualization, data management, mobile security, cloud computing, IoT, behavioral analysis, and tools. The topics of training, cybersecurity research methods, and research datasets are also raised.

21. *Machine Learning and Security: Protecting Systems with Data and Algorithms* (Chio and Freeman 2018)
This is a practical guide on the application of machine learning to a range of security issues. Topics covered include intrusion detection, malware classification, and network analysis. Data and model quality are addressed as focused topics. Adversarial machine learning, the analysis of vulnerabilities, and attacks on machine learning systems are raised and discussed.

22. *Data Science for Cybersecurity (*Heard et al. 2018*)*
    This work compiles papers submitted to an eponymous workshop held at the
    Imperial College London in 2017. The focus is on the application of statistical
    and data science techniques to cyber defense. Particular topics covered include
    feature selection, threat detection, anomaly detection, botnet clusters, red team-
    ing, extracting labels for machine learning, and data-driven decision-making. A
    major dataset for cybersecurity data analysis research is presented from the Los
    Alamos National Laboratory (LANL).

23. *Guide to Vulnerability Analysis for Computer Networks and Systems (*Parkinson
    et al. 2018*)*
    This edited collection of contributions covers a range of topics related to the
    application of artificial intelligence to vulnerability analysis. Followed by an
    overview and survey of domain approaches, key sections examine the areas of
    vulnerability assessment frameworks, case-based applications, focused meth-
    ods, and visualizations. Vulnerability analysis is discussed across a range of
    platforms and architectures, including computer networks, file systems, cloud
    computing, SCADA systems, telecommunications, and mobile devices.

24. *AI in Cybersecurity (Leslie F. Sikos 2018a, b)*
    A collection of authored chapters, this work covers emerging methods for pro-
    active cyber defense. Topics covered include ontological modeling of cyberse-
    curity knowledge, adversarial machine learning, software vulnerability
    detection, AI methods for network attack detection, machine learning for intru-
    sion detection, and mobile application analysis.

25. *Malware Data Science: Attack Detection and Attribution (*Saxe and
    Sanders 2018*)*
    Focused on applying data science to malware detection and analysis, this work
    explores topics including machine learning, statistics, social network analysis,
    and data visualization. Neural networks and deep learning methods are covered.
    Guidance is offered to those seeking to advance professionally as a security
    data scientist.

26. *Mastering Machine Learning for Penetration Testing (*Chebbi 2018*)*
    This is a practitioner-oriented guide to the application of machine learning for
    penetration testing. With examples focused on Python, this work covers phish-
    ing, malware, botnet, and advanced persistent threat detection approaches. Also
    addressed are techniques for evading intrusion and machine learning detection
    systems, as well as adversarial machine learning methods. In addition to intro-
    ducing the basics of machine learning, best practices in machine learning and
    feature engineering are covered.

27. *Hands-On Machine Learning for Cybersecurity (*Halder and Ozdemir 2018*)*
    This is a practitioner-oriented introduction to machine learning addressing a
    range of security use cases. A variety of both statistical and machine learning
    approaches are demonstrated through hands-on Python examples. A range of
    security threats are introduced and addressed through detection and mitigation
    techniques, for example, network intrusion, phishing, and malware.

28. *Machine Learning for Cybersecurity Cookbook (*Tsukerman 2019b*)*
    This hands-on guide collects and demonstrates a series of applied machine learning algorithms and techniques, including deep learning. Utilizing Python, focused examples address topics such as malware, social engineering, and network intrusion detection. Other subjects addressed include penetration testing, password attacks, password security, encryption, and data privacy.

29. *Hands-On Artificial Intelligence for Cybersecurity* (Parisi 2019*)*
    Per the title, this practitioner-oriented work demonstrates the application of a variety of machine learning approaches to cybersecurity. Utilizing Python and including deep learning approaches, examples address a variety of use cases in the areas of threat detection (spam, malware, network anomalies), data and asset protection, and model management.

30. *Machine Learning for Computer and Cybersecurity (*Gupta and Sheng 2019*)*
    An edited collection of curated papers, this work focuses on challenges and applications of machine learning and data mining to computer and cybersecurity. Topics covered include deep learning, phishing detection, machine learning use cases, mobile security, and feature selection. Organizationally focused topics such as incentive systems in IT security and teaching security best practices are also covered.

31. *Deep Learning Applications for Cybersecurity (*Alazab and Tang 2019*)*
    This compellation of contributed works focuses on the application of deep learning and machine learning to a variety of cybersecurity use cases. The application of deep learning to cyber threat detection, modeling, monitoring, analysis, and defense is addressed. Focused use cases include intrusion detection, infrastructure security, botnet detection, malware analysis, adversarial methods, and forensic approaches.

32. *Data Science in Cybersecurity and Cyber Threat Intelligence* (Sikos and Choo 2020*)*
    This compellation of contributions addresses the application of data science methods and approaches to cybersecurity and cyber threat detection. Focused topics covered include automated reasoning for cyber threat intelligence, information fusion, semantic engineering, malware detection, big data processing for machine learning, IoT security for mobile health networks, and pitfalls associated with applying data science to cybersecurity.

33. *Cybersecurity Analytics (*Verma and Marchette 2020*)*
    Commencing with a foundational introduction to both the analytics and cybersecurity domains, this well-researched book offers a thorough overview of the CSDS field. The work blends statistical and machine learning perspectives, addressing both theory and practice. Topics covered include statistics, machine learning, text mining, natural language processing, and big data techniques. A number of examples, supplemented with links to relevant datasets, are provided for practitioners.

2.6.4 Result 4: CSDS Literature Gap Analysis

Being focused in the MIS domain, this research effort is interested in integrated technical, process, and organizational perspectives on CSDS. It has been asserted that there is a lacuna of organization and theoretical content in CSDS literature. In that this research effort intends to address this lacuna, it is important to be able to substantiate this claim more formally.

As noted in the development of the CSDS corpus in *Result 3* (Sect. 2.6.3), engineering-focused and applied implementation topics are well-addressed in CSDS research. Noting that there were several thematic meta-studies of journal-based CSDS technical research (Berman et al. 2019; Buczak and Guven 2016; Sarker et al. 2020; Spring et al. 2019; Stamp 2018), it was determined that an MIS-focused review of CSDS literature, being a gap, would be a useful contribution.

A comparative literature gap analysis was undertaken to substantiate the assertion that organizational factors are less represented in CSDS literature. To motivate the importance of surfacing this gap, organizational factors have been noted as being central to successful analytics implementations (Kiron et al. 2014; LaValle et al. 2010). Raising awareness of the need for greater focus on management and organizational orchestration topics in CSDS literature thus aligns with the objective of advancing the effectiveness of the field.

Having extrapolated and reviewed a CSDS corpus of manuscript-length works, per *Result 3*, it was of interest to understand and profile the central themes raised in these works more rigorously. It was of particular interest to understand the relative representation of specific themes to identify potential thematic gaps.

As this inquiry focuses on CSDS from an MIS context, it was of interest to determine to what degree organizational topics, along with focused theoretical topics, are or are not covered in CSDS literature. An understanding of key CSDS literature themes and gaps provides a reference point to assess and validate CSDS themes and gaps identified through other means (i.e., in *Phase II* practitioner interview research). More generally, a substantiation of key CSDS themes from literature helps inform sensitizing concepts as a basis for interpreting interview results (per *Result 5* in Sect. 2.6.6 following).

A staged, semi-grounded process was followed to identify and analyze comparative themes across the CSDS corpus of 33 book-length works. The main analysis stages included (1) identification of sources, (2) initial thematic coding content review, (3) refinement of themes, (4) focused thematic content review, (5) summarization of theme frequency across all sources, (6) addition of new works to the analysis, and (7) theme frequency and gap analysis. The research was conducted over a 2-year period between May of 2018 and April of 2020.

The approach applied observed content analysis research best practices as framed by Krippendorff (2019). Regarding the thematic extrapolation process, a structured process for qualitative thematic coding was applied based on guidance from

Polkinghorne (2015). As the designation of themes was sensitized by concepts raised during literature analysis and from prior research, the approach can be considered semi-grounded (van den Hoonaard 1996).

In terms of reinforcing rigor and reproducibility, details of the staged thematic analysis process followed will be enumerated. Starting with stage 1, identification of sources, 20 initial book-length CSDS-related works were identified and procured. Details of this effort are profiled in *Result 3* in Sect. 2.6.3. As a result, an initial corpus of 20 books was identified, validated, selected, and reviewed.

In stage 2, initial thematic coding content review, the works were reviewed and notes were taken concerning key topics raised. A detailed subject-matter review was undertaken for each, and core topics were documented in structured notes. The analysis resulted in a preliminary list of 27 topics raised across all works. The initial set of 27 topics helped to drive the identification of a refined set of key themes. In order to validate and refine the themes raised, a comparison was made to prior CSDS topic models raised in cybersecurity literature-based research: (a) a security analytics taxonomy derived by Grahn et al. (2017) and (b) a cybersecurity data science topic model extrapolated by Bechor and Jung from research literature via text mining (Bechor and Jung 2019).

Progressing to stage 3, refinement of themes, themes were progressively consolidated and resolved as the effort progressed, observing a process similar to *initial open coding* applied in grounded interview research. At the point of having reviewed 12 works, the list of themes reached what is termed *coding saturation*, a point where the topics identified seem to be stable and mature. At this point, a stable set of themes was designated. Cross-referencing topical analysis results with prior CSDS topic models from research resulted in a refined set of CSDS topics raised in the corpus:

1. Focused use cases
2. Risk quantification
3. Decision support
4. Data management
5. Data collection
6. Scientific methods
7. Feature engineering
8. Statistical methods
9. Anomaly detection
10. Machine learning
11. Model management
12. Visualization
13. Adversarial methods
14. Organizational management

Having established a stable set of themes, the content analysis process progressed to stage 4, focused thematic content review. In this phase, a more focused and targeted review of all works in the corpus was conducted. This involved a three-step process: (a) reviewing the table of contents for indications of the key themes, (b) reviewing the index, and (c) having determined the location of target themes in the work, going to the indicated chapter, section, or page to assess the depth and quality of the related content.

In this respect, a brief mention of a theme such as "machine learning" or "data management" was not considered enough for the work to have covered the related theme in detail. For verification of thematic content, there needed to be a substantial discussion of the theme in the body of the work. For instance, for the theme "model management," a work was judged as having contained the theme if there was at least a two-page discussion that covered the topic in methodological depth, or otherwise raised specialized examples, conceptual diagrams, and/or best practices.

In this way, each work was progressively profiled in terms of the amount of substantial expert content covered related to the list of themes. Completion of this process for the entire corpus composed step 5, summation of thematic frequency across all sources. To record the results, a table was composed which cross-indexed each work with the canonical themes. The presence of a theme was recorded as a simple binary "yes" or "no"—no gradations of partial thematic coverage were tracked or documented.

Stage 6 involved the addition of new works and refreshing the analysis. A reality of the fast-developing CSDS domain was that during the 2-year review process, a non-trivial number of additional new works were published. Also, new recommendations were received from CSDS practitioners concerning older works that were worthy of adding to the corpus. As the initial review of 20 works was socialized amongst the CSDS practitioner community via the researcher's professional blog and via LinkedIn (Mongeau 2020a), new recommendations were raised. Also, new recommendations resulted from practitioners reacting to presentations by the researcher at cybersecurity industry events (Mongeau 2020b, c). As explained in *Result 3*, via these channels, 13 new works were reviewed, validated, and added to the corpus for a total of 33 works. Analysis conducted in stages 4 and 5 was repeated for the additional 13 works.

The completion of the thematic coding exercise across the corpus resulted in a final summary of thematic frequency across the corpus per Table 2.15. As a result of having extrapolated key themes across the corpus, it was of particular interest to understand the relative frequency of topics to be raised in comparison to others across the corpus. This constitutes the final stage, stage 7, theme frequency and gap analysis. A comparative accounting of the topics raised across all works was undertaken. Results of the topical cross-analysis are recorded in the footer of Table 2.15.

Table 2.15 CSDS topic coverage across central literature Result 5: CSDS sensitizing concepts

		Focused use cases	Risk quantification	Decision support	Data management	Data collection	Scientific methods	Feature engineering	Statistical methods	Anomaly detection	Machine learning	Model management	Visualization	Adversarial methods	Organizational management
1	Computer Intrusion Detection and Network Monitoring	✓				✓		✓	✓	✓	✓	✓	✓	✓	
	Marchette (2001)														
2	Statistical Methods in Computer Security	✓	✓	✓					✓	✓				✓	✓
	Chen (2005)														
3	Machine Learning and Data Mining for Computer Security	✓	✓	✓	✓	✓			✓	✓	✓	✓			
	Maloof (2006)														
4	Intrusion Detection: A Machine Learning Approach	✓							✓	✓	✓			✓	
	Yu and Tsai (2011)														
5	Data Mining and Machine Learning in Cybersecurity	✓		✓	✓			✓	✓	✓	✓	✓	✓		
	Dua and Du (2011)														
6	Network Anomaly Detection: A Machine Learning Perspective	✓		✓		✓		✓	✓	✓	✓	✓	✓	✓	
	Bhattacharyya and Kalita (2013)														
7	Applied Network Security Monitoring	✓	✓	✓	✓	✓		✓	✓	✓			✓		✓
	Sanders and Smith (2013)														
8	Network Security Through Data Analysis	✓		✓	✓	✓		✓	✓	✓			✓	✓	
	Collins (2014)														

Table 2.15 (continued)

		Focused use cases	Risk quantification	Decision support	Data management	Data collection	Scientific methods	Feature engineering methods	Statistical methods	Anomaly detection	Machine learning	Model management	Visualization	Adversarial methods	Organizational management
9	Data Analysis for Network Cyber-Security, Adams and Heard (2014)	✓		✓		✓		✓	✓	✓	✓	✓	✓	✓	
10	Data-Driven Security, Jacobs and Rudis (2014)	✓	✓	✓	✓	✓	✓	✓	✓	✓	✓	✓	✓	✓	✓
11	Fraud Analytics Using Descriptive, Predictive, and Social Network Techniques, Baesens et al. (2015)	✓	✓	✓	✓	✓	✓	✓	✓	✓	✓	✓	✓	✓	✓
12	Essential Cybersecurity Science, Dykstra (2016)	✓	✓	✓		✓	✓	✓	✓	✓	✓	✓	✓	✓	
13	Dynamic Networks and Cyber-Security, Adams and Heard (2016)	✓	✓			✓	✓	✓	✓	✓		✓	✓	✓	
14	Cybersecurity and Applied Mathematics, Metcalf and Casey (2016)			✓				✓	✓	✓			✓		
i5	How to Measure Anything in Cybersecurity Risk, Hubbard and Seiersen (2016)		✓	✓			✓		✓			✓	✓		✓

(continued)

Table 2.15 (continued)

		Focused use cases	Risk quantification	Decision support	Data management	Data collection	Scientific methods	Feature engineering	Statistical methods	Anomaly detection	Machine learning	Model management	Visualization	Adversarial methods	Organizational management
16	Data Analytics and Decision Support for Cybersecurity	Carrascosa et al. (2017)	✓		✓	✓	✓		✓	✓	✓	✓	✓	✓	✓	
17	Introduction to Machine Learning with Applications in Information Security	Stamp (2017)	✓						✓	✓	✓	✓	✓	✓	✓	✓
18	Information Fusion for Cyber-Security Analytics	Alsmadi et al. (2017)	✓	✓	✓	✓	✓		✓	✓	✓	✓	✓	✓	✓	
19	Research Methods for Cyber Security	Edgar and Manz (2017)	✓	✓	✓	✓	✓	✓	✓	✓	✓	✓	✓	✓	✓	
20	Big Data Analytics in Cybersecurity	Savas and Deng (2017)	✓	✓	✓	✓	✓	✓	✓	✓	✓	✓	✓	✓	✓	✓
21	Machine Learning &Security	Chio and Freeman (2018)	✓		✓	✓	✓		✓	✓	✓	✓	✓	✓	✓	
22	Data Science for Cybersecurity	Heard et al. (2018)	✓	✓	✓		✓		✓	✓	✓	✓	✓	✓	✓	
23	Guide to Vulnerability Analysis for Computer Networks and Systems	Parkinson et al. (2018)	✓	✓	✓		✓		✓	✓	✓	✓	✓	✓	✓	

Table 2.15 (continued)

		Focused use cases	Risk quantification	Decision support	Data management	Data collection	Scientific methods	Feature engineering	Statistical methods	Anomaly detection	Machine learning	Model management	Visualization	Adversarial methods	Organizational management
24	AI in Cybersecurity	Sikos et al. (2018)	✓		✓	✓			✓	✓		✓	✓		✓	
25	Malware Data Science: Attack Detection and Attribution	Saxe and Sanders (2018)	✓				✓		✓	✓	✓	✓	✓	✓	✓	
26	Mastering Machine Learning for Penetration Testing	Chebbi (2018)	✓				✓		✓	✓	✓	✓	✓	✓	✓	
27	Hands-On Machine Learning for Cybersecurity	Halder and Ozdemir (2018)	✓				✓		✓	✓	✓	✓	✓	✓	✓	
28	Machine Learning for Cybersecurity Cookbook	Tsukerman (2019a, b)	✓				✓		✓	✓	✓	✓	✓	✓	✓	
29	Hands-On Artificial Intelligence for Cybersecurity	Parisi (2019)	✓				✓		✓	✓	✓	✓	✓	✓	✓	
30	Machine Learning for Computer and Cyber Security	Gupta and Sheng (2019)	✓	✓	✓		✓		✓	✓	✓	✓	✓		✓	

(continued)

Table 2.15 (continued)

		Focused use cases	Risk quantification	Decision support	Data management	Data collection	Scientific methods	Feature engineering	Statistical methods	Anomaly detection	Machine learning	Model management	Visualization	Adversarial methods	Organizational management
31	Deep Learning Applications for Cyber Security	Alazab and Tang (2019)	✓				✓				✓	✓	✓		✓	
32	Data Science in Cybersecurity and Cyberthreat Intelligence	Sikos and Choo (2020)	✓		✓	✓	✓		✓		✓	✓	✓		✓	
33	Cybersecurity Analytic	Verma and Marchette (2020)	✓		✓	✓	✓	✓	✓	✓	✓	✓	✓	✓	✓	
		94%	42%	70%	45%	82%	24%	85%	94%	94%	82%	82%	76%	85%	21%

Based upon the frequency analysis, it was of interest to understand which topics were less commonly raised across the corpus, themes that represented gaps. This would indicate a lacuna in coverage—topics that are significant, yet which are less often covered. Per Table 2.15, the topical cross-analysis suggests that the following topics are addressed relatively less frequently, being represented in 50% or less of the works spanning the corpus:

- Risk quantification: 42%
- Data management: 45%
- Scientific methods: 24%
- Organizational management: 21%

These findings are tangentially reinforced by the broader observation, made previously, that security journal and conference research focus heavily on engineered solutions and technical case studies (Longstaff 2012; Meushaw and Landwehr 2012). Very little published journal research related to CSDS addresses organizational, methodological, and/or (scientifically oriented) theoretical perspectives. Per the results here, although these topics appear in manuscript-length works, the topics are less well represented.

As a result of the thematic gap analysis, guidance concerning organizational factors and scientific methods can be considered the most pronounced CSDS research lacunas. The importance of addressing these gaps is highlighted by practitioner observations that organizational factors are a key shortcoming and persistent challenge in security implementations (Ponemon Institute 2017). Also, many cybersecurity researchers bemoan the absence of fundamental scientific methods and perspectives in research literature (Craigen 2014; Longstaff 2012; Maxion et al. 2010; Meushaw and Landwehr 2012; Pavlovic 2012; Tardiff et al. 2016). These gaps observed in literature and practice will be interrogated in more detail in the subsequent research phase, *Phase II*, through interviews with CSDS practitioners.

2.6.5 Result 5: CSDS Sensitizing Concepts

A key goal of this phase has been to extrapolate a set of foundational CSDS concepts. In synthesizing literature significant to the emergence of the CSDS profession, a set of *sensitizing concepts* has been refined. Sensitizing concepts are primary concepts derived to orient qualitative research efforts (van den Hoonaard 1996). The intention has been to establish a foundation for qualitative interview and gap analysis research. The concepts derived provide central guidance and context for the qualitative interview research undertaken subsequently.

While cybersecurity has a high level of relative professional maturity, the field is confronted by the specter of rapid change. Technological evolution has ushered in a range of new and growing vulnerabilities. The exponential proliferation of networks and networked devices are overwhelming traditional cybersecurity approaches. The implication is that traditional techniques associated with cybersecurity defense require extension and adaptation.

CSDS inherently addresses a range of challenges facing the cybersecurity profession. However, greater methodological rigor and clarity of scope are required for CSDS to emerge as a robust cybersecurity sub-domain. Institutional facilitation would facilitate and improve this process. A complication is that the rapidity of change underlying the CSDS domain transcends traditional institutional timeframes associated with established professions. Certificate and standards bodies are bureaucratic and do not adapt easily to exponential or rapidly emergent phenomena. This is symptomatic of a larger phenomenon as occupations generally are mutating rapidly, especially those inherently tied to digitization (World Economic Forum 2016, 2018).

The underpinning technologies to which cybersecurity professionals dedicate themselves evolve at a speed increasingly beyond the scope of human organizational adaptation and change. One need not look far to identify cybersecurity threats that are evolving faster than human knowledge systems can seemingly address. An array of new technologies challenge traditional definitions of network and device interaction, creating new classes of challenges—both vulnerabilities and threats: IoT attack vectors and platforms (Schneier 2018), machine learning driven attacks (Brundage et al. 2018; Forrester 2020; Gopalakrishnan 2020), adversarial machine learning (Barreno et al. 2006; Chebbi 2018, 2019; Chio and Freeman 2018; Munoz-Gonzalez and Lupu 2018; Samuel 2019; Strout 2019; Yin et al. 2019), cloud attacks and attack platforms (Vacca 2017), and container and microservice vectors and threats (Chaturvedi 2019).

CSDS is marked by interdisciplinary complexity and involves a domain heavily subject to continual technological change. Shifting technologies affect the domain in multiple ways:

- **Objectives**: protecting evolving technical devices and infrastructures
- **Countermeasures**: guarding against rapidly adapting, technically sophisticated adversaries
- **Threats**: adversaries continually developing and utilizing innovations to attack new targets
- **Methods**: applying fast-evolving data science methodological and technical innovations in the mission to safeguard shifting infrastructure against evolving adversaries

While data science is recognized as a fast-growing and popular field, gaps in methodological maturity and rigor have been raised. However, previous examples of the hybridization of data science with focused professional domains suggest that cybersecurity domain challenges have the potential to ground and focus data science efforts. CSDS frames a paradigm shift that addresses a range of central challenges facing the cybersecurity domain. Likewise, focused cybersecurity problems provide tangible use cases for data science methods to establish clarity and rigor. This proposes CSDS as a symbiotic hybridization whereby gaps in the parent fields hybridize to address their respective gaps. Figure 2.25 highlights *body of theory* (BoT) gaps in the parent domains and their synthesis in CSDS.

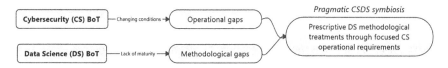

Fig. 2.25 CSDS as symbiotic hybridization addressing domain gaps

In this context, CSDS can be viewed as being both pulled and pushed towards specialization and hybridization.

A structured assessment of CSDS professional maturity vis à vis the root parent domains was undertaken. A summary of key sensitizing concepts focused on CSDS professional maturity extrapolated from literature analysis:

1. CSDS in a management research context

   (a) *Growing pressure to improve cybersecurity automation and effectiveness*

   - Centrality of digital infrastructure to modern organizations frames cybersecurity as general enterprise risk.
   - Growing digital exposure from the expansion of networks and proliferation of devices.
   - Growing digitization scale and complexity generates overwhelming amounts of cybersecurity data.
   - Attackers have asymmetric advantages of methods, opportunities, time, and choice.
   - Threats are growing in scale and sophistication.
   - Traditional rules-based and castle-and-moat cybersecurity paradigm is increasingly ineffective.
   - Challenges of differing regulatory approaches and uneven global oversight.

   (b) *Managerial pressure to mediate costs and risks*

   - Economic trade-offs must be made between profit and access versus risk and exposure.
   - Inadequate human resources to surveil expanding infrastructure.
   - Control and assurance of risk require quantification and measurement.
   - Better approaches needed to quantify, triage, and remediate cyber risks.
   - Need for analytical decision support methods to focus and automate.
   - Commercial hype proliferating distortions regarding analytics capabilities.
   - Specter of cyber threat actors utilizing machine learning mechanisms.

   (c) *MIS as a natural disciplinary locus for systematic inquiry into CSDS*

   - CSDS hybridizes a broad set of cybersecurity needs with a diverse set of data science methods.
   - MIS frames CSDS across organizational, process, and technical contexts.
   - MIS addresses a lacuna in CSDS literature concerning organizational and management factors.

- CSDS frames cybersecurity demands with the supply of data science prescriptions.

(d) *Assessment framework for professional maturity and emergence*

- CSDS is a hybrid emergent professional specialty in the midst of rapid change.
- Two models concerning the maturity of professions, the trait approach and professional emergence, support an assessment of CSDS with respect to its parent domains.
- A developed body of theory (contextualizing and situating practice) and institutional facilitation are key aspects of maturing professions.

2. Cybersecurity as a profession

(a) *Definition of the domain*

- Traditional definitions of cybersecurity focus on technical and tactical factors.
- A broader definition of cybersecurity encompasses organizational and strategic aspects.
- Due to the proliferation of the cybersphere, cybersecurity has increasingly been framed in organizational, legal, regulatory, economic, social, and political contexts.
- It is useful to position the array of professional cybersecurity functions as combinations of tactical versus strategic and technical versus organizational concerns.
- Traditional cybersecurity professional practice is straining to accommodate the growing expansion of digital infrastructure, leading to prolific exposure and broadening context.
- Continuity assurance, beyond audit, is increasingly a key management concern.

(b) *Professional maturity*

- Cybersecurity as a profession can be considered relatively mature.
- Key aspects include an array of facilitating institutions, academic programs and degree offerings, certifications, professional training programs, and conceptual frameworks.
- Challenges include weak authority, rapid technical proliferation and change, and expansion of disciplinary breadth.
- CSDS offerings address a range of emerging gaps in the cybersecurity profession: challenges integrating fragmented systems and data, data overload, lack of resources, overwhelming volumes of alerts, difficulties triaging events, and challenges assessing risk.

3. Data science as a profession

(a) *Definition of the domain*

- Data science has multiple frames depending on context: academic, professional, organizational-programmatic, and/or commercial-technological.

- There are unresolved disciplinary debates, such as the relative importance of statistical (explanatory and diagnostic) versus predictive methods (machine learning algorithms).
- Presents as a collaborative role with implicit stewardship over scientific rigor.
- The breadth of the data science domain pressures practitioners towards specialization.
- There are standing examples of hybrid data science specialty domains emerging professionally, such as financial analytics and marketing analytics.

(b) *Professional maturity*

- Despite its popularity and rapid growth, data science is overall less mature as a profession than the cybersecurity field.
- The breadth of the domain challenges the ability to focus and delineate boundaries.
- There is a central practitioner emphasis on engineering and IT solutions (technê).
- Commercial enthusiasm (technê-driven) dominates to the dearth of theoretical rigor (logos-driven epistêmê), which leads to distortions and confusions concerning domain scope.
- Data science body of theory, overarching context and rigor, is comparatively immature as goal-oriented practice and engineering (technê: teleological implementation) dominates theoretical context (logos-driven epistêmê: ontological and epistemological clarity).
- Debates concerning academic focus persist despite many recent degree offerings and a range of practice-oriented research journals, forestalling body of theory maturation.
- Professional organizations and certifications are of varying reach and uneven quality.
- Data scientists are responsible for scientific rigor, but often lack resources and decision-making rights, creating incentive-based conflicts and the risk of decision capture.
- A lack of clear and well-socialized ethical codes calls stewardship into question.

4. CSDS as an emerging hybrid profession

(a) *Definition of the domain*

- Fundamentally a data-focused domain, the field applies quantitative, algorithmic, and probabilistic methods; attempts to quantify risk; focuses on producing focused and efficacious alerts; promotes inferential methods to categorize behavioral patterns; and ultimately seeks to optimize cybersecurity operations.
- Represents a partial paradigm shift from traditional cybersecurity approaches, which are rule-and-signature-based and focus on boundary protection.
- Seeks situational awareness and assumes persistent and prolific threats which may be human, automated, or "cyborg" in origin.

- Goals connect historically with cybersecurity continuous monitoring and forensics functions in particular.
- Has emerged from two parent domains which themselves are undergoing rapid transformation. As such, the "body of theory" surrounding CSDS is still evolving.
- Has evidenced pragmatic successes using analytics and machine learning in focused use cases such as spam filtering, phishing email detection, malware and virus detection, network monitoring, and endpoint protection.
- Applied CSDS involves addressing cybersecurity challenges with data science prescriptions and implies a gap analysis is conducted (i.e., focused cybersecurity challenges frame particular data science prescriptions for applied CSDS treatments).
- Hybrid human and machine teaming is a growing central organizational goal.

(b) *Professional maturity*

- Growth currently issues from practical efficacy in addressing growing cyber challenges.
- The field currently derives professional trappings from the root parent domains.
- Expansion of purview in the cybersecurity field extends the range of challenges to be addressed and thus risks overextension and diffusion (e.g., focus on assuring core network security versus addressing broader challenges such as addressing "fake news").
- Relative immaturity in the data science domain and confusions concerning methods hinder the ability to strengthen methodological rigor and to standardize best practices.
- Practitioner methods currently rarely address organizational and management perspectives, focusing on engineering solutions and applied methods.
- There is a lacuna in domain literature concerning the topics of organizational management, scientific methods, data management, and risk quantification.
- There is currently a lack of organizational bodies to foster domain development and to standardize best practices.
- There is an underlying need for a more refined and codified body of theory to support domain knowledge individuation.
- The problem of rigor is preeminent and requires stronger efforts to frame cybersecurity challenges as scientific processes and problems (i.e., CSDS as a scientific process).
- Need for more systematized approaches to data management, to the degree that preparation and framing of data embed methodological and model assumptions.
- Need for focus and specification in model framing to manage model complexity.

2.6.6 Result 6: Methodological Concepts

A number of key concepts were framed, summarized, or extrapolated during the course of the literature analysis. These concepts were raised as being of interest to CSDS practitioner body of theory. A number of concepts raised methodological

Table 2.16 Key CSDS analytical-methodological concepts

#	Reference[a]	Description	Summary
1	Figure 2.10	Analytics methods model	A summary representing the range of analytics methods available. Identifying the goal of an inquiry helps focus the selection of methods
2	Figure 2.11	Analytical methods per model scope framework	Cross-referencing analytics methods (per Fig. 2.10) with model scope (technical, economic, behavioral) helps identify use cases and facilitate analytical model development
3	Figure 2.14	Key CSDS value propositions	Identifies central CSDS use cases resulting from a matching of cybersecurity demands with data science offerings
4	Figure 2.17	Cyclical data-focused cybersecurity detection and prevention	Framing of a cyclical, self-reinforcing cybersecurity detection learning process based on continual feedback and refinement
5	Figure 2.21	CSDS analytical, use, temporal context model	An extension of Fig. 2.11 adding a temporal dimension, this further assists in model framing and development efforts, particularly as an approach to target model specificity and to control complexity

[a]See relevant figure for a full description along with an overview of the source(s) and/or description of the derivation. Of note, these items are framed as general concepts rather than rigorous theoretical vehicles

frameworks derived from analytics-focused practice. Analytical-methodological and foundational concepts of interest to CSDS were framed by visual figures. As these key concepts will be re-examined in *Phase III* (design problem-solving), they are summarized in Table 2.16 as notable outcomes of the literature analysis.

2.7   Summary of Literature Analysis Results

2.7.1   Research Relevance

CSDS, being an emergent hybrid profession, embodies new organizational phenomenon which currently lacks focused coverage in MIS research literature. In aggregate, *Phase I* results address a research lacuna by producing a set of MIS-grounded organizational research perspectives concerning the emerging CSDS profession.

This literature effort pursues Verschuren and Doorewaard's framing of diagnostic research via background analysis, suitable for complex and emergent phenomena (Verschuren and Doorewaard 2010). Best practices concerning general and MIS-specific literature analysis were observed and applied (Hart 2000; Levy and Ellis 2006; Onwuegbuzie et al. 2012; Sammon et al. 2010).

The multifaceted literature review synthesizes concepts from several domains to deliver a structured account of CDSD as an emerging field. Literature reviewed triangulates academic, practitioner, and commercial literature, appropriate given the rapidly developing nature of the field and gaps in research.

The specific outcomes achieved, profiled in Sects. 2.6.1–2.6.6, are submitted as novel research contributions:

1. Specific gaps forestalling the process of CSDS professionalization are profiled.
2. Discrete forces propelling growth of the CSDS domain are isolated in a summary model.
3. A corpus of key CSDS literature is identified and summarized.
4. The assertion of gaps in CSDS literature is substantiated.
5. Sensitizing concepts are framed to support CSDS interview research conducted in *Phase II*.
6. Data science methodological concepts are positioned to support *Phase III* problem-solving.

The resulting six outcomes serve to ground and motivate the local research effort herein. As a broader contribution to the MIS discipline, the outcomes jointly intend to orient and motivate research efforts towards integrated technical, process, and organizational assessments of CDSD as a new phenomenon.

The CSDS organizational research gap, substantiated (*Result 4*), mirrors a lack of codified body of theory in CSDS practice (*Result 1*). The lack of theoretical foundations hinders professionalization, compromises the effectiveness of practitioner efforts, and forestalls programmatic efforts. It is asserted that CSDS must develop domain body of theory to advance as a profession (as reinforced in *Result 1*).

To circumscribe the domain and support literature gap analysis, a corpus of manuscript-length CSDS-associated works is framed and summarized (*Result 3*). In order to understand the practical drives impelling growth in the CSDS field, a functional model is framed (*Result 2*) based on Looijen and Delen's information systems management model (Looijen and Delen 1992). This posits a supply-and-demand relationship between data science best practices and cybersecurity challenges, respectively.

In order to support subsequent steps in the local research effort, two particular outcomes are framed. In *Result 5*, a key set of CSDS sensitizing concepts are framed to drive *Phase II* interview research. As framed by van den Hoonaard, a sensitizing concept is a starting point providing an initial guide to orient qualitative inquiry (van den Hoonaard 1996). Sensitizing concepts are especially of use in orienting research in complex and novel domains, as is the case with CSDS. Finally, *Result 6* surfaces a set of key data science methodological approaches of interest to CSDS practice. These methods are utilized to support design problem-solving in the concluding phase, *Phase III*.

2.7.2 *Meta-theoretical Implications*

A practical objective of this research effort is to advance CSDS body of theory to promote professional advancement. A crucial but less developed element of CSDS body of theory concerns *meta-theory* for assuring scientific rigor in CSDS practice. Put another way, there is a need for more refined CSDS body of theory concerning approaches for instilling scientific rigor in CSDS practice. It is asserted that a macro-outcome of this literature analysis has been to create a foundation for this discussion to advance. To summarize this foundation and frame-associated implications, the meta-theoretical implications of *Phase I* outcomes will be summarized.

Currently, CSDS, per the detailed maturity assessment in Sect. 2.5.6, can be considered as being in the process of knowledge individuation and body of theory codification through applied practice. Prescriptive data science treatments are actively being applied to an array of cybersecurity challenges. Maturation of body of theory requires iterating domain knowledge to specify the theoretical foundations (as testable assumptions) underpinning actions in practice. This amounts to asserting the importance of theory (*logos*—how and why) in guiding the design and implementation of solutions (*technê*—practical treatments).

A shortcoming in pursuing CSDS body of theory individuation relates to the relatively irresolute professional maturity of the data science parent domain as compared to cybersecurity. While cybersecurity faces a range of growing challenges that are reframing the domain, the like-to-like comparison reveals that the data science field is overall less mature in comparison (see *Result 1* per Sect. 2.6.1). A central challenge identified is that data science offers a potentially overwhelming array of methods, techniques, and technologies to address cybersecurity challenges.

Data science has been positioned as being a supplier of prescriptions to accommodate cybersecurity challenges through the vehicle of CSDS (*see Result 2* per Sect. 2.6.1). As data science is a still evolving and rapidly shifting profession, CSDS encounters struggles to resolve ad hoc prescriptions into best practices. This relative lack of maturity affects the ability of CSDS to clarify its value proposition in tension with cybersecurity. As CSDS is an emerging domain, this results in a lack of clarity as to which methods and techniques should be applied to meet focused cybersecurity objectives.

As a goal for development of the CSDS domain, the prescription of data science techniques to meet cybersecurity challenges ideally tracks through codified body of theory. Theory-codified and validated diagnostic approaches suggest prescriptions for treatments, much as in the medical profession. The implication is that developing CSDS body of theory requires specifying and advocating standard data- and science-driven processes (theory-driven *logos*), as opposed to the isolated prescription of techniques and technologies (practice-driven *technê*).

Technê-driven approaches are otherwise the default in the cybersecurity profession (de Bruijn and Janssen 2017; Foroughi and Luksch 2018). However, data science problem-solving, similar to medical diagnostics, espouses a process of contextual discovery prior to prescribing technical treatments (INFORMS 2013; Kelleher and Tierney 2018; Voulgaris 2017). In the context of CSDS, this principle is interpreted and summarized in Fig. 2.26, positioning CSDS body of theory as serving an assurance function, as a steward of rigor, in overseeing problem-solving diagnostics prior to prescribing data science treatments.

This is otherwise to recognize that engineering problem-solving in isolation is not sufficient to ensure rigor in CSDS practice. CSDS, absent efforts to instill rigor in diagnostic problem-solving, falls into the trap of achieving focused practical solutions while potentially missing key aspects of problem context. As an example, a random password generator to create complex passwords ignores the problem that

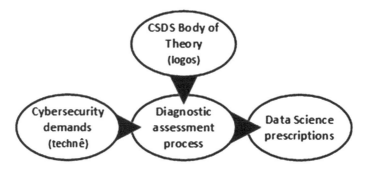

Fig. 2.26 CSDS body of theory context

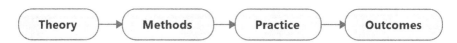

Fig. 2.27 Theory-driven CSDS

the results cannot easily be memorized, and thus users write their passwords on post-it notes. With an improper diagnosis and treatment, the cure becomes worse than the disease, so to speak.

The promotion of CSDS occupational closure initially rests upon appeals to the field of data science to deliver demonstrative practical methods for addressing challenges in the host cybersecurity profession. One path to improving this situation would be to sensitize CSDS stakeholders and managers regarding the need for scientific processes to diagnose cybersecurity challenges and to prescribe data science treatments.

This suggests the growing importance not of technical prescriptions, but of improved scientific understandings of the cybersphere allied with clear processes for systematic problem diagnosis. Whereas there are no lack of analytical methods to apply, and a fast-expanding set of challenges to address in the cybersecurity domain, the next evolutionary step for CSDS demands structured improvements to the scientific diagnosis and classification of cybersecurity phenomena. In the engineering-focused status quo, practice specifies methods absent theoretical context. An improvement occurs, per Fig. 2.27, when general theory, particularly as a guide to problem diagnostics, is interposed as a primary steward for the application of methods (which then orient implementation practice).

This implies the lacuna in CSDS professionalization, vis à vis data science, is not deciding on better methods but involves retrenching to improve understandings of cybersecurity challenges as scientific phenomena susceptible to research processes and methods. As Greenwood notes concerning the importance of the scientific method to extracting body of theory:

> The importance of theory precipitates a form of activity normally not encountered in a nonprofessional occupation, viz., theory construction via systematic research. To generate valid theory that will provide a solid base for professional techniques requires the applica-

tion of the scientific method to the service-related problems of the profession. Continued employment of the scientific method is nurtured by and in turn reinforces the element of rationality. (1957)

2.7.3 Managerial Relevance

In mapping CSDS boundaries and gaps, this analysis offers guidance to practitioners, managers, and researchers interested in advancing CSDS professional practice and general effectiveness. This includes those planning operational programmatic implementations and/or MIS-based action research initiatives (Baskerville and Pries-Heje 1999; De Villiers 2005).

CSDS is asserted as a hybrid domain and nascent profession. The field has emerged from the fusion of cybersecurity needs and data science methods in response to growing economic and organizational pressures associated with the effects of rapid digital expansion and evolution. From a managerial perspective, a structured understanding of the broad dynamics shaping the emergence of CSDS provides an ability to control and foster this transformation.

Being a practitioner-led discipline, managerial stewardship is essential to the maturation of the CSDS field. This analysis seeks to inform managerial actors concerning the landscape and issues resident in the current-versus-idealized state of the field. Citing Greenwood's framework for professionalization, next steps in the professional emergence of the CSDS domain requires knowledge individuation through operationalization and institutionalization of a *body of theory* (1957). Managers, as institutional mediators, are essential to iterating this evolution by facilitating the codification of a clear *body of theory*, which in practitioner terms constitutes operational *best practices* united with *theoretical context*. This mirrors the classical distinction between *technê*, knowing how to do, and *epistêmê*, knowing what to do and why.

The background analysis herein frames the major theoretical, practical, and structural elements necessary for transitioning from CSDS *body of theory* knowledge individuation to *occupational closure* via institutional codification. Specifically, institutional facilitation via professional training providers and certification authorities would be a key next step in achieving *occupational closure*, whereby the CSDS specialty can assert professional specialty status. As Beer and Lewis remark, the evolution of science in an age of increasing complexity will see a growing need, indeed a burgeoning discipline, for managerial specialists who understand the science of managing science (Beer and Lewis 1963). Understanding the path to professionalization outlined herein is central to arming such managerial "science stewards" to facilitate CSDS *occupational closure*.

Adjacent exemplars of similar successful analytics-associated *occupational closure* transitions are instructive. The nascent CSDS discipline has similar characteristics to other analytically focused interdisciplinary fields that have successfully emerged from sub-specialty to varying degrees of specialized professional status.

Operations research is a distinct professional field that emerged from military operations, economics, and computer science (Sodhi and Tang 2010). Financial Planning and Analysis (FP&A) emerged from accounting and computer science via Decision Support Systems (DSS) (Alter 1980; Power 2007, 2008). Both now are subjects of focused specialized professional training, education, and certification programs, partial earmarks of *occupational closure*.

More recently, the field of customer analytics has individuated professionally from marketing by way of business intelligence (BI) and data analytics. This creates a precedent for CSDS as a data analytics allied specialty to individuate professionally within the cybersecurity domain. Next steps are for a *body of theory* to be strengthened and codified, following the process and criteria extrapolated from Greenwood's model. Beyond axiomatic methodological prescriptions to pragmatic challenges, broader theoretical guidance on the application of scientific methods and processes are essential.

The next phase pursues qualitative interview research to chart the current and perceived future state of CSDS *body of theory* from the perspective of active practitioners.

2.7.4   Research Questions Addressed

See Table 2.17.

Table 2.17 Research questions addressed—literature analysis

Management Problem 1	**What gaps challenge the effectiveness of the CSDS field and impede professionalization?** • Immature CSDS body of theory hinders professionalization • CSDS lacks clear standards for driving scientific rigor in practice • Uneven institutional sponsorship forestalls professionalization • Data science as a profession lacks clear disciplinary boundaries • Cybersecurity as a profession is increasingly challenged by the rapid expansion and evolution of threats • Cybersecurity practice is engineering-focused and lacks strong theoretical foundations
Research Objective 1	*Analyze* **the as-is state of the CSDS field based on comparative literature** • The CSDS field is in the process of body of theory individuation, a prerequisite for professional emergence • Gaps in CSDS literature include risk quantification, data management, scientific methods, and organizational management • There is a need for body of theory codification through institutionally sponsored training and certification

Research Question 1	**What is the basis for asserting CSDS as a nascent professional domain?** • Active named and specialized practitioners are employed globally • CSDS derives sanction through its parent domains and demonstrates efficacy in focused use cases • The field represents a paradigm shift from traditional approaches to security assurance • CSDS is presently demand-driven as security professionals face growing challenges which are addressed by data science methods
Research Question 2	**What are the disciplinary boundaries of CSDS as a hybrid professional domain?** • The locus of the CSDS field is most logically situated within the cybersecurity professional domain • Applied CSDS involves addressing cybersecurity challenges with data science prescriptions and implies a gap analysis is conducted • There are early indications of recognition of the CSDS domain as a focused specialty from a range of institutions (commercial, professional, governmental, academic)
Research Question 3	**Where is CSDS in a process of professionalization?** • CSDS is in the process of professional emergence through body of theory codification • There is a need for clearer guidelines concerning the application of scientific rigor in practice • Institutional facilitation is necessary to achieve professional individuation, particularly professional training and certification • Institutions are beginning to frame CSDS as a unique sub-specialty within the security profession • There is a growing body of focused CSDS literature
Research Question 4	**What challenges face the CSDS domain on the path to professionalization?** • Both the security and data science domains are undergoing rapid technology-driven change • The methodological permissiveness of the data science domain challenges the ability to codify CSDS body of theory • The security domain lacks rigorous scientific theoretical foundations, focusing on applied engineering solutions • Data management is a key challenge requiring focused solutions • Commercial hype creates informational distortions compromising clarification of CSDS body of theory • Paradoxically, there is a need for institutional sponsorship, although uneven body of theory forestalls institutional adoption

Chapter 3
Phase II: CSDS Practitioners—Diagnostic Opinion Research and Gap Analysis

3.1 Research Objectives

Cybersecurity data science (CSDS) has been explored as an emerging field in the early stages of professionalization. The results of *diagnostic background analysis* undertaken through integrated literature analysis in *Phase I* assert that CSDS body of theory must advance to facilitate the professional emergence of the new domain. A range of sensitizing concepts were derived from literature analysis (*Result 5* per Sect. 2.6.5) to provide a foundation for interview and gap analysis undertaken in this phase. Key themes resident in CSDS literature were also summarized (*Result 4* per Sect. 2.6.4).

To promote the development of the CSDS profession, a systematic understanding of the current state of CSDS practice from the perspectives of active practitioners is useful. Deriving practitioner views of perceived challenges and best practices in the CSDS domain provides a basis for gap analysis to frame design requirements.

Supported by sensitizing concepts raised, *Phase II* combines qualitative and quantitative research methods. For the qualitative component, *diagnostic opinion research* is undertaken in the form of interviews with CSDS practitioners. This is followed by a quantitative component supporting *diagnostic gap analysis*, undertaken to surface and substantiate CSDS gap themes raised in interviews.

In the qualitative research component, 50 global CSDS practitioners were interviewed between June 2018 and April of 2019. The results profile shared practitioner perceptions of the CSDS domain. Perceived challenge and best practice themes raised in the interviews are coded and analyzed. Text analytics is undertaken on interview transcripts to provide supplementary substantiation of themes raised. Themes are cross-referenced with topics surfaced in CSDS literature.

In the quantitative research component, a range of multivariate statistical pattern analysis methods are applied to coded interview results. The goal of quantitative analysis is to extrapolate and substantiate patterns across interviews in the form of macro-themes. Results suggest shared practitioner themes representing key CSDS

© Springer Nature Switzerland AG 2021
S. Mongeau, A. Hajdasinski, *Cybersecurity Data Science*,
https://doi.org/10.1007/978-3-030-74896-8_3

gaps. Interpretation and framing of patterns surfaced through quantitative analysis are reinforced by mixed qualitative findings, bridging literature analysis, interview memoing, interview thematic analysis, and text analytics methods.

In terms of quantitative methods utilized, factor analysis is applied to extrapolate latent themes resident across coded interview results. Factor-to-factor linear fitting and correlation analysis results in a set of key CSDS thematic challenges and matching best practice prescriptions. Advancing a detailed understanding of practitioner prescriptions to address CSDS challenges supports *diagnostic gap analysis*. Shared themes advocate practical treatments of interest to a range of CSDS stakeholders.

Triangulating mixed methods across qualitative and quantitative components reinforces rigor. Variant inductive and deductive approaches are combined to substantiate central themes surfaced related to CSDS professionalization. The newness and complexity of the domain advocates the application of triangulated mixed methods aimed at exploring and extrapolating hidden patterns, as opposed to pursuing hypothetical-deductive approaches. There being no target variable or causal model to test, the methodological approaches applied focus on pattern extraction and substantiation. As such, no epistemological claims to empirical causality are asserted, the mixed methods being focused solely on rigorous latent pattern extrapolation and analysis.

In extrapolating practitioner views on key CSDS gaps and prescriptions, a foundation for subsequent design science problem-solving is established. Results frame problem-solving design requirements to be addressed comprehensively in the concluding phase (*Phase III*).

Table 3.1 summarizes the key methods and outcomes of this research phase. Table 3.2 lists the focused research objectives addressed.

Table 3.1 Summary of research and results: opinion and gap analysis

Chapter	Phase	Method	Result
3. CSDS Practitioner Interviews & Gaps	II. Diagnostic opinion research	Qualitative interview research	Key challenge & best practice themes
	II. Diagnostic gap analysis	Quantitative analysis of themes	Diagnosis of CSDS gap-prescriptions

Table 3.2 Summary of methodology research questions

Management problem 2	What prescriptive treatments address categorical CSDS gaps?
Research objective 2	*Diagnose* gaps impeding professionalization based on qualitative research
Research question 5	What gaps can be diagnosed in the emerging CSDS field from practitioner perspectives?
Research question 6	What treatments are prescribed to address gaps based on practitioner input?

3.2   Qualitative Component: Diagnostic Interview Research

3.2.1   Interview Research Preparation and Planning

3.2.1.1   Establishing Context

Titled CSDS professionals are active globally in a growing number of institutions. Practitioners formulate and apply a range of CSDS methodological approaches to address focused security challenges. However, as raised in literature analysis, CSDS lacks many institutionalized aspects of a mature profession, including standards bodies, certifications, training programs, and a well-codified body of theory. Cohesive CSDS body of theory has not yet been fully syncretized and articulated between the parent cybersecurity and data science domains. This evidences as a lacuna between CSDS professional practice objectives and outcomes.

The lack of well-codified body of theory at the junction of data science and cybersecurity forestalls the staged process of professionalization. This lacuna results in difficulties envisioning strategic and tactical plans, developing and delivering standardized training curricula, applying and managing recognized methods and processes, and assessing individual and organizational maturity.

Given the nascent nature of the CSDS field, qualitative interviews of active practitioners were undertaken to extrapolate an aggregated assessment concerning perceived gaps and best practices in the field. Extrapolating a practitioner understanding of the CSDS domain offers benefits to managers, researchers, educators, planners, and analysts alike.

By improving an understanding of current perceived challenges and opportunities in the field, prescriptive design requirements can be framed. Connecting practitioner-perceived challenges with best practice guidance provides a structured understanding of where the CSDS discipline is currently *as-is*, where it is evolving in terms of dominant trends, and where organizations need to foster future development to address *to-be* goals.

3.2.1.2   Qualitative Research Overview

Diagnostic opinion research was undertaken to understand perceived challenges and advocated best practices in the CSDS field based on shared stakeholder opinions. A qualitative interview approach was taken to clarify and diagnose perceived CSDS practitioner challenges and gaps. As an emergent practitioner-led domain, direct interaction with CSDS professionals provided an opportunity to understand the status of professionalization, to gauge the maturity level of current practice, and to identify gaps in the form of perceived impediments and concerns.

Between June 2018 and April of 2019, 50 interviews were conducted with titled CSDS practitioners. The format was semi-structured, guided by a set of standard open-ended questions. Allowing flexibility in replies encouraged detailed, ad hoc responses intended to reduce researcher bias. There was a desire not to impose set views, to allow the participant to self-identify and emphasize themes they personally viewed as being important. The format allowed interviewees to extrapolate and associate as they saw fit among the themes important to them, meaning some topics were embellished upon throughout the interview.

Interviews were recorded, transcripts were produced, and memos of researcher impressions were taken following the interviews. Transcripts were refined and structured to improve interpretability. Interview transcripts were coded in an open coding phase to identify central topics and themes. A final ordered codebook was derived and applied to code all interview transcripts. Final coded results were analyzed using a variety of quantitative methods to extrapolate shared patterns.

Given the newness and complexity of the CSDS domain, a range of themes were raised in interview feedback. Quantitative methods applied to the qualitative inquiry surfaced and substantiated shared latent themes. Latent themes concerning perceived CSDS domain gaps and prescriptions across all respondents were derived through factor analysis of coded interview data. Quantitative methods applied were focused on pattern extrapolation. Thus, no epistemic claims to empirical outcomes were asserted.

Combined, triangulated results from mixed literature analysis, qualitative, and quantitative components were summarized and simplified through root cause analysis. Results were framed to support concluding design requirements analysis undertaken in the subsequent phase (*Phase III*).

3.2.1.3 Interview Research Process

The goal of the structured interview research process was to facilitate the extrapolation of insights from practitioner interviews in the form of patterns. A structured overview of the qualitative interview research process undertaken:

1. *Preparation and Planning*

   (a) Structuring interview format and approach
   (b) Review sensitizing concepts from domain literature (van den Hoonaard 1996)
   (c) Develop and frame interview questions
   (d) Determination of target sample
   (e) Outreach/invitations to CSDS professionals

2. *Initial Coding and Refinement*

   (a) Interviews with participant volunteers with memoing
   (b) Transcription of interviews
   (c) Open coding
   (d) Coding saturation reached ($n = 22$)
   (e) Ordering of codebook

- Ordering families with sensitizing concepts (Glaser 1978, 1998; Glaser and Strauss 1967)
- "Coding mall" classification (Corbin and Strauss 2015)

3. *Structured Coding and Results*

   (a) Selective axial coding from ordered codebook

   - Recode initial interviews ($n = 22$)
   - Complete and code remaining interviews ($n = 28$)

   (b) Final code cleaning and merging
   (c) High-level results

   - Demographic analysis
   - Frequency and basic statistical analysis

   (d) Complementary text analytics content analysis of transcripts

4. *Quantitative Component: Mixed Methods Gap Analysis*

   (a) Logistic regression fitting
   (b) Exploratory factor analysis
   (c) Factor fitting and extrapolation
   (d) Factor-to-factor linear fitting and correlation analysis
   (e) Gap-prescription model extrapolation
   (f) Root cause analysis of gaps
   (g) Final interpretations and guidance

3.2.1.4   Methodological Underpinnings

Given the emergent nature of CSDS, it was determined that a loosely inductive approach influenced by, yet not rigorously adherent to, grounded theory was desirable to elicit insights directly from practitioners. Grounded research is an inductive methodology and advocates allowing impressions and insights to emerge from the research process. Although surveys would support data collection at greater scale, it was felt that survey design would impose a priori assumptions and structure upon responses. To avoid this and to stimulate breadth in responsiveness, a semi-structured interview format was pursued.

This approach was deemed desirable given the emergent, relatively new, complex, and methodologically diverse nature of the CSDS domain. The semi-structured approach departs from strict grounded theory principles while still allowing for flexibility in responses to support inductive discovery. As the CSDS field is new and rapidly changing, this approach supported flexible discovery, as opposed to a survey tool, which would impress and constrain responses to a much greater degree.

In undertaking the qualitative inquiry and subsequent analysis, best practices were observed concerning general management research methods and processes (Baarda 2010; Lee and Lings 2008; Saunders et al. 2009), grounded theory research

(Glaser and Strauss 1967; Mason 1994; Miles and Huberman 1984; Mortelmans 2019; Myers and Newman 2007; Strauss 1987), content analysis (Krippendorff 2019), interviewing (Bernard and Ryan 2010; Richards and Morse 2007), codebook ordering (Glaser 1978, 1998; Goffman 1961; Polkinghorne 2015; van den Hoonaard 1996), and interview transcript coding (Bryman and Burgess 1994; Burnard 1991; Corbin and Strauss 2015; Saldana 2016).

The interview research approach, while not being strictly adherent, was influenced by grounded theory in facilitating the extrapolation of interpretations from raw results (Charmaz 2014; Corbin and Strauss 2015; Glaser and Strauss 1967). The semi-structured interview format allowed interviewees to freely interpret and frame their own responses from a core set of interview questions. The raw results produced were transcript text artifacts capturing the free verbal interpretations and impressions of the interviewees. While prompted to discuss particular topics, interviewees were permitted to range freely in their interpretive responses.

Given the complexity of the domain under scrutiny, an allowance was made for the use of sensitizing concepts from preliminary literature analysis (van den Hoonaard 1996). Sensitizing concepts are permitted within the bounds of grounded research, although their application is considered by some to be a departure. A key sensitizing conceptual model concerning the relationship between CSDS trends, challenges, and best practices guided the preparation of interview questions. The questions were framed as explicitly open-ended to encourage respondents to interpret and provide feedback per their own priorities, preferences, and perceptions.

The initial interviewing phase applied open coding, a bottom-up, emic approach. Questions were framed as intentionally broad and open-ended, allowing the respondents to provide their own interpretations and to surface issues and topics most of interest to them. This provided opportunities for respondents to free-associate and raise their own interests and observations as they saw fit. The intention was to support broader and more flexible assessments of practitioner views and concerns across the interviewee sample.

At coding saturation ($n = 22$), the codebook was ordered and consolidated into a "coding mall" (Corbin and Strauss 2015). Topics were extrapolated into notable themes. This incorporated rough sensitizing concepts (Glaser and Strauss 1967) and led to axial coding from the final ordered codebook. In composite, progressive codebook structuring and ordering reflected a process of *iterative recursive abstraction* as advocated by Polkinghorne (2015). Subsequently, the original transcripts were recoded, and the remaining interviews were completed and coded ($n = 28$). Coding and structuring of the codebook was managed utilizing the NVivo 12 software tool.

Within grounded theory, Goffman advocates organizing *family of terms* to consolidate empirically related concepts (1961). Van den Hoonaard details methodological approaches to employing sensitizing concepts, including establishing family resemblances to connect and organize empirical observations (1996). In complex domains, of which CSDS is an example, there is a precedent for deriving sensitizing concepts to facilitate codebook ordering from initial open coding results.

To support inductive interpretation via triangulated mixed methods, thematic content analysis was applied to transcript corpora utilizing natural language processing and text analytics. This provided a supplemental inductive approach to extrapolate and frame topical themes from the transcript text artifacts (Bechor and Jung 2019; Krippendorff 2019; Onwuegbuzie et al. 2012). The result was a set of sensitizing themes to support the final analysis and interpretation of results across the sample.

3.2.1.5 Sensitizing Concepts

Given the complexity of the CSDS domain, initial sensitizing concepts were derived from background literature to focus and orient the preparation of interview questions. A central sensitizing concept was that the CSDS parent domains, cybersecurity and data science, are both undergoing rapid evolution due to technical and methodological innovations (Buczak and Guven 2016; de Bruijn and Janssen 2017; Lehto 2015; Mahmood and Afzal 2013; Verma et al. 2015). Evolution in this context may lead to both negative (threatening) and positive (fortifying) security outcomes. For example, the proliferation of cloud-based storage and services creates new security vulnerabilities while also enabling new platforms and mechanisms for cybersecurity event monitoring and detection.

Based on the notion that CSDS is subject to both constructive and destructive forces, it was posited that innovations can be framed as cybersecurity challenges (in the form of security threats, new vulnerabilities, or impediments to assuring security) and/or data science treatments (in the form of improved methods for event monitoring and/or detection). In composite, an emerging innovation, in the form of a trend, might frame challenges and/or best practices. Challenges and best practices may, in turn, create or avert a security threat, respectively. This observes that technical and methodological innovations lead variously to cybersecurity symptoms and/or CSDS data science treatments. This reciprocal relationship between CSDS challenges (gaps) and best practices (prescriptions) is represented in a sensitizing conceptual model in Fig. 3.1, below.

Central to this sensitizing conceptual model is the recognition that practitioners may potentially frame CSDS themes as challenges and/or best practices. For instance, a value-neutral theme highlighting *the aggregation of distributed security data sources* might be framed as a challenge, inhibiting CSDS practice when not addressed properly, or as an enabler, when implemented as a best practice. The conceptual model thus anticipates and allows for an interviewee to raise CSDS themes as issues requiring attention (challenges) and/or as facilitating enablers (best practices).

A more general principle underlying the observation of dualism resident in CSDS innovations is evident: namely, that interview subjects may raise topics with a positive or negative valence. This has an aspect of the more general phenomenon of positive versus negative framing. The central subject matter, perceptions of the CSDS field, thus may conceivably invoke responses that identify the following: (1) a challenge as a required best practice or (2) the absence of a best practice as a standing challenge. The sensitizing concept of duality is thus a general observation

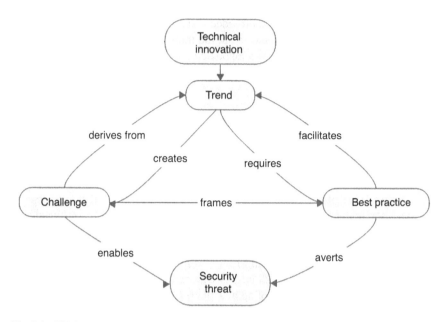

Fig. 3.1 CSDS challenge-best practice reciprocity conceptual model

beyond the CSDS domain: that interviewees frame base, positive (scientific) assertions and observations with normative judgments. Of interest in interpretation is the
ability to abstract, where needed, normative from positive assertions in order to
refine interpretation.

3.2.1.6   Interview Questions and Format

Based on sensitizing concepts derived from literature analysis, there was a desire to
understand practitioner perceptions regarding challenges and best practices in the
CSDS domain. There was an interest in evoking related feedback concerning methods applied in practice and adversarial security trends. A standard set of interview
questions was framed:

1. How did you become involved in cybersecurity data science (CSDS)?
2. What key trends are emerging in CSDS?
3. What are top challenges in the CSDS domain?
4. Are there CSDS best practices you feel should be socialized more broadly?
5. Do you draw upon adjacent domains in applying CSDS?
6. What adversarial trends do you see emerging?
7. Do you have any colleagues you can recommend to participate in this research?

The questions were explicitly communicated to the participants before interviews as open response questions to be discussed within a 30-min interview format.

In pursuing dramaturgical interview guidelines advocated by Myers and Newman (2007), question 1, a grounding question on interviewee background with relevance to the CSDS domain, was framed at the outset of the interview. In addition to collecting demographic details, this encouraged role-based identification as a professional CSDS interviewee and allowed for the establishment of communicative rapport and inclusion through mirroring.

Question 2 began the substantive interview and was intended as a value-neutral query to elicit interviewees to raise central CSDS themes of importance to them as practitioners. This question was explicitly intended to be value-neutral to stimulate open feedback on respondent thematic interests. It was assumed respondents might choose to raise and emphasize negative (challenge) and/or positive (best practice) themes in equal measure according to their own preferences and perspectives.

Once interviewees were primed to free-associate on key CSDS themes of unique interest to them as a practitioner and according to their own value valences, subsequent questions pursued perceived negative and positive themes specifically. Questions 3 and 4 raised the core interview questions regarding perceived CSDS challenges and best practices. The question on challenges was framed first and was left open-ended such that the respondent could raise adversarial, technical, methodological, and/or organizational issues. Similarly, the subsequent question on best practices was left quite broad for the respondent to interpret according to their own thematic interests and perceptions.

Two closing substantive questions were framed to pursue feedback on specific positive and negative themes. Question 5 inquired as to whether there were particular CSDS methods derived from external domains that the practitioner advocated. As the emergence of the CSDS field constitutes a relatively recent phenomenon, the intention was to identify methodological advocacy in the form of practices applied from more established domains.

Subsequently, a question on perceived emerging cybersecurity adversarial threats was posed. This allowed for the identification of particularly worrying adversarial innovations emerging. An implicit implication was that such emerging challenges were significant and notable in that they constitute a detection "blind spot" or otherwise imply the need for novel CSDS treatments.

The interview sequence was closed with an open question as to whether the respondent might be willing to recommend colleagues or professional associates for participation in the research. This allowed for general closure of the dramaturgical interview dialogue but also was of practical interest in perpetuating the research effort.

3.2.1.7 Interviewee Selection

In total, 50 global CSDS practitioners were interviewed between June of 2018 and April of 2019. Interview candidates were identified and invited through the online networking website LinkedIn. LinkedIn, a professional networking website and Microsoft subsidiary, is recognized as the largest online global professional network with 610 million members across 200 countries and territories (LinkedIn 2019).

As the objective was to identify a focused list of only those professionals currently practicing as named *cybersecurity data scientists*, a refined LinkedIn search was conducted. The targeted search sought to identify only those professionals with a current job title explicitly including the words "cyber" with either "data scientist" or "analytics." The search allowed for "analytics" in place of "data scientist" in the job title.

A filtering specifier was added to exclude any professionals in non-practitioner roles, such as marketing, sales, or recruiting (LinkedIn 2018). The formalized LinkedIn search logic applied was as follows: job title = (cyber *and* data scientist) *or* (cyber *and* analytics) *and not* (marketing *or* sales *or* recruiter *or* recruiting). Subsequent manual review removed several candidates if the job title listed was judged as not being accurate or was aspirational in character (e.g., a student intending to enter the field but with no current role or professional work experience). This resulted in a final set of 348 global professionals as interview candidates.

Over a period of 9 months beginning in June of 2018, all 348 professionals on the initial candidate list were contacted via the LinkedIn messaging function and were invited to participate in an interview in support of academic research on the emerging CSDS field. Approximately 70 respondents indicated interest, and a final 37 were able to commit to scheduling and completing the interview. Thirteen additional interviewees were identified and added through recommendations from the core interviewees. In total, 43 cybersecurity data scientists were interviewed verbally through web conferencing for 30 min. Due to scheduling challenges, seven interviewees opted to provide written feedback via email in lieu of a discussion. Given the structured set of interview questions and the text-focused interpretive approach, written feedback was deemed acceptable for these exceptions.

To address sample integrity, LinkedIn profiles for the entire interviewee pool ($n = 348$) were profiled to identify demographic aspects, including gender, estimated age, years employed, educational level, and academic specialty. Statistical tests were conducted to substantiate equivalency between the group of 50 interviewed individuals and the remaining group of 311 candidates not interviewed (bearing in mind 13 interviewees were identified as referrals from outside the original LinkedIn target candidate group of 348). Comparing several key measures, statistical tests suggested that the two groups were not significantly different.

Analysis of variance (ANOVA) tests indicated that the average estimated age and years employed for the group of 50 respondents did not differ significantly from the group of 311 candidates remaining. Linear test assumptions were confirmed concerning normality, robustness, and uniformity in standard deviations. Examining estimated age between the interviewee group and the remaining candidate group, the null hypothesis H_0 of "significant difference between group means" was rejected, indicating a lack of statistically significant differences in sample means with test statistics $F(1, 358) = 0.970$, MSE $= 89.75$, P-value $= 0.7196$, and $\eta2 = 0.0004$.

Concerning selection bias, one potential biasing factor is that the individuals identified and contacted had chosen to maintain a LinkedIn profile and thus to promote their professional background publicly. It should be considered that some practicing CSDS professionals might choose not to publicize their employment or activity in this field, for instance, to maintain secrecy, to avoid scrutiny, or because they are not seeking networking or external employment opportunities. However, as 13 of the 50 final interviewees (26%) were identified through references from initial interviewees and thus were not directly identified on LinkedIn, this, in principle, partially ameliorates potential selection bias.

3.2.1.8   Interviewee Demographics

A demographic profile of the interviewees was conducted, as extrapolated from LinkedIn details (Table 3.3). The results are summarized in Tables 3.3 and 3.4 below.

The age estimate was inferred as the date of beginning undergraduate studies less 18 years. When no undergraduate start date was recorded, a general estimate was made based on work history (i.e., first recorded employment, less 18 years). The relatively high variation in ages logically mirrored a high variation in years of employment experience. Striking was the relatively low number of comparative years of explicit CSDS work experience (mean 2.9; standard deviation 1.9; median 2.5), validating that CSDS is indeed a relatively new job designation.

Active region was aligned with the region of origin in 78% of the interviewees ($n = 39/50$), measured as an alignment of the past university and regional employment locations with the current location. Concerning migration trends, 22% ($n = 11/50$) moved from a different region of origin to their present regional locale.

Table 3.3 Demographic profile of interviewees

Gender	n	%
Male	43	86
Female	7	14

Age	Years	
Mean	36.8	
StdDev	9.1	
Median	35.0	

Current region	n	%
North America	35	70
Western Europe	10	20
Eastern Europe	2	4
Middle East	2	4
South America	1	2

	Years employed	Years CSDS	Years data sci	Years cybersecurity
Mean	14.2	2.9	3.6	5.3
StdDev	9.5	1.9	3.4	5.2
Median	12.0	2.5	3.0	4.0

Of these, 18% ($n = 9/50$) moved specifically to the USA, with 10% ($n = 5/50$) moving specifically from the Asia Pacific region to the USA.

Prior to moving into a cybersecurity data scientist or allied role, 68% ($n = 34/50$) of respondents had both cybersecurity and data science work experience. Twelve percent of respondents ($n = 6/50$) had no data science *or* analytics experience before moving into a CSDS role. Eight percent of respondents ($n = 4/50$) had no cybersecurity experience prior to moving into a CSDS role. Eight percent ($n = 4/50$) lacked both data science/analytics *and* cybersecurity experience. The average age of the four respondents with no experience in either domain was 31, indicating that some of these professionals may have recently completed post-graduate or certificate training and undertook a CSDS role based on this criterion. This observation suggests that some employers appear willing to accept a lack of professional work experience in one or both of the CSDS parent domains, likely with on-the-job training provided after moving into the role.

Table 3.4 summarizes interviewee job role, industry, and educational background, as extrapolated from self-reporting during interviews and cross-referenced with LinkedIn profiles. Concerning academic field, as noted in Table 3.4, more than one field was possible per respondent, where multiple degrees were held: 34% ($n = 17/50$) had degrees in two different academic domains; 16% ($n = 8/50$) held degrees in three fields. Twenty-four percent ($n = 12/50$) of respondents specifically held both "science/quantitative" and "IT/engineering" degrees (this being a notable pattern).

At the outset of the interview, respondents were asked whether they had become involved in CSDS intentionally or if they had found themselves in a CSDS role more serendipitously. A notably high 38% ($n = 19/50$) reported having found themselves entering the CSDS domain by serendipitous circumstances or happenstance (Table 3.5).

Table 3.4 Employment and educational profiles of interviewees

Current job role	n	%
Data scientist	33	66
Manager/team lead	9	18
Engineer	6	12
Consultant	2	4

Current industry	n	%
Software and services	28	56
Consulting	7	14
Finance/financial svcs/ insurance	7	14
Government/military	3	6
Consumer products	2	4
Academics/research	2	4
Telecom	1	2

(continued)

Table 3.4 (continued)

University study	Years	
Mean	6.8	
StdDev	3.1	
Median	6.0	

^ Degree	n	%
PhD	13	26
MA	18	36
BA/BSci	15	30
None	4	8

Academic field	n	%a
CompSci/IT/engineer	36	72
Science/quantitative	20	40
Humanities	2	4
None	4	8

Certificates	n	%
No certs	21	42
Cyber & DS	11	22
Cyber cert	9	18
DS cert	9	18

aSeveral interviewees held degrees from multiple fields

Table 3.5 Entry into CSDS domain by interviewees

Entry into CSDS domain	n	%
Intentionally pursued and planned to specialize in CSDS	31	62
Serendipitous, unplanned entry into CSDS role	19	38

3.2.1.9 Logistics and Responses

All participants were provided the central list of questions for consideration prior to the interview. Participants were provided an interview research consent agreement, to which they agreed and which is held on record. The consent agreement and associated aspects, especially concerning confidentiality, were summarized verbally at the outset of the interview. This served to stage the formal interview and addressed a key topic of potential concern related to the inherent sensitivity of the domain itself, reassuring respondents that all responses were to be held in strict confidence.

Interviews were conducted through web conferencing (WebEx), with the interviewer establishing an amenable date and time via email and sending a link for the web-based conference via an emailed invitation. In some cases, appointments needed to be rescheduled due to conflicts. During the interview, the interaction was voice only (no video), which was a practical consideration to maintain audio quality but was also intended to reduce the introduction of distracting visual cues.

Of note, a minority of respondents, 14% ($n = 7/50$), did not wish to be recorded, and in these instances, detailed notes were taken on key themes raised. A subset, 14% ($n = 7/50$), for various reasons, preferred to provide feedback in free-form textual responses via email. Both exceptions were allowed in the context of the research, and allowances were made to carefully treat the textual results like-to-like with transcripts, although there was a reduction of textual richness involved in these cases. A majority, 72% ($n = 36/50$), of interviews were recorded, and full transcripts were produced. All interviews resulted in textual artifacts embodying responses to the interview questions: summary transcripts, formatted textual notes, or textual replies.

Committing and carrying out the research interviews involved substantial logistical efforts. Scheduling interviews often involved multiple attempts and communications. As a rule, professionals in the domain reported being intensely busy such that scheduling a 30-min interview was often difficult (and in the cases of those who provided textual responses, not possible).

Of note, requests for participation also resulted in a number of volunteers from allied backgrounds, but who lacked indications of hands-on experience with CSDS work. These professionals were involved in allied roles, principally marketing, business development, consulting sales, product sales, or recruiting. These professionals were not included in the research effort as target interviewees were intended to be explicitly active, hands-on CSDS practitioners.

In staging the interviews and structuring researcher communications during the interviews, focused best practices were applied based upon guidance from Myers and Newman concerning qualitative interviews in IS research (Myers and Newman 2007). Observing the dramaturgical model advocated, the following guidelines were followed, as noted:

1. *Situating the researcher as actor*: Initial questions framed the effort clearly as a research initiative, and it was communicated that a number of other CSDS professional interviewees were involved. Initial questions related to the background of the interviewee, encouraging the participant to identify in the role of a professional CSDS interviewee.
2. *Minimize social dissonance*: An effort was made to make the interviewee feel comfortable concerning the research process. The level of in-/formality in communication was balanced to the perceived comfort level of the interviewee. Applying mirroring, in general a more formal, structured approach was taken with those in military, managerial, or corporate roles, whereas a more flexible, informal disposition was taken with those in a start-up or independent contractor role, depending on their indicated communicative disposition.

3. *Represent various voices*: Respondents were from a range of different CSDS professional roles, from individual contributors to practitioner-managers. Professional backgrounds varied, ranging from military to management consulting. There was a gender imbalance (14% female participants), and efforts were made to facilitate input concerning gender disparity in CSDS practice. Age ranges were relatively broad (mean 36.8; standard deviation 9.1; median 35), and an effort was made to reflect experience and status as befitting the indicated desires of the respondent. Education levels varied, and framing of the research effort was balanced to address varying comfort levels with formal academic research practices.

4. *Everyone is an interpreter*: The interview transcript was viewed as a central artifact, but a provision was made to incorporate variant texts. Memos regarding general themes and impressions were recorded alongside the transcript to provide alternate views concerning the perceived context and meaning of the topics discussed. Further, in support of inductive grounding, text analytics was applied to extrapolate semantic themes directly from structured versions of the transcripts.

5. *Mirroring in questions and answers*: A conscious effort was made to reflect terminology and themes raised by the interviewees, to speak in a lexicon comfortable to them and to encourage elaboration on topics of particular interest to them personally.

6. *Flexibility*: While there was a list of seven open-ended questions, and the goal was to address all questions within 30 min, there were otherwise no restrictions on responses. Interviewees were given the flexibility to respond as they desired and to raise topics as they saw fit and in any order they wished. The main structural constraint concerned attempting to move the questions forward and to address all questions as time progressed in the set 30-min format.

7. *Confidentiality of disclosures*: Confidentiality and research ethics guidelines were established and agreed with a formal academic consent agreement, which was assented to via an explicit email acknowledgment.

3.2.2 Initial Coding and Refinement

3.2.2.1 Open Coding and Refinement Process

In the initial interview phase, 22 interviews were conducted, textual results were refined and structured, resulting textual artifacts were open coded, and a standard ordered codebook was derived. Structured versions of the initial transcripts, notes, and research memos were produced. Textual replies were segmented to align with the key interview questions to facilitate interpretation.

As a semi-structured interview approach was taken, interviewees at times would circle back and return to previous questions. In these cases, an effort was made to aggregate textual responses into the structured question categories for clarity of interpretation. Further to improve clarity in the textual artifacts, interviewer questions and commentary were extracted from the text, with a label added to specify which particular questions were being addressed.

The result of the refinement process was focused, categorized textual artifacts upon which open coding was conducted. The resulting textual artifacts were clearly structured to identify responses to the key topics corresponding to the main interview questions (questions 2 through 6): trends, challenges, best practices, adjacent domains, and adversarial trends.

3.2.2.2 Open Codes and Categories

Phase I open coding pursued extrapolation of key themes raised in the refined interview textual artifacts. The goal was to achieve a standard set of codes in the form of an ordered codebook to support *Phase II* selective axial coding. Progressing through the initial 22 responses, codes evolved, aggregating and bifurcating as emerging relative frequency and thematic overlaps dictated. This observed an *iterative recursive abstraction* approach as advocated by Polkinghorne (2015). Phase 1 open coding reached code saturation within 22 interviews (44% of the final set of 50 interviews).

The codebook ordering process was partially guided by a sensitizing ordering family (Glaser 1978, 1998; Glaser and Strauss 1967) or "coding mall" (Corbin and Strauss 2015). To categorize extrapolated themes-as-codes, the classical MIS tripartite frame of organization, process, and technology was applied. This scheme oriented CSDS thematic codes into categorical topics that aligned with either organizational aspects (i.e., management, training), process factors (i.e., methods and modeling), or technology topics (i.e., engineering and adversarial technical vectors).

The saturated initial open coding scheme extracted prior to final codebook ordering is shared as it may be of inherent interest to practitioners and researchers. The initial open coding scheme gives insights into the raw themes observed in the interview artifacts prior to the abstraction of final codebook ordering, consolidation, and refinement:

CSDS Trends: 17 codes in 3 categorical themes
- Organizational and management: 7 codes

  - Cyber risk becoming general enterprise risk
  - Cyber resilience as a structured decision-making process
  - Data-driven paradigm shift
  - Persistent threats and vulnerabilities assumption
  - Outsourcing to managed security service providers (MSSPs)
  - Increasing regulatory pressure
  - Cyber data science as efficiency driver

- Methodological and modeling: 7 codes

  - Facilitating knowledge and decision management
  - Deep learning and decision automation
  - Automation of vulnerability testing

- – Anomaly detection automation (including UEBA)
- – Hosting and pushing detection to endpoints
- – Preventative threat intelligence
- – Unsupervised machine learning pattern detection

- Data engineering and management: 3 codes

  - – Data quality, exploration, and preparation as a process
  - – Cloud and container-based tools and data storage
  - – Emergence of open source technology stacks

CSDS Challenges: 29 codes in 5 categorical themes
- Organizational and management: 6 codes

  - – Unclear ownership, decision-making, and processes
  - – Marketing hype creating unrealistic expectations
  - – Shadow IT and proliferation of exposure
  - – Inherent cost overhead of hosting operations in house
  - – Uncertainty and confusion (infrastructure) leads to reactive, over proactive stance
  - – Unanticipated overhead in deploying full open source platforms

- Skills and training: 5 codes

  - – Over focus on engineering and IT
  - – Lack of training in data science methods
  - – Lack of cybersecurity domain experience
  - – Qualitative impressions dominate quantitative and risk-focused decision-making
  - – Breadth and complexity of the cyber domain

- Methodological and modeling: 9 codes

  - – Limitations of rule-based methods giving too many false positives
  - – Contextual nature of normal versus anomalous phenomena
  - – Application of scientific research processes and methods
  - – Deriving probabilistic and risk models
  - – Real-time detection difficult
  - – Transparency and interpretation of machine learning results
  - – Deep learning requires very large datasets and data prep
  - – Designing and developing new methods
  - – Over focus on predictive machine learning

- Data engineering and management: 4 codes

  - – Privacy and regulatory concerns
  - – Data access, volume, quality, preparation, integration, and feature engineering
  - – Standardizing data and context models to "map" local environment
  - – Lack of labeled data

- Nature of cyber vulnerabilities: 5 codes

  - – Logical and physical infrastructure have inherent vulnerabilities

- Time to detection and reconnaissance advantage
- Unpredictable nature of adversarial human behavioral motivations
- Expanding attack surfaces
- Securing the cloud and increasing exposure from cloud services/storage

CSDS best practices: 23 codes in 4 categorical themes
- Organizational management: 6 codes

  - Upper management buy-in and support
  - Clear decision processes with a quantitative risk focus
  - Process focused cross-functional collaboration
  - Adopt IT service and quality management standards
  - Patching and updating outdated systems
  - Segment risk programmatically and outsourcing components

- Skills and training: 6 codes

  - Balance between specialization and general skills
  - Train cybersecurity staff in data science methods
  - Train data scientists in cybersecurity foundations
  - Address AI hype and unrealistic expectations directly
  - Coding/programming skills and capabilities
  - Train general staff on vulnerabilities and threats

- Methodological and modeling: 7 codes

  - Providing interpretability and context for model derived alerts
  - Framing data science as an IT process and service
  - Focus on implementation efficiency and agility
  - Designing for perpetuity in the face of constant technical change
  - Survey academic methods and techniques
  - Human-in-the-loop solutions and approaches (as opposed to automation)
  - Exploratory process to understand data and derive models

- Data engineering: 4 codes

  - Data preparation and feature engineering as a process
  - Track and label incidents and exploits
  - Standard data model frameworks (data structures and ontologies)
  - Data-sharing consortiums and initiatives

Adversarial trends: 13 codes
- Adversaries constantly adapting to countermeasures and innovating methods
- Machine learning mechanisms, vectors and threats, including adversarial ML
- Attackers exploiting vulnerabilities and vectors in new tech. (i.e., cloud, mobile, IoT, ICSs)
- Growing sophistication of social engineering and identity hacking
- Kits and tool-based automation of penetration and attacks
- Exploiting inherent vulnerabilities of logical and physical infrastructures

- State actors hiring data scientists
- As-a-service hacking with cryptocurrency mediation
- Reverse engineering or leveraging known detection and prevention mechanisms
- Time to detection and reconnaissance advantage
- Growing internal threats and corporate espionage
- Information warfare, fake news
- Hijacking systems for cryptocurrency mining (cryptojacking)

Adjacent domains: 8 codes
- Fraud analytics, forensic and criminology
- Deep learning—computer vision and acoustics
- Medical, epidemiological, ecological, and bioinformatics
- Network graph analytics
- Social and behavioral sciences, including (micro-)economics and game theory
- Natural language, semantic, and knowledge engineering
- Risk management
- Forecasting and time series analysis

3.2.2.3 Memoing Insights

Notes were captured as memos following each interview, observing the interviewing practice of memoing. Memoing recorded high-level thematic insights and impressions supplementary to the transcript. This observed Myers and Newman's advocacy to support multiple interpretive vehicles beyond raw transcripts (Myers and Newman 2007).

Concerning subject matter, a general impression gained across memos concerned the dominance of the following themes:

1. *Data management*: Data management as both a challenge and an advocated best practice was an overwhelmingly dominant theme. Typically, the topic was raised in terms of the activity being time-consuming but highly important, and that this was prone to causing conflicts with management concerning perceived priorities.
2. *Organizational factors*: A majority of interviewees raised organizational and management factors at some point in the interview, both in the context of challenges and best practices.
3. *Scientific inquiry and discovery*: Although the scientific method was not always raised explicitly, it was referenced frequently in terms of allied activities, such as gathering evidence, exploring data in a structured manner, discovering emerging phenomena, developing contextual environmental understandings, and building and testing quantitative models.

Concerning the interview process, a key impression noted in memos concerned the degree to which the intentionally open-ended interview questions revealed judgments based on unique practitioner outlooks and backgrounds. This was an important observation given the relatively new, dual-disciplinary, and still evolving nature of the CSDS domain. Memoing observed interviewee judgment patterns in the following areas:

1. *Interpretive value judgments*: In framing themes, respondents chose to variously emphasize critiques of or positive aspects of the CSDS domain. Inherent to the open-ended interview model was the notion that CSDS trends might be framed as advocated best practices for domain improvement or that best practices might be framed as challenges when perceived to be lacking. The relative emphasis on one or the other was resident in responses. It appeared that value judgments stemmed from experience in professional practice. It was observed that a challenge was surfaced when the respondent observed roadblocks experienced in practice, whereas best practices were advocated when efficacious approaches allowed practitioners to overcome impediments. As examples were often cited in interview responses, it was posited that value judgments were indicators of hands-on practitioner experience, rather than proxy indicators of emotional or personality traits.

2. *Parent domain emphasis*: Per the professional demographic overview, respondents varied in terms of their skills and background. There were varying levels of exposure to the two CSDS parent domains, with not all respondents having a balanced exposure to both fields. This was reflected in responses, with some interviewees choosing to emphasize one or the other parent CSDS domain topically to the dearth of the other. For instance, a cybersecurity professional relatively new to data science and analytics would use terminology and raise topics more strongly oriented with traditional cybersecurity professional practice, and vice versa. While 86% ($n = 43/50$) of respondents had experience in both cybersecurity and data science, the extent of relative experience was typically not equal. In aggregate, cybersecurity professionals pursuing and adding analytical methods to their professional practice were more common than focused data scientists building competency in the cybersecurity domain. This was observable in the professional demographic measure of 5.3 years on average of cybersecurity professional practice experience versus 3.6 years on average with data science. Otherwise, this also recognizes that data science itself is a relatively new professional field as compared to the cybersecurity field, having emerged as a job title only in the last 5–6 years.

3. *MIS tripartite classification*: Recognizing that the interview questions did not explicitly prompt for organizational, procedural/methodological, or technical topics, respondents chose to select and emphasize their own perceived frames. Some respondents were more or less attuned to managerial and organizational topics. Similarly, some respondents chose to focus more or less on technical and engineering topics. Raising the topic of processes and methods was variable, with some respondents highlighting and emphasizing the need for processes and methods and others ignoring this terminology and situating their procedural observations in the terminology of management and/or engineering factors. This otherwise observed than not all respondents were necessarily attuned to the tripartite MIS conceptual framework (organization, process, technology). This led

eventually to the diffusion of clarity concerning these formal distinctions in aggregate responses, although in some cases the respondent applied this conceptual terminology explicitly.

Recognizing that the above factors influenced individual responses, it was determined that a simplified, aggregated coding scheme would reduce the potential distortion of interpretive bias. An effort was made to review and streamline the codebook in order to focus and balance themes outside potential value, domain, or framing contexts.

3.2.2.4   Final Ordered Codebook

Coding saturation denotes the point in the open coding process when no new codes emerge, indicating a threshold has been reached, whereby the majority of focused themes raised by respondents have been identified (within the purview of the interview questions). Coding saturation was achieved at the 22nd interview, following which a systematic review was conducted to refine and structure a standard ordered codebook of themes raised. A key goal for refinement was conceptual and quantitative economy while not abandoning focused themes. Related themes were aggregated or disassociated as guided by relative frequencies in response. Initial frequency analysis revealed that there were too many codes, particularly concerning highly specific codes with few responses. Several codes had less than 10% representation (i.e., one or two responses out of 22). This indicated the need for thematic code consolidation with allied themes where possible. In these cases, opportunities were sought to roll-up concepts into overarching thematic codes.

In select cases, there were indications that the code was too general, and attempts were made to bifurcate the code conceptually. One specific theme was quite frequent in both challenge and best practice responses: "data management." However, it was difficult to subsegment this topic adequately in interpreting the interview responses. The theme had several allied subtopics, but the majority of respondents spoke of data management as a general end-to-end process. This implied that the range of data management subtopics were essentially linked together in the view of practitioners and that breaking, for instance, "data cleaning" from "reducing and consolidating variables" would violate the key theme of "end-to-end data management as a process." Despite the prevalence of this theme, it was determined to retain it as an aggregate as segmenting it into subthemes would violate the central observation that "data management" was being framed by respondents as an integrated, end-to-end process.

The "CSDS trends" question category (interview question 2) typically elicited responses that could be framed as challenges and/or best practices, as anticipated. Detailed analysis of responses indicated that "trend" was a proxy and was not a thematic family in-of-itself. The "trend" question acted as an open invitation for the respondents to raise themes they intended to subsequently develop as challenges and/or best practices. For instance, as a trend response, the topic of "cloud computing" was framed as either a challenge (e.g., introducing new vulnerabilities and attack vectors) or as a best practice (e.g., providing a platform for aggregating security data and conducting data analytics).

Observing that trend responses were typically a proxy for framing challenges and/or best practices, a thematic model was extrapolated to segment the relevant "trend" themes typically raised. This model provided a guide to extrapolate and segment trend interview responses into suitable matching challenge and/or best practice areas. Applying this guide, the refined transcripts "trend" responses were segmented into either challenge or best practice areas. Furthermore, it was determined to merge and subsume the initial open coding "trend" codes into the relevant challenge and/or best practice codes in the final ordered codebook. The following categorical segmentation model describes the logic which guided the repurposing and integration of trend codes (Fig. 3.2).

Following an effort to consolidate and reduce codes thematically through aggregations and roll-ups, "challenge" themes were greatly reduced from 29 to 11 codes. Family categories were dispensed in "challenge" code reduction. Best practice themes expanded slightly from 23 to 26 codes as a result of integrating trend-derived codes. "Trend" responses suggested several new best practice topics, per the conceptual model in Fig. 3.2. Sensitizing best practice family categories were refined to align with the MIS tripartite classification scheme of organization, process, and technology.

The final ordered codebook, following consolidation and refinement, is documented below in Table 3.6. Of note, the original ordered codebook was unnumbered and unordered. The code numbering scheme represented here emerged post-completion ($n = 50$) and is ordered by the most to the least frequent responses to assist with interpretation.

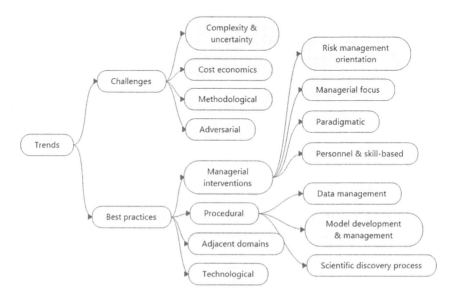

Fig. 3.2 Categorical segmentation of perceived CSDS trends

Table 3.6   Final ordered codebook

Challenge codes
CH1: Data preparation (access, volume, integration, quality, transformation, selection)
CH2: Unrealistic expectations proliferated by marketing hype
CH3: Contextual nature of normal versus anomalous behavioral phenomenon
CH4: Lack of labeled incidents to focus detection
CH5: Own infrastructure, shadow IT, and proliferation of exposure
CH6: Uncertainty leads to ineffective reactive stance
CH7: Traditional rule-based methods result in too many alerts
CH8: Program ownership, decision-making, and processes
CH9: Resourcing, developing, and hosting in house
CH10: Expanding breadth and complexity of cyber domain
CH11: Policy, privacy, regulatory, and fines

Best practice codes[a]	
BP1: Structured data preparation, discovery, engineering process	Proc
BP2: Building process focused cross-functional team	Org
BP3: Cross-training team in data science, cyber, engineering	Org
BP4: Scientific method as a process	Proc
BP5: Instill core cyber domain knowledge	Org
BP6: Vulnerability, anomaly and decision automation to operational capacity	Tech
BP7: Data normalization, frameworks and ontologies	Tech
BP8: Model validation and transparency	Proc
BP9: Data-driven paradigm shift away from rules and signatures	Org
BP10: Track and label incidents and exploits	Proc
BP11: Cyclical unsupervised and supervised machine learning	Proc
BP12: Address AI hype and unrealistic expectations directly	Org
BP13: Understand own infrastructure and environment	Org
BP14: Cloud and container-based tools and data storage	Tech
BP15: Distinct exploration and detection architectures	Tech
BP16: Participate in data-sharing consortiums and initiatives	Tech
BP17: Deriving probabilistic and risk models	Org
BP18: Upper management buy-in and support	Org
BP19: Human-in-the-loop reinforcement	Proc
BP20: Survey academic methods and techniques	Org
BP21: Cyber risk as general enterprise risk and reward	Org
BP22: Segment risk programmatically and outsource components	Org
BP23: Adding machine learning to SIEM	Tech
BP24: Preventative threat intelligence	Org
BP25: Hosting and pushing detection to endpoints	Tech
BP26: Honeypots to track and observe adversaries	Tech

(continued)

Table 3.6 (continued)

Adversarial trend codes
ADV1: Adversaries constantly adapting to countermeasures and innovating methods
ADV2: Machine learning mechanisms, vectors and threats (adversarial ML)
ADV3: Attackers exploiting vulnerabilities and vectors in new tech. (i.e., cloud, mobile, IoT, ICSs)
ADV4: Growing sophistication of social engineering and identity hacking
ADV5: Kits and tool-based automation of penetration and attacks
ADV6: Exploiting inherent vulnerabilities of logical and physical infrastructures
ADV7: State actors hiring data scientists
ADV8: As-a-service hacking with cryptocurrency mediation
ADV9: Reverse engineering or leveraging known detection and prevention mechanisms
AD10: Time to detection and reconnaissance advantage
ADV11: Growing internal threats and corporate espionage
ADV12: Information warfare, fake news
ADV13: Hijack systems for bitcoin mining

Adjacent domain codes
ADJ1: Fraud analytics, forensic and criminology
ADJ2: Deep learning—computer vision and acoustics
ADJ3: Medical, epidemiological, ecological, and bioinformatics
ADJ4: Network graph analytics
ADJ5: Social and behavioral sciences, including (micro-) economics and game theory
ADJ6: Natural language, semantic, and knowledge engineering
ADJ7: Risk management
ADJ8: Forecasting and time series analysis

[a]Includes rough classification family of MIS focus area: *Org* organization, *Proc* process, *Tech* technology

3.2.3   *Structured Coding and Results*

3.2.3.1   Selective Axial Coding Process

The final ordered codebook resulted in 58 CSDS theme codes across 4 question categories: 11 challenges, 26 best practices, 13 adversarial trends, and 8 adjacent domains. Subsequent to ordering the final codebook, the original 22 interviews were recoded. An additional 28 interviews were then conducted and coded utilizing the final codebook, resulting in 50 total coded interviews.

Concerning the coding process, following each interview, a raw transcript was derived from the audio recording. In a minority of cases where recordings were not allowed, email responses ($n = 7$) or researcher notes ($n = 7$) were prepared as text-based equivalents. Transcripts resulting from the interviews were manually reviewed for quality and clarity. In particular, when responses strayed across questions (i.e., raising a best practice during a discussion on challenges), a focused effort was made to manually group the relevant text with the related question category.

Per the procedure previously developed, trend responses were manually segmented into either challenge or best practice responses, depending on respondent framing. Content related to adversarial trends and adjacent domains were segmented, respectively. As a result, post-processing of raw transcripts resulted in treated transcripts which explicitly segmented text related to the four core interview question categories, with trend responses allocated to either challenges or best practices.

Treating the transcripts to segment text clearly into the four question categories aided interpretation and coding. The segmented transcripts were subsequently manually coded specific to the related question categories (challenges, best practices, adversarial trends, adjacent domains). The NVIVO qualitative data analysis software package was used to manage the coding process.

All 50 interviews were coded across the four categories, composed of 58 total codes. The final resulting dataset was thus a 50-by-58 matrix. Each of the 58 codes across four categories was marked with a binary indicator: 1 = theme was raised by the respondent during the interview, or 0 = theme was not raised by the respondent during the interview. Demographic measures associated with interviewees were also captured along with codes raised in the interviews.

The resulting matrix supported frequency as well as correlation and logistic fit analysis. The latter two methods were used to assess the propensity of particular responses to co-occur across all interviews. Given the complexity and newness of the domain, this supported the interpretation of multi-thematic insights, revealing the relations between themes.

Of note, the frequency of response per respondent per theme was also captured and considered. However, in aggregate, examining response weight per respondent per theme (a weighted measure per theme) did not enhance clarity of analysis. Frequency and related analysis thus examine a simple count of the number of individual respondents from the sample of 50 that mentioned a theme at least once during the interview, each code noting a binary measure per respondent (1 = response; 0 = non-response).

3.2.3.2 Challenge Codes and Results

Coding of CSDS challenge themes (question 3) was conducted on the refined transcripts resulting from each interview. Eleven challenge codes from the ordered codebook were utilized to log responses. This resulted in the frequency distribution displayed in Table 3.7, indicating the percentage of respondents who raised the related challenge theme at least once in their responses ($n = 50$).

The theme "CH1: Data preparation (access, volume, integration, quality, transformation, selection)" was the most frequent response, with 84% of respondents raising this topic as a challenge ($n = 42/50$).

As this theme was so prevalent, it was of interest to determine to what degree the theme correlated with other of the challenge themes. Correlation analysis and a nominal logistic fit were run on "CH1: Data preparation" against the other challenge themes (codes). Logistic fit allowed for a quantitative assessment of significance to accompany correlation analysis.

Table 3.7 Frequency of "challenge" response codes ($n = 50$)

Coded responses: Perceived challenges	N	$\%$
CH1: Data preparation (access, volume, integration, quality, transformation, selection)	42	84
CH2: Unrealistic expectations proliferated by marketing hype	35	70
CH3: Contextual nature of normal versus anomalous behavioral phenomenon	30	60
CH4: Lack of labeled incidents to focus detection	28	56
CH5: Own infrastructure, shadow IT, and proliferation of exposure	27	54
CH6: Uncertainty leads to ineffective reactive stance	25	50
CH7: Traditional rule-based methods result in too many alerts	25	50
CH8: Program ownership, decision-making, and processes	20	40
CH9: Resourcing, developing, and hosting in house	16	32
CH10: Expanding breadth and complexity of cyber domain	16	32
CH11: Policy, privacy, regulatory, and fines	15	30

Table 3.8 Challenge responses co-occurring with "CH1: Data preparation"

Challenge responses relating to CH1: data preparation	Correlation	P-val
CH4: Lack of labeled incidents to focus detection	0.38	0.002
CH7: Traditional rule-based methods result in too many alerts	0.22	0.016
CH8: Program ownership, decision-making, and processes	0.02	0.048

Correlation and logistic fit analysis signified a positive relation between theme CH1 to themes CH4, 7, and 8, per Table 3.8. The "challenge" logistic model fit to CH1 indicated a significant whole model at a P-value of 0.0017 (<0.01) and an $R2$ of 0.64 (judged as being reasonably strong for nominal regression). The related challenge themes thus co-occurred positively and significantly with "CH1: Data preparation." Conceptually, these themes indicate potential allied issues that relate to the challenge of data management. Table 3.8 summarizes results from quantitative analysis of topics allied with challenge "CH1: Data preparation":

Interpreting this observation through sensitizing concepts extrapolated in literature analysis, the correlating themes indicate that data preparation is viewed across respondents as being a manifestation of a larger programmatic challenge. There is an indication that breakdowns in managing data overlap with challenges in organizational programmatic ownership, decision-making, and incident tracking. An impetus is indicated, namely, that rule-based methods are increasingly inadequate, but that overcoming this limitation requires a more concerted and systematic approach to collecting and analyzing data. A recognition of the need to label incidents in order to advance probabilistic and machine learning-based detection models is clearly indicated.

Concerning sympathies among other responses, the frequency of challenges 3, 5, and 6 reinforces the notion that many practitioners feel there is generalized confusion in organizations regarding what and how to track and assure cyber continuity in the face of growing complexity. One interpretation is that many CSDS practitioners perceive that fundamental questions of *what to track*, *what data to collect*, *what is normal*, and *how to analyze* are inadequately defined in many environments and that this is viewed as being an organizational challenge as much as it is a technical one.

Drawing from an understanding of the CSDS domain derived from literature analysis, themes surfaced as challenges observe that for CSDS prescriptions to provide value, analytical (scientific) questions need to be carefully framed and validated. In aggregate, practitioners appear to surface a central underlying challenge: that organizations are seemingly confused concerning how to programmatically implement CSDS and that this manifests principally in not having a concerted vision concerning what phenomena to track, what data is required, which baselines to establish, and which analytical methods to apply. Further, to the degree analytical (scientific) processes are required, the implication is that these questions must be addressed iteratively, indicating that multiple, staged treatments are required.

3.2.3.3   Best Practice Codes and Results

Best practice themes (question 4) were similarly examined for response frequencies across the sample. As noted previously, the range of best practice codes variously invoke themes related to the key MIS focus areas. Advocacy surfaced in best practices are noted in Table 3.9 as aligning with the tripartite MIS areas of organization (Org), process (Proc), or technology (Tech):

While the MIS categories are rough sensitizing designations, it is of interest to observe that respondents raised conceptual themes fairly evenly across the range of MIS context frames: 46% *Org* = organization, 23% *Proc* = process, 31% *Tech* = technology.

Organizational themes, composing 46% of the codes, were slightly more common, indicating that despite being a technically and methodologically intensive domain, CSDS practitioners were highly attuned to organizational dependencies in CSDS efforts. This may indicate that organizational integration and sponsorship of CSDS is perceived as a sensitive gap requiring greater accommodation. This observation is reflected in literature analysis as organizational perspectives were found to be a relative gap in CSDS literature, with only 21% of the corpus addressing this topic (7 out of 33 works).

Similar to challenge responses, the theme of "data management" was the most frequently raised best practice, with 84% ($n = 42/50$) advocating "BP1: Structured data preparation, discovery, and engineering process." In this case, practitioners framed the theme as a positive facilitator of CSDS practice as opposed to something which is observed as being a gap in practice. Thirty-eight respondents raised data

Table 3.9 Frequency of "best practice" response codes ($n = 50$)

Responses: advocated best practices	Family	N	%
BP1: Structured data preparation, discovery, engineering process	Proc	42	84
BP2: Building process focused cross-functional team	Org	38	76
BP3: Cross-training team in data science, cyber, engineering	Org	37	74
BP4: Scientific method as a process	Proc	34	68
BP5: Instill core cyber domain knowledge	Org	33	66
BP6: Vulnerability, anomaly and decision automation to operational capacity	Tech	33	66
BP7: Data normalization, frameworks and ontologies	Tech	32	64
BP8: Model validation and transparency	Proc	31	62
BP9: Data-driven paradigm shift away from rules and signatures	Org	29	58
BP10: Track and label incidents and exploits	Proc	28	56
BP11: Cyclical unsupervised and supervised machine learning	Proc	25	50
BP12: Address AI hype and unrealistic expectations directly	Org	23	46
BP13: Understand own infrastructure and environment	Org	23	46
BP14: Cloud and container-based tools and data storage	Tech	22	44
BP15: Distinct exploration and detection architectures	Tech	22	44
BP16: Participate in data-sharing consortiums and initiatives	Tech	21	42
BP17: Deriving probabilistic and risk models	Org	20	40
BP18: Upper management buy-in and support	Org	16	32
BP19: Human-in-the-loop reinforcement	Proc	14	28
BP20: Survey academic methods and techniques	Org	13	26
BP21: Cyber risk as general enterprise risk and reward	Org	12	24
BP22: Segment risk programmatically and outsource components	Org	9	18
BP23: Adding machine learning to SIEM	Tech	5	10
BP24: Preventative threat intelligence	Org	4	8
BP25: Hosting and pushing detection to endpoints	Tech	4	8
BP26: Honeypots to track and observe adversaries	Tech	2	4

Table 3.10 Best practice responses co-occurring with "BP1: Structured data preparation"

Best practice responses relating to BP1: structured data prep.	Correlation	*P*-val
BP13: Understand own infrastructure and environment	0.18	<0.0001
BP9: Data-driven paradigm shift away from rules and signatures	0.40	<0.0001
BP7: Data normalization, frameworks and ontologies	0.24	0.1

management both as a challenge and as an advocated best practice during the interviews, with eight respondents raising this topic only as a challenge OR a best practice. This indicated that the majority viewed data management both as a common pain point and a key facilitator when implemented correctly.

It was also of interest to examine which other themes were raised frequently along with BP1. Correlation and logistic model fitting indicated a significant relationship between BP1, as the most frequent response, and best practice themes BP 7, 9, and 13, per Table 3.10. The logistic fit model to BP1 indicated a significant whole model at a *P*-value of 0.01 (=0.01) and an *R*2 of 1.0. Collectively, this combination of best practices advocates structured data gathering accompanied by a contextual understanding of data in terms of the environment being protected.

Again, results appear to reinforce the observation that implementing data preparation as a process necessitates answering more fundamental questions. Fundamental questions emerge concerning what phenomena to analyze with which data and which methods. These questions can be associated with a drive to frame positive (scientific) problems in CSDS practice, rather than to pursue normative assertions (assertions of "should" or "good," which are not discretely measurable).

Addressing fundamental framing questions of context and purpose ties to the notion that a *scientific process* necessitates a programmatic effort to build systematic consensus within the organization prior to implementing technical and procedural solutions. Invoking the importance of rigorous framing implies that CSDS practice requires careful planning and scoping prior to attempting to pursue treatments. By association, the collective themes observed as correlating with "BP1: Structured data preparation" tangentially highlight the importance of BP4, advocacy for pursuing the scientific method as a process.

Cross-functional team composition and cross-training, BP2 and 3, were the next most frequent responses, followed by BP4, advocacy for framing the scientific method as a process. Organizational advocacy concerning the need for training and cross-functional coordination in BP2, 3, and 5 observes the inherent breadth and complexity of the CSDS domain. The preponderance across these responses implies a general recognition that deep expertise in both cybersecurity and data science is rare to find in single individuals. The implication is that cross-training and coordination should be undertaken systematically and programmatically to inculcate CSDS organizational competence in and across teams.

3.2.3.4 Adjacent Domains Codes and Results

Interview question 5 inquired as to which adjacent domains CSDS practitioners drew upon to inform their CSDS work. Fifty percent indicated that fraud analytics, including forensics and criminology, centrally informed their CSDS work. The range of the various adjacent domains cited provides insight into the sheer breadth of analytical methods resident across CSDS practice. The range of implied methods is broad and ambitious, encompassing inferential statistics, econometric approaches, diagnostics, quantitative analysis, specialized techniques (i.e., graph analytics and natural language processing (NLP)), and the full range of machine learning approaches. Table 3.11 captures the frequency of responses to question 5 on adjacent domains.

Responses to question 5 elicited a range of responses from general subject-matter domains to focused methods. ADJ1, 3, 5, and 7 relate to broad subject-matter areas, whereas ADJ2, 4, 6, and 8 are aligned with particular methods (* denoted and *italicized* in Table 3.11). Concerning subject-matter areas, CSDS practitioners cited an interest to embrace external paradigms ranging across the fields of fraud analytics (ADJ1), medicine (ADJ2), social science (ADJ5), and risk management (ADJ7). Methodologically, a common element across these fields is multivariate inferential statistics.

A range of focused analytical methods were also cited: deep learning (ADJ2), graph analytics (ADJ4), NLP (ADJ7), and time series analysis (ADJ8). While these methods are not exclusive to the subject-matter areas raised (e.g., fraud analytics frequently applies network graph analytics), their explicit invocation in interviews signifies that CSDS professionals view these isolated methods to have particular relevance to evolving practice.

A general observation related to the breadth of domains cited concerns the impression that the CSDS field appears in search of methodological and paradigmatic guidance. Given the breadth of responses, it appears that there were fundamental differences of opinion concerning the paradigmatic and methodological

Table 3.11 Frequency of "adjacent domain" response codes ($n = 50$)

Coded responses: adjacent domains	N	%
ADJ1: Fraud analytics, forensic and criminology	25	50
ADJ2: *Deep learning—computer vision and acoustics**	19	38
ADJ3: Medical, epidemiological, ecological, and bioinformatics	17	34
ADJ4: *Network graph analytics**	14	28
ADJ5: Social and behavioral sciences, including (micro-)economics and game theory	13	26
ADJ6: *Natural language, semantic, and knowledge engineering**	8	16
ADJ7: Risk management	7	14
ADJ8: *Forecasting and time series analysis**	7	14

underpinnings of the CSDS domain. This reinforces the assertion that CSDS is in the process of body of theory development. Results also reinforce a finding from literature analysis that the security domain lacks a strong theoretical and scientific tradition, being largely engineering-driven.

3.2.3.5   Adversarial Trends Codes and Results

The final substantive question, question 6, inquired concerning perceptions regarding trends in adversarial cybersecurity threats. Table 3.12 captures the frequency of responses.

The most frequent response 76% ($n = 38/50$) involved the general observation that adversaries are constantly adapting and innovating. This was not surprising and is broadly remarked upon in the professional and popular press, as surfaced in literature analysis. Explanatory background raised in several memos indicated the perception that organizations struggle to move at the speed and scale of adversaries and thus that defenders are often thrust into a highly reactive stance.

The second most frequently cited theme, raised by 50% of interviewees ($n = 25/50$), concerned the perception that adversaries were actively developing methods utilizing machine learning techniques to target and orchestrate attacks, as

Table 3.12 Frequency of "adversarial trend" response codes ($n = 50$)

Coded responses: adversarial trends	N	$\%$
ADV1: Adversaries constantly adapting to countermeasures and innovating methods	38	76
ADV2: Machine learning mechanisms, vectors and threats, including adversarial ML	25	50
ADV3: Attackers exploiting vulnerabilities and vectors in new tech. (i.e., cloud, mobile, IoT, ICSs)	20	40
ADV4: Growing sophistication of social engineering and identity hacking	20	40
ADV5: Kits and tool-based automation of penetration and attacks	19	38
ADV6: Exploiting inherent vulnerabilities of logical and physical infrastructures	11	22
ADV7: State actors hiring data scientists	9	18
ADV8: As-a-service hacking with cryptocurrency mediation	8	16
ADV9: Reverse engineering or leveraging known detection and prevention mechanisms	6	12
AD10: Time to detection and reconnaissance advantage	5	10
ADV11: Growing internal threats and corporate espionage	4	8
ADV12: Information warfare, fake news	3	6
ADV13: Hijack systems for bitcoin mining (cryptojacking)	3	6

well as to evade machine learning-driven detection mechanisms. Although the degree to which adversaries might presently be using machine learning to stage attacks was a subject of some dispute, it is a growing topic of general interest in the professional press and among practitioners (Chebbi 2018; Drinkwater 2018; Forrester 2020; Open Data Science 2018).

Memos indicated some respondents felt that the utilization of machine learning to perpetrate attacks, as well as machine learning systems as targets of attacks, is an inevitability in a domain framed by some as undergoing a technological "arms race." There was a suggestion that this is a topic of special interest among state-level actors, those with generous support and resources to pursue advanced developments. This ties to ADV7, the assertion that state actors are increasingly employing data scientists. The ADV2 theme includes the topic of adversarial machine learning, which encompasses methods intended to interrupt, trick, or otherwise interfere with the operation of machine learning mechanisms. This concern is also of increasing interest in recent research and professional press articles (Brundage et al. 2018; Chebbi 2018, 2019; Forrester 2020; Gopalakrishnan 2020; Samuel 2019).

Concerning ADV5, 38% ($n = 19/50$) of respondents surfaced a perceived trend toward increasing automation in attacks. It was remarked in memos that several respondents felt that "white hat" automation tools, such as those applied to automate penetration testing and red teaming exercises, often seemingly quickly end up being repurposed for "black hat" purposes. White hat cybersecurity actors are those seeking to surface and address the vulnerabilities of security systems in order to communicate said vulnerabilities and thus to improve security integrity. Generally, this category of actor is regarded as being benevolent, as opposed to black hat actors. Black hat actors are those driven by personal, financial, and/or political motives that impel them to destroy or otherwise interfere with systems, services, and/or data.

The perception, substantiated in memos, was that there is a worrying degree of permeability between white hat defensive tools and adversarial techniques and approaches. This gives rise to the notion of "grey hat" actors as intermediaries, those who operate fluidly between benevolent and self-serving frames. There is further evidence that a market for specialized hacking tools is burgeoning, propelled by demand from state-sponsored and associated actors (The Economist 2019c).

To the degree that statistical and machine learning features are being increasingly added to penetration testing automation toolkits, it was observed in memoing that this may be an additional path by which machine learning methods find their way into the hands of adversarial actors. An implication is that greater restrictions should be pursued regarding the availability and use of defensive tools which can be rapidly repurposed for adversarial means.

The implied motivations of attackers were surfaced across adversarial trends. A 2019 Verizon report on data breaches based on over 40,000 incidents attributed 23% of incidents to state actors and 71% as being financially motivated (Verizon 2019).

While ADV2, 5, and 7 responses implied sophisticated state-level activities, other themes recognized adversarial actors motivated by financial rewards. This distinction may be nonexclusive, as in the case of contracted third-party state actors, yet is helpful to demarcating gross motivations, means, and targets. Codes ADV8 and 13 raised profit mechanisms and mercenary motivations. ADV8 noted an increase in "contracted hacking," adversarial actions undertaken for a fee under some type of agreement. Both of these trends identified untraceable cryptocurrencies and anonymous darknet facilitated communications as enablers.

Memos noted that the ADV8 category encompassed ransomware activities, hijacking and encrypting data and systems until a ransom is paid, also in the form of cryptocurrencies. ADV13 noted a growing trend toward cryptojacking or cryptomining—hijacking computing resources to mine bitcoins or other cryptocurrencies. For perspective, the 2019 Verizon data breach report cited 24% of malware incidents as being tied to ransomware, whereas cryptojacking encompassed 2% (Verizon 2019).

Together, the ADV8 and 13 trends note that financially motivated actors respond to risks and rewards by actively innovating methods and mechanisms to facilitate the full scope of extortionary or fraudulent transactions. The year 2018 saw a relative decline in ransomware incidents and a balanced increase in cryptojacking, a phenomenon that coincided with the peaking of the 2018 bitcoin valuation bubble. This indicates a preference on the part of adversarial actors to actively optimize rewards against perceived proportional risk (Schwartz 2018). Cryptojacking is less aggressively visible and disruptive, in that it seeks to commandeer computing resources surreptitiously, and as such, hypothetically may be seen as a less risky undertaking.

As with several of the other adversarial trends, it was noted that logistical and economic anonymity is a powerful enabler. An accompanying observation is that technological advancements and trends, such as powerful encryption and cryptocurrency, while not initially intended to facilitate adversarial cybersecurity actors, have indeed become a central enabler. The implication is that, observing ADV1, adversaries are constantly innovating using a range of logistical and technical developments, not necessarily only those with a purely exploitative or adversarial scope.

An infrequently cited theme, ADV12, concerned information warfare and fake news, topics which are garnering increasing public attention (Booz Allen Hamilton 2019). This theme was notable as it is an emerging phenomenon outside the traditional purview of the cybersecurity professional domain, dealing as it does with the veracity and integrity of news and media. A topic of increasing geopolitical interest, the phenomenon of orchestrated disinformation and information warfare campaigns, particularly motivated at the state actor level, raises urgency. Implicit is the observation that public discourse is regionally and nationally sensitive but that the cybersphere is transnational and loosely monitored in those countries where this is becoming a central topic of concern, i.e., Western democratic nations.

Interview memos indicated that respondents viewed the origination and perpetuation of fake news as both a technical and a societal problem. While broad-based digital literacy is an essential component to curtailing disinformation contagions, the prospect of applying algorithmic methods to automate the detection of fake news and orchestrated disinformation campaigns has garnered increasing research interest (Bara et al. 2019). As CSDS offers a range of methods for systematizing the detection of dubious digital information and substantiating progenitorship, this topic will inevitably grow as a specialized area of CSDS attention.

In aggregate, responses to interview question 6 on adversarial trends reinforced the literature analysis finding that adversarial phenomena are constantly adapting as underlying behavioral and technical factors evolve. The primacy of human motivations and behavioral factors in themes surfaced in responses draws a sharp contrast with the engineering focus resident in the security domain. This highlights CSDS practitioners' propensity to defer to external domains that focus specifically on social and behavioral factors, per responses to question 5 on adjacent domains. Ostensibly the practitioner deference to the fraud analytics and social science fields indicates an impetus to shift the security domain toward methods for analyzing and assessing human behavior, particularly as an explicit aspect of criminology.

3.2.3.6 Demographic and Professional Experience Observations

Responses were compared to demographic background factors to determine the potential presence of significant influential relationships. Sample size was an issue in analyzing some factors, for instance, correlating professional industry with responses, as subsamples were of inadequate size. Grosser demographic and professional background measures and characteristics gave clearer, more rigorous indications.

Concerns regarding sample size were partially addressed earlier through ANOVA substantiation of similar demographic group means between the 50 interviewees and the larger original sample of 348 prospects. Thus, there are grounds for asserting that the sample size of 50 is demographically representative of the larger original target sample.

Several key demographic aspects had no discernable or weak indications of influence. Concerning gender, there was a strong imbalance in distribution, with only 14% of respondents being female-identified ($n = 7/50$). However, gender identification had no notable significance in response propensity, with a whole model categorical chi-square test failing to reject the null hypothesis of no gender effect for all responses at 0.1 significance (chi-square = 0.829).

Age and work experience similarly failed to indicate significant influence in responses. The number of years employed had a 0.93 correlation with estimated age, which was unsurprising. However, there was only a weak correlation between years employed and years of cybersecurity professional experience (0.33), indicating many respondents had moved into the cybersecurity field from other backgrounds, perhaps pursuing opportunities due to labor shortages.

There was a lack of strong correlation between years employed and other professional background measures, including the number of years at university (-0.17), years in the CSDS field (0.027), and years in data science/analytics (0.024). This otherwise serves to underscore the relative newness, cross-generational, and interdisciplinary nature of the CSDS field. Additionally, as interviewees reported their entry into the CSDS domain as either being intentional/planned or accidental/serendipitous (e.g., an opportunity presented while studying or working in a different domain), no significant effects were observed between reported motivation of entry and responses.

As CSDS is a relatively new field (mean 2.9 years CSDS experience across the interviewee sample, 1.9 years standard deviation), it was of interest to determine whether relative CSDS experience had significant effects. Demographically there was little correlation between CSDS experience and age or years of total employment, both measures having only a 0.02 correlation with years of CSDS experience. This attests to CSDS as having a wide representation of entrants in terms of age and prior work experience. As age and work experience were not statistically significant to responses, the indication was that level of prior experience did not appear to influence opinions concerning the CSDS field. Again, the field being relatively new, CSDS gives indications of undergoing a process of professional emergence as a new domain with unique theoretical foundations.

After ruling out the influence of other demographic factors on aggregate CSDS experience, it was of interest to determine whether challenge and/or best practice responses were more prevalent with greater relative CSDS work experience. Applying a least squares model of independent responses to the number of years of CSDS experience as the dependent variable produced a significant model, indicating a specific set of themes which correlated strongly and positively with greater amounts of CSDS work experience, per Table 3.13.

The interpretation of the results is that greater amounts of experience working as a CSDS practitioner increases awareness concerning the operational importance of structured data management and engineering. This general observation is mirrored in CSDS practitioner survey research, with the Ponemon Institute noting that 65% of 651 respondents surveyed identified "data challenges" as being the biggest issue preventing the implementation of security analytics (Ponemon Institute 2017).

There is also an indication that there is a growing awareness of the inherent challenges posed by the sheer breadth of the CSDS discipline. CSDS professionals demonstrate a need for organizational focus and cross-disciplinary collaboration, as opposed to a "go-it-alone" mentality. Responses suggest that an awareness of the themes noted strengthen with increasing tenure in practical operational settings.

Table 3.13 Least squares model for years of CSDS experience to responses

# of years employed in CSDS domain (y) against challenge and best practice responses (x)

Least squares model	LogWorth	Prob > ChiSq
BP16: Participate in data-sharing consortiums and initiatives	2.397	0.0036
BP7: Data normalization, frameworks and ontologies	2.315	0.0048
BP2: Building process focused cross-functional team	1.854	0.014
CH10: Expanding breadth and complexity of cyber domain	1.660	0.0219
CH2: Unrealistic expectations proliferated by marketing hype	1.395	0.0403
BP1: Structured data preparation, discovery, engineering process	1.219	0.0604
BP14: Cloud and container-based tools and data storage	1.071	0.085

Significant at 0.05: Whole model test P-value 0.04, $R2$: 0.89

Additional influences related to professional experience observed that greater data science/analytics work experience correlated with a greater propensity to highlight two particular best practice themes in interview responses:

- BP6: Vulnerability, anomaly and decision automation to operational capacity (correlation 0.32)
- BP7: Data normalization, frameworks and ontologies (correlation 0.31)

Although moderate, the correlation measures were judged as reasonably notable for categorical responses. The interpretation of the correlation between greater analytics professional experience and BP6 and 7 responses is that experienced data scientists are potentially more sensitive to practical organizational and operational constraints with attention to data quality and context.

Similarly, holding a professional certificate in analytics or data science correlated with raising the challenge themes:

- CH3: Contextual nature of normal vs. anomalous behavioral phenomenon (0.33)
- CH4: Lack of labeled incidents to focus detection (0.36)

This implies that those professionals with a stronger background in data science were sensitized to practical operational issues, again raising the importance of the topics of data quality and context.

In contradistinction, examining those with greater proportional cybersecurity professional experience, a moderate correlation was noted among those with a technical background in "networking, systems administration, and security" and the challenge-related response:

- CH10: Expanding breadth and complexity of the cyber domain (0.38)

Additionally, holding one or more cybersecurity certifications also correlated with the response:

- BP16: Participate in data-sharing consortiums and initiatives (0.34)

This implies that those with a deeper prior cybersecurity background are more sensitized to the security domain's practical challenges, particularly that the field is rapidly expanding in breadth and complexity.

Concerning years of university study, a weak but notable categorical correlation was revealed between the number of years of university study and the response advocating:

- BP20: Survey academic methods and techniques (0.42)

Likewise, a moderate correlation with the same measure was observed for those holding Ph.D. (0.34) or MA/MSci/MBA (0.38) degrees. Similarly, a moderate correlation (0.49) on the same measure for those currently working in an academic setting was noted. The implication is that greater exposure to higher academic studies or working in an academic setting predisposed respondents to advocate surveying academic research to support CSDS efforts.

Whereas the small sample size restricted the ability to determine the targeted effects of deeper professional industry and educational background, this may be a fruitful avenue for future inquiry. A larger response sample from a survey-based inquiry would support an examination of the degree to which CSDS professional and academic backgrounds might predispose practitioners to differing stances on central themes.

3.2.3.7 Thematic Insights via Text Analytics

Text analytics was utilized as a supplementary inductive approach to surface themes implicit across the structured interview transcript artifacts. Observing Myers and Newman's advocacy (2007) to utilize multiple interpretive approaches in qualitative research, text analytics provided a supplemental interpretation of themes surfaced.

Themes extrapolated from text analytics were viewed as an additional guide to reinforce thematic interpretations. In addition to the results from literature analysis, text analytics provided an alternate approach to triangulate major themes identified in the CSDS domain. Text analytics offered an independent inductive mechanism to offset the risk of researcher bias in the thematic interpretation of interview transcripts.

Text analytics was not used to modify or reinterpret the structured codebook nor to recode the transcripts. The structured codebook reached saturation at 22 interviewees, and text analytics was undertaken after all transcripts were complete and coded ($n = 50$). Recoding and reinterpreting the transcripts based on text analytics was considered out of scope, although results ultimately served to validate and reinforce the themes identified in coding.

Content analysis through text mining or text analytics is an increasingly recognized research approach to extract semantic themes from text artifacts (Krippendorff 2019; Onwuegbuzie et al. 2012). Substantiating this approach in the information technology domain, Bechor and Jung have applied text analytics specifically to data science and cybersecurity research corpora to extract a CSDS intersectional topic model (Bechor and Jung 2019).

As text analytics is a machine-driven extrapolative method that identifies statistical-structural patterns in language and text, the approach is inductive in scope. Growing guidance is available to researchers on the practical methods and tools to undertake computer-aided text and content analytics (Bechor and Jung 2019; Jockers 2014; Miner et al. 2012).

Text mining was applied to extract key topics from segmented versions of the interview transcripts. A tool-driven sequential process was followed to extrapolate themes from refined versions of the transcripts. Figure 3.3, adapted from "Text Mining and Analysis" (Chakraborty et al. 2013), summarizes a high-level representation of the applied text mining process.

Topical text analytics was performed to identify thematic patterns among the structured transcripts. The SAS Text Miner text analytics tool was utilized to facilitate the process of analyzing raw text from the prepared and segmented transcript text corpora. SAS Text Miner facilitated the steps of text parsing, text filtering, transformation, and text mining.

In the first step, core text documents were prepared for analysis. Having been segmented and structured to align with key questions, the transcripts were reduced to versions specifically detailing feedback associated with challenges or best practices. Additionally, a combined corpus of challenge and best practice responses was prepared for comparison. Transcript text responses related to trends had been previously extracted and aggregated into related challenge and/or best practice areas.

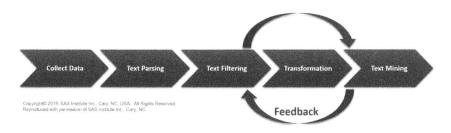

Fig. 3.3 Text mining process flow

Interviewer commentary had been previously removed from the textual record in order to focus on core respondent feedback. This resulted in structured challenge, best practice, and combined corpora, one document per interviewee created for each type. Content related to other topics, namely, professional background, adversarial trends, and adjacent domains, was removed. Preliminary exploratory text analytics was applied to these topics but did not result in new findings.

Natural language processing (NLP) was undertaken in sequence to extract a matrix of key terms and topics from the corpora. Tasks such as spell-checking and English grammatical segmentation were driven through this process. Settings and configurations to optimize results were managed through the SAS Text Miner tool. For instance, specifications were set on stop words and grammatical terms to ignore, such as articles and prepositions, and the indication to allow for bi- and trigrams (key terms composed of common two- or three-word sequences such as "machine learning" or "training model validation"). Proper nouns and other associated entities were identified and ignored, and synonyms and variant spellings of common nouns and verbs were aggregated via an embedded thesaurus.

Text corpora were transformed to key term-by-document matrices through the application of latent semantic analysis (LSA) (Landauer et al. 1998), latent semantic indexing (LSI) (Deerwester et al. 1990), and vector space modeling (Chakraborty et al. 2013; Hull and Li 1993). These methods support discrete matrix-based analysis of thematic patterns in textual content, in particular addressing challenges with large, sparse matrices.

A model of relative term frequency and semantic relationships between terms was thus derived. This created a model of key terms specific to the corpora based on frequency and linked terms, with statistical proximity indicating semantic relationships. Table 3.14 gives an example of key terms (noted in the top row) as identified with statistically associated sub-terms as extrapolated from the entire corpus of combined challenge and best practice transcript content. Key topics relevant to CSDS are clearly visible:

Key terms were thus extracted and associated in matrices encoding semantic relationships based on proximity and frequency. As a result, key terms were also exposed for parsing via spider diagrams to explore networks of connected terms and concepts. As an example, Fig. 3.4 is a representation of a spider diagram extrapolated from the combined corpora for the key term "data scientist".

In the final text analytics step, singular value decomposition (SVD) was applied to reduce term dimensionality and to consolidate the matrices for analysis. A single results matrix was the result, surfacing the most frequently framed themes as agglomerated collections of statistically interconnected terms. Thus, key topics were extrapolated through relative proximity and frequency in the structured transcripts and surfaced as summarized topical results.

Three text analytics assessments were run sequentially on the three corpora, each corpus an aggregation of extracted transcript text (one document per respondent):

1. Aggregation of combined challenge and best practice responses corpus
2. Challenge-only response corpus
3. Best practice-only response corpus

Table 3.14 Examples of key term associations resulting from text analytics

Algorithm	+ risk	event	+ incident	+ vulnerability
• big data	• focus	• big data	• relevant	• quickly
• people	• first	• successful	• artificial	• guess
• Potentially	• expertise	• + create	• feedback	• avoid
– + impact	• engineering	– + approach	• right	• + trust
– + spend	• + manage	– + type	• artificial	– behavioral
– necessarily	– + talk	– + detection	intelligence	– moment
– discussion	– + focus	– + sort	• + opinion	– + discovery
– specific	– + several	– good	– + level	– social
– + set	– + engineer	– + miss	– + raw	– Internal
– Metrics	– + place	– + human	– + high	– + job
• + quality	– + build	• + bad	– + action	– Financial
– + push	– + cloud	– Model	– + reason	– + compromise
– + structure	• +	– + result	– + advance	• + address
– + step	recommendation	– + end	• + exist	– + down
– + clean	– + operation	– + impact	– + job	– + implement
– + explain	– + lose	– + book	– + apply	– + operation
– + complexity	– + advantage	• + endpoint	– + create	– + allow
– interesting	– random	– + evolve	• + book	– + move
• + bring	– basically	– + implement	– nature	– + approach
– + common	– + depend	– + develop	– successful	• + user
– + level	– + compare	– Internal	– + area	– malicious
– + challenge	• + pick	– + cloud	– + hard	– + impact
– + high	– + researcher	• + set	– + involve	– + directly
– usable	– domain	– + potentially	– +	– + deal
• + design	knowledge		unsupervised	– + automate
– + build	– + deep		technique	– + collect
– end	– + difference			• + client
– development	– + fall			– + best practice
– + system	• + business			– + well
– throw	– artificial			– compromise
– false	– artificial			– + late
– number one	intelligence			– money
	– + project			• + talk
	– Basic			– + lot
	– + job			– next
	– quickly			– + paradigm
				– + instance
				– + group
				– + mean
				– right

Beginning with analysis of the combined corpus, the key themes resulting from text analytics are documented in Table 3.15 as statistical term aggregations, along with high-level thematic interpretations based on CSDS domain sensitizing concepts gleaned from literature analysis.

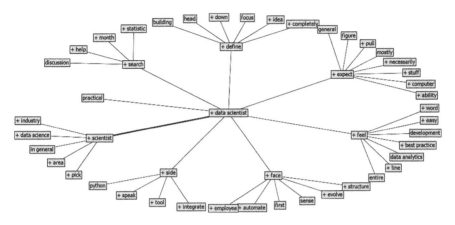

Fig. 3.4 Term linkage spider diagram for "data scientist"

Table 3.15 Text analytics—general topical themes extracted

Combined text topics	Thematic domain interpretation
preparation, +product, research, +application, +spend	Invest in tools to facilitate processes
+sample, +constraint, program, literature, +obtain	Implement a programmatic process based on research and samples of known incidents
+answer, +normalize, normalization, +understand, +format	Normalize and structure data based on an understanding of core challenges
+matter, subject, incident, +language, +analyst	Ensure analysts have subject-matter expertise to properly triage incidents
capture, +organization, +increase, +tool, sophistication	Embed knowledge of own infrastructure in analytics and tools
+owner, +database, IP, +vulnerability, +feed	Track vulnerabilities and utilize threat feeds
+risk, ai, perimeter, soc, critical	Risk-based focus in data science efforts
production, +label, +engineer, performance, malware	Track and label incidents in an operational process
model, adversarial, adversarial machine learning, +team, +law	Adversarial focus in model validation and consciousness of machine learning vulnerabilities
prevention, +client, +identify, +attacker, +anomaly	Preventative focus in anomaly detection

In interpreting the results from topic extraction across the combined responses corpus, several CSDS themes of note are apparent (per the list following):

- Importance of implementing programmatic organizational processes
- Data management/structuring and analytics as processes
- Embedding risk, prevention, vulnerability, and incident context in data
- Comprehensively track incidents through structured data collection
- Contextual configuration and implementation of technical tools
- Ensure staff are cross-trained and are knowledgeable concerning local security context

Table 3.16 Text analytics—challenges topic themes extracted

Challenge text topics	Thematic domain interpretation	Codes
analysis, +risk, +explain, +standard, +clean	Challenge processing and analyzing data to drive risk analysis	CH1
+feature, +communicate, domain knowledge, extent, +rely	Domain knowledge required for selecting features and communicating results	CH3
+system, critical, unsupervised, +machine, +human	Need to integrate human and machine assessments of incidents	CH8
performance, +engineer, +research, +team, +product	Need for research to adopt performance-enhancing tools	CH2
internal, +client, +company, +vulnerability, +prevent	Inherent vulnerabilities and preventing internal threats	CH5, 10
model, +customer, +alert, +result, false	Challenges refining models to reduce false alerts	CH7
+data science, +science, knowledge, cyber cyber	Lack of combined cybersecurity and data science knowledge	CH8, 9
+language, +enterprise, +incident, +mitre, +create	Need for standard data frameworks for tracking and assessing incidents	CH1, 3
+database, robust, +run, +analyst, +result	Challenges accessing, gathering, and analyzing data	CH1, 4
+solution, +environment, +answer, +space, perspective	Difficulties embedding operational context in models	CH5, 6

At a high level, these themes serve to validate and reinforce general topics identified in coding. Themes are reinforced by literature analysis sensitizing concepts. Further, the results mirror themes raised subsequently in factor analysis.

To refine insights from text analytics, structured challenge and best practice corpora were subsequently analyzed separately. In addition to offering thematic interpretations, relevant matching challenge and best practice codes were proposed to link thematic text extrapolation with the standard codes derived through the transcript coding process. Results are presented in Table 3.16.

In assessing topics extrapolated in text analytics of the structured challenge transcripts, the following high-level themes were observed:

- General challenges associated with data management
- Lack of structured data models to frame environments in context and to capture incidents
- Too many alerts and lack of consensus concerning how to reduce alerts in context
- Challenges framing data and analysis to identify discrete risk
- Issues related to the breadth of combined cybersecurity and data science domain knowledge

Linking extrapolated text topics with codes surfaced in codebook ordering served to reinforce and validate central themes identified.

Lastly, best practice structured transcripts were analyzed via the same process, with results summarized in Table 3.17. Greater diversity was observable in

Table 3.17 Text analytics—best practice topic themes extracted

Best practice text topics	Thematic domain interpretation	Codes
+application, internal, +company, +lack, +network	Understanding own infrastructure	BP13
domain knowledge, +field, communicate, knowledge, extent	Integrating cross-domain expertise	BP2, 3, 5, 18
+environment, +big, +risk, management, +step	Integrating big data with risk management	BP14, 17, 21, 22
analysis, +structure, +algorithm, +standard, +challenge	Structured process to design and validate standard algorithms	BP8, 9, 10, 25
+system, unsupervised, critical, +supervise, +label	Integrating unsupervised and supervised methods to surface and label incidents	BP11, 15
privacy, +feature, engineering, feature engineering, +event	Feature engineering focusing on events and classification	BP10, 26
+engineer, +research, +team, python, +organize	Integrated team of engineers, data scientists, and cybersecurity experts	BP2, 3, 4, 20
+database, management, +vulnerability, +analyst, robust	Focused engineering effort to aggregate and integrate data for situational awareness	BP1, 7, 12, 13, 19, 23
model, +team, +result, +operation, +investigation	Model validation integrating operational investigative results	BP4, 6, 10, 11, 15, 19, 24
+language, +mitre, +community, +event, +common	Applying standard data schemas and frameworks to track and share events	BP1, 7, 10, 16, 24, 26

associating derived topics with best practice codes. The following high-level best practice themes were observable in text analytics derived results:

- Advocacy for understanding own infrastructure in context
- Structured efforts to refine data models, variables, and representations
- Need for risk-focused approaches to surfacing threats
- Apply a structured process to design and validate algorithmic models
- Integrated-but-distinct discovery and detection processes
- Need for cross-domain cybersecurity and data science training and collaboration

In aggregate, text analytics provided valuable inductive reinforcement of themes surfaced across multiple methods: literature analysis, memoing, codebook extraction, and frequency analysis. In particular, themes extrapolated aligned with and reinforced codebook codes. Results also provided an additional interpretive vehicle to sensitize and orient subsequent quantitative pattern extrapolation.

3.2.3.8   Summary of Mixed Method Qualitative Insights

Interviews were interpreted through memoing, frequency analysis, demographic measures, and text analytics. Notable demographic impressions gained across 50 interviewees included the short relative tenure of titled CSDS professionals, averaging 2.9 years, with a median of 2.5 years. Furthermore, the diversity of backgrounds and experience levels of professionals working in the CSDS field was notable. An additional insight was that demographic and professional background appeared to have a very weak influence on themes surfaced in interviews, which implies that responses reflected shared impressions of the emerging domain rather than of disparate professional origins.

It was clear across all interpretations that challenges managing and processing data was a dominant reoccurring interview theme. The centrality of data management was the most frequent topic raised as both a challenge and best practice. Eighty-four percent ($n = 42/50$) of respondents raised the challenge theme "CH1: Data preparation, volume, and processing overhead." Eighty-four percent ($n = 42/50$) raised the best practice theme "BP1: Structured data preparation, discovery, and engineering process." Seventy-six percent ($n = 38/50$) of the interviewees raised both themes in their responses, indicating a strong sentiment that data management is both a challenge and an essential element of driving CSDS practice. Data management was surfaced not only as an engineering topic but also as an organizational undertaking requiring cross-domain collaboration and programmatic focus.

In summary, coding, memoing, frequency analysis, and text mining surfaced the following dominant themes across challenges and best practices:

1. *Data management as a process*: There was a recognition that data management as a process is not only a central CSDS enabler but also a complex multidomain challenge. More than a pure technical or engineering challenge, data management was framed across interviews as encompassing a complex of topics including scoping and problem framing; data aggregation, refinement, structuring, selection, and transformation; analytical modeling; and data science methods. The implication was that data management was viewed as an end-to-end process that necessitates the integration of organizational, methodological, and technical factors. Further, the activities associated with data management gave indications of requiring context. The key questions of *what to track*, *what data to collect*, *what is normal*, and *how to analyze* were raised. These prerequisites indicate the need for strategic organizational introspection and a concerted effort to frame data management as an integrated systematic and programmatic activity.

2. *Scientific analysis as a process*: There was a strong recognition of the need to embed scientific inquiry as a set of processes. The plurality recognizes that scientific problems do not tie to a single methodological approach or process but that CSDS presents a range of scientific problems, each requiring specialized process-driven methods and data. Indications from memos differentiate the specialized methodological approaches associated with the distinct challenges of CSDS problem framing, identifying hidden patterns, statistical diagnostics, segmentation, categorization, data reduction/engineering, framing and substantiating causal models, testing and selecting correlative models, model validation, and building a foundation upon which optimization models can be staged.

3. *Risk management orientation*: Interviewees emphasized the need to frame risk contextually, both in terms of external adversarial threats and as related to unique vulnerabilities associated with the local infrastructure being protected. It was noted that challenges encountered with leveraging data inherently hinder efforts to measure and frame risk. Scientific methods were presented as a facilitator to the degree quantitative analysis and models support approaches to measure and compare risks.

4. *Organizational orchestration*: A challenge observed by interviewees was that growing CSDS domain complexity has led to confusion concerning organizational priorities and ownership. This lack of strategic clarity hinders focus and the ability to orchestrate programmatic solutions. Sources of confusion are variously attributed to the growing breadth of the security domain, challenges determining organizational ownership, commercial market noise, continual technical evolution, the rapid expansion of exposure, growing environmental complexity, and continually shifting threats from adversarial actors.

5. *Cross-domain training and coordination*: There was a clear recognition that the CSDS domain is one of inherent breadth and complexity. This resulted in a perceived need for the concerted coordination of multidisciplinary teams and cross-training. In particular, there was a recognition of the need for a healthy mix of specialists and generalists. The notion of facilitating communication across disciplinary divides was also raised.

Several niche topics were raised infrequently by only two or three respondents during the interviews. In some cases, these were supplementary or adjacent topics that were aggregated with a larger theme during codebook ordering. However, in select cases, niche topics were raised that aligned with emerging themes in research and professional literature. For instance, adversarial machine learning, the rise of cryptojacking, and the topic of fake news were commented upon. While noted as less frequently cited interview themes, they were retained due to the growing prevalence in literature.

3.3 Quantitative Component: Diagnostic Gap Analysis

3.3.1 Overview of Analytical Methods

Having undertaken qualitative diagnostic background analysis and opinion research, interview results were next analyzed utilizing quantitative methods to drive diagnostic gap analysis (invoked in research objective RO2 and research question RQ5, per Sect. 1.3). Quantitative pattern analysis methods were applied to extrapolate latent themes across coded interview data. The intention was to reinforce literature review and qualitative findings through triangulated methods.

Triangulating mixed methods with quantitative pattern analysis was motivated by the newness and complexity of the CSDS domain. Given the breadth and complexity of themes raised, quantitative pattern analysis provided a rigorous supplementary approach to extrapolate latent thematic patterns. Quantitative mixed methods supported interpretive rigor in diagnostic gap analysis as a supplement to, and validation of, qualitative findings.

Indications of key themes in the CSDS domain were previously surfaced through qualitative mixed methods. Several key themes were raised in mixed literature analysis. Thematic patterns were also surfaced across interview responses through basic statistical approaches, namely, frequency, correlation, and logistic fit analysis. Additionally, text analytics provided inductive substantiation of themes raised. Based upon structure indicated through these triangulated approaches, there were grounds to apply focused quantitative methods to reinforce the interpretation of thematic patterns across interview responses.

Specifically, given the original sensitizing conceptual model concerning the reciprocal relationship between CSDS challenges and best practices (per Fig. 3.1), there was a desire to determine if particular challenge and best practice codes statistically co-occurred across interviews. Per the model, it was anticipated that significant affinities might be detected between CSDS challenges and best practices. This might be taken to indicate latent advocacy of practitioner prescriptions (best practices) to address key gaps (challenges).

Of central interest was to surface possible significant connections between challenges and best practices raised across interviews to inform diagnostic gap analysis. While triangulated literature analysis, interview memos, interview code frequency analysis, and text analytics all raised a key set of master themes, there was a desire to substantiate findings through quantitative pattern analysis to provide an additional layer of interpretive rigor.

In terms of the quantitative pattern analysis methods applied, logistic regression was pursued to assess significant relationships between core challenge and best practice responses across all interviews. This resulted in a set of significant gap-prescriptions, matching CSDS best practices statistically to focused challenges.

Additionally, given the relatively large number of overall challenge and best practice themes, there was a desire to determine if higher-level latent themes were resident individually within the existing challenge and best practice responses. The intention was to ease interpretability by aggregating higher-level latent themes. Explanatory factor analysis was applied separately to challenge and best practice interview responses to derive aggregated latent constructs for the two categories individually. This resulted in a compressed set of challenge and best practice macro-theme factors.

Lastly, to determine whether there were statistical connections between the two sets of latent macro-themes surfaced for challenges and best practices, statistical associations between the two sets of factors were assessed. As factor analysis resulted in two rotated factor matrices, statistical relationships between the two matrices were assessed. Linear regression and correlation analysis between extrapolated challenge and best practice factors was undertaken to frame statistically co-occurring latent factor relationships. These were interpreted to represent latent high-level treatments prescribed by practitioners to address identified gaps. Based upon the results, root cause analysis was undertaken to aid the interpretation of the derived insights and to offer practical guidance.

3.3.2 Methodological Underpinnings

Quantitative analysis of coded results was undertaken by applying mixed multivariate methods (Bartholomew et al. 2008; Harlow 2005). Factor analysis was applied to derive latent challenge and best practice themes across coded interview data (Kim and Mueller 1978a, b; O'Rourke and Hatcher 2013; Osbourne and Banjanovic 2016). Latent thematic factors were extrapolated, and factor-to-factor linear fit and correlation analysis were assessed. Factor-to-factor analysis was undertaken to diagnose latent relationships between challenge and best practice themes raised across interviews.

Quantitative analysis was pursued to substantiate and reinforce qualitative findings surfaced in both literature analysis and thematic interpretation of interviews. The central intention of applying quantitative analysis was to extrapolate and substantiate thematic patterns resident across interviews. Given the range of themes raised in interview coding, reflecting both the newness and complexity of the CSDS domain, quantitative methods provided a clear and rigorous approach to reduce and aggregate macro-themes.

Applying mixed quantitative methods supported rigorous pattern extrapolation. Combined results from quantitative analysis of interview code data (via frequency analysis, logistic regression, factor analysis, and correlation) served to reinforce qualitative patterns surfaced in both literature analysis and thematic interpretation of interviews (via memoing, frequency analysis, and text analytics).

Factor analysis, in particular, was central as it identifies latent themes—underlying commonalities between and among topics. As noted by Kim and Muller, "factor analysis assumes that the observed variables are linear combinations of some underlying (hypothetical or unobservable) factors" (1978b). The interpretability of factors in exploratory factor analysis was judged efficacious as factors separated cleanly both quantitatively and conceptually. Interpretation of factors was reinforced by sensitizing concepts derived in literature analysis and memoing, and inductive patterns surfaced in text analytics and thematic frequency analysis. The methodological issue of treating dichotomous data with factor analysis was addressed directly by cross-referencing results with variant methods supported in research.

Factor analysis, along with subsequent factor-to-factor linear fit and correlation analysis, served to surface and clarify macro-themes resident across the broader and more complex range of themes raised in interview coding. The final results were interpreted through the lens of triangulated mixed methods, including sensitizing concepts surfaced from literature analysis, interview memos, thematic frequency analysis, and text analytics.

In the larger research context, mixed methods were applied to triangulate interview findings utilizing approaches ranging from the more inductive (text analytics, thematic frequency analysis) to the more deductive (literature analysis, sensitizing concepts, memoing). Factor analysis provided a rigorous methodological bridge between inductive and deductive approaches.

Factor analysis is an inductive method for extrapolating latent concepts that must ultimately be validated deductively within the context of the conceptual domain under scrutiny. Factor analysis thus provided an efficacious and rigorous method to surface macro-themes across interviews and to insinuate findings from adjacent mixed methods through factor interpretations. Combined, the methods were applied to extrapolate and substantiate perceived patterns. Thus, no empirical claims to a causal model or hypothetical proof were intended or implied.

3.3.3   Logistic Fitting of Challenges with Best Practices

Having previously identified indications of significant relationships within challenge and best practice responses and having surfaced common cross-categorical themes through literature analysis and text analytics, a quantitative assessment of challenge-to-best practice responses was undertaken. Each individual challenge theme was assessed against the range of best practice responses to determine which collection of best practices were most commonly advocated by respondents along with specific challenges. The datasets utilized were two binary matrices capturing themes raised across 50 interviews: an 11-by-50 matrix for challenges and a 26-by-50 matrix for best practices. The binary response for each coded theme indicated either 1 = theme code raised by the interviewee or 0 = theme code not raised by the interviewee.

A nominal logistic regression model fitting approach was applied to assess significance between challenge (CH) and best practice (BP) responses across all interviews. All 11 individual challenge response themes were treated in sequence as dependent (Y) factors against all possible best practice response codes as independent (X) factors.

The intention was to assess grounds for statistically valid models across all responses, asserting a significant tendency to surface a particular challenge with a subset of particular best practice themes. Significance implied that, across all responses, single challenges, as identified gaps, tended to co-occur with a subset of advocated best practices, as treatment prescriptions. The overriding implication is that challenges associate singularly with unique subsets of best practices, constituting practitioner-based advocacy of combined prescriptions (best practices) to treat singular perceived gaps (challenges).

Four significant models, identifying with 4 of the 11 challenge themes, were derived utilizing the logistic model-fitting approach. Only positively significant best practice themes were reported from the resulting models. Negative significance indicated a pronounced lack of the best practice response being raised along with the related challenge theme.

Additionally, although "CH1: Data preparation" was identified as being highly prevalent in frequency analysis, it did not surface a significant subset of advocated best practices. The relative frequency of this response across respondents diluted its association broadly across best practices. In terms of interpretation, while this was the most frequent challenge raised, best practices associated with addressing data management challenges appeared to range broadly across all themes, indicating this is a complex and wide-ranging challenge.

Challenge-Best Practices Model 1 (CH2)   Among the four challenge codes correlating significantly with particular subsets of best practices, the first logistic fit model extrapolated was for CH2 "unrealistic expectations proliferated by marketing hype." Table 3.18 identifies those best practice codes (BPs) which demonstrated significant logistic regression whole model worth to CH2, indicating that interviewees tended to raise these particular best practices along with challenge 2 (CH2).

In terms of interpretation, best practice prescriptions for challenge theme CH2 address organizational challenges associated with framing realistic CSDS operational goals. This can be considered to address the problem of *what to analyze* as well as how. As was observed in literature analysis, the cybersecurity solutions market has led to informational distortions that have confused organizational expectations and goal setting.

The prescriptive advocacy identified in matching best practices suggests that organizations adopt a structured and systematic stance. The implication is for organizations to develop formal criteria for designing and validating data science solutions. This includes advocacy for cross-training and for implementing structured analytical processes focused on model validation and operational reinforcement.

The recommendations for implementing tracking honeypots and for tracking the results of investigations to build models (human-in-the-loop reinforcement) indicate a focus on evidentiary data collection. These mechanisms connect data science methods and solutions to a "real" phenomena. The implication is that by focusing

Table 3.18   Gap-prescription 1 (CH2)—refining expectations and capabilities

CH2: Unrealistic expectations proliferated by marketing hype		
Logistic regression effect likelihood ratio tests	LogWorth	Prob > ChiSq
BP3: Cross-training team in data science, cyber, engineering	60.809	<0.0001
BP8: Model validation and transparency	24.045	<0.0001
BP19: Human-in-the-loop reinforcement	6.606	<0.0001
BP26: Honeypots to track and observe adversaries	4.922	<0.0001
BP12: Address AI hype and unrealistic expectations directly	4.564	<0.0001

Significant at 0.01: Whole model test *P*-value 0.0001, $R2$: 1.0

Table 3.19 Gap-prescription 2 (CH3)—understanding phenomenon in operational context

CH3: Contextual nature of normal versus anomalous behavioral phenomenon		
Logistic regression effect likelihood ratio tests	LogWorth	Prob > ChiSq
BP3: Cross-training team in data science, cyber, engineering	6.472	<0.0001
BP11: Cyclical unsupervised and supervised machine learning	2.553	0.0028
BP2: Building process focused cross-functional team	1.888	0.013

Significant at 0.01: Whole model test P-value < 0.0001, $R2$: 1.0

on evidentiary data, the pitfalls of attempting to chase after phantom anomalies are avoided (particularly as promulgated in marketing distortions).

Challenge-Best Practices Model 2 (CH3) A second gap-prescription was associated with challenge theme 3 "CH3: Contextual nature of normal versus anomalous behavioral phenomenon." The model addresses the question of *what is normal*. This identifies the difficulty of disassociating normal from "abnormal" behavior in settings of increasing technical, organizational, and adversarial complexity. Results are summarized in Table 3.19. Best practices associated with CH3 advocate interdisciplinary collaboration, namely approaches to unite specialized cross-domain cybersecurity and data science expertise.

The best practice of cyclical unsupervised and supervised machine learning observes that data science does not offer magic bullets. Distinct data science methods need to be applied to address differing cybersecurity challenges. As a key example, unsupervised methods are exploratory in scope and can be used to discover emerging patterns or to segment entities into distinct categories. Supervised methods are used to detect (or predict) the presence of potentially adversarial phenomena but require hard evidence in the form of prior examples to train models.

The need to integrate cross-domain expertise advocates connecting CSDS treatments to well-defined challenges. By matching focused treatments to well-defined challenges, unique contextual aspects of cybersecurity phenomena are addressed. This includes systematic approaches to establish baselines, an understanding of what is statistically expected within a particular environment. Framing a notion of "normal" for an environment is a crucial foundation for staging the detection of "abnormal" phenomenon through a process of categorical reduction and elimination.

As a gap-prescription, this model highlights organizational management's importance in promoting training and collaboration as a foundation for cross-disciplinary exchange. Both the challenge and treatments raise the danger of oversimplifying CSDS solutions. Implicit is the notion that an instinctive pitfall of inherent complexity is to oversimplify responses. In aggregate, prescribed best practices emphasize the need to integrate organizational and process-based responses closely.

Table 3.20 Gap-prescription 3 (CH4)—capturing incidents as discrete representations

CH4: Lack of labeled incidents to focus detection		
Logistic regression effect likelihood ratio tests	LogWorth	Prob > ChiSq
BP10: Track and label incidents and exploits	415.564	<0.0001
BP4: Scientific method as a process	202.121	<0.0001
BP20: Survey academic methods and techniques	91.303	<0.0001

Significant at 0.01: Whole model test P-value < 0.0001, $R2$: 1.0

Challenge-Best Practices Model 3 (CH4) Prescriptions to address challenge theme 4, "CH4: Lack of labeled incidents to focus detection," advocate programmatic tracking and labeling of incidents and exploits to address the need for phenomenological evidence. Best practice responses statistically associated with CH4 are summarized in Table 3.20.

CH4-associated best practices also emphasize the need for the application of process-driven scientific methodologies. This recognizes that a question exists concerning how to label incidents through representations in data. Diagnostic and feature engineering methods support a focused approach to specify *what data to collect*. Advocacy for formal scientific methods is further emphasized in highlighting the importance of accessing academic research. This recognizes that organizations should not seek to "reinvent the wheel" when there may well be prior guidance available in research literature.

Challenge-Best Practices Model 4 (CH5) Gap-prescriptions addressing challenge theme 5, "CH5: Own infrastructure, shadow IT, and proliferation of exposure," associates a range of best practice treatments to address the challenge of expanding vulnerabilities and risks. Table 3.21 summarizes best practices associated with CH5.

Challenge CH5 observes that centralized IT in many organizations has lost oversight concerning what devices are attached to the network. As other departments and employees increasingly attach their own devices to the organizational network, there is a growing lack of control and accompanying expansion of exposed entry points, attack vectors, and resources at risk. The question of *what to track* become difficult to address in environments where it is increasingly uncertain what devices are attached to and active on a particular network.

A broad number of prescribed best practices are associated with CH5, recognizing the complex, multi-domain nature of this challenge. Prescriptions range across the MIS frames of organization, process, and technology. At the organizational level, there is an observation concerning the need for management support, a risk and probability focus, and guidance to avoid the pitfalls of marketing hype. Furthermore, there is advocacy to promulgate a paradigm shift to data-focused methods, to adopt a risk management stance, and to ensure cybersecurity domain knowledge is instilled in the organization.

Table 3.21 Gap-prescription 4 (CH5)—understanding what is being protected

CH5: Own infrastructure, shadow IT, and proliferation of exposure		
Logistic regression effect likelihood ratio tests	LogWorth	Prob > ChiSq
BP18: Upper management buy-in and support	486.684	<0.0001
BP5: Instill core cyber domain knowledge	430.839	<0.0001
BP6: Vulnerability, anomaly and decision automation to optimize operations	343.513	<0.0001
BP22: Segment risk programmatically and outsource components	150.152	<0.0001
BP21: Cyber risk as general enterprise risk and reward	142.937	<0.0001
BP7: Data normalization, frameworks and ontologies	138.086	<0.0001
BP14: Cloud and container-based tools and data storage	97.298	<0.0001
BP9: Data-driven paradigm shift away from rules and signatures	41.342	<0.0001
BP13: Understand own infrastructure and environment	6.358	<0.0001

Significant at 0.01: Whole model test P-value < 0.0001, $R2$: 1.0

Concerning processes, it is advocated that an effort be made to map and track the existing environment and represent this environment in structured data. It is also advocated to automate vulnerability identification, anomaly detection, and decision-making processes. Lastly, technical aspects are raised in advocating the application of cloud and container-based tools and approaches to support CSDS efforts.

3.3.4 Exploratory Factor Analysis and Interpretations

The identification of significant relationships between challenge and best practice interview responses through logistic regression suggested the potential presence of additional thematic patterns. With an interest in surfacing macro-thematic patterns across responses, factor analysis was applied to identify possible latent constructs. Exploratory factor analysis was undertaken to extrapolate factor solutions and to offer interpretations based upon sensitizing concepts. Extrapolated factors with explanatory integrity were identified, interpreted, and framed as macro-themes.

Factor analysis was undertaken on coded challenge and best practice interview responses separately in order to extrapolate latent constructs specific to the two categories. Factor analysis of a combined dataset was trialed but did not result in a valid factor model. This was otherwise evidence of the distinct conceptual difference between challenge and best practice responses. As factor analysis identifies latent variables, the combined dataset was unable to resolve shared factors across the conceptually distinct categories. A secondary interpretation was that the small sample size impeded cross-response factor extrapolation.

The same response matrix datasets utilized in logistic regression were utilized for factor analysis. These two binary matrices captured themes raised across 50 interviews, an 11-by-50 matrix for challenges (CHs) and a 26-by-50 matrix for best practices (BPs), with 1 = theme raised by the interviewee or 0 = theme not raised by the interviewee. The goal was to identify clusters of responses suggesting macro-themes within each of the two categories, challenges and best practices. The analysis resulted in two sets of significant thematic factor constructs, one for challenges and one for best practices.

As factors are statistical assertions of latent constructs, interpretation is nuanced and must defer to theoretical concepts and domain knowledge. Extrapolated factors were interpreted by triangulating concepts surfaced in literature analysis along with themes identified in interview research memoing, frequency analysis, and text analytics.

Two methodological aspects of the application of factor analysis must be caveated. Concerning sample size, larger samples are generally considered better for factor analysis. Advocacy varies, but 50 cases is generally considered a quite small sample. However, there are examples in research where a sample of 50 is supportable (Barrett and Kline 1981). The conceptual interpretability of factors is considered a key threshold for sample sufficiency. In addition to meeting diagnostic thresholds for significance, both sets of factors extrapolated were conceptually interpretable.

The second methodological caveat concerns the treatment of binary or dichotomous data, with factor analysis typically requiring interval or ratio data. There are specific exceptions to this mandate, which are addressed comprehensively subsequent to the analysis. Based on the data meeting a set of acceptable exceptions, the utilization of binary data was judged to be acceptable.

Methodological best practices were observed in the assessment and extraction of factors, for instance, regarding determining the number of factors and rotating factors to assist in interpretation (Bartholomew et al. 2008; Cudeck and MacCallum 2007; Jöreskog 1977; O'Rourke and Hatcher 2013; Osbourne and Banjanovic 2016; Rennie 1997). A maximum likelihood (ML) factor analysis approach was taken, which iteratively identifies factors that maximize the likelihood of the population correlation matrix (Jöreskog 1967; Jöreskog and Lawley 1968; Lawley and Maxwell 1971; Osbourne and Banjanovic 2016).

The SAS JMP data analysis software tool was used to perform factor analysis. Several specifications were determined to configure the factor analysis appropriately. A maximum likelihood (ML) factoring method was applied, as opposed to a principal axis approach. The former led to greater separability and interpretability of factors.

Variance scaling was configured based on correlations, which imposes assumptions on the diagonal of the extracted correlation matrix. Concerning prior communality estimates, the amount of variation explainable by common factors, a common factor analysis squared multiple correlations (SMC) approach was applied. This approach uses communality estimates based on diagonal measures to capture the variation in elements explainable by common factors. The common factor analysis SMC prior communality approach supported clearer interpretability in resulting factors (as opposed to the principal components diagonals method which sets diagonals to 1 by default).

Table 3.22   Challenge (CH) theme factor loadings

Factor loadings: Challenges (CH)	F1	F2	F3	F4	F5	F6
CH6: Uncertainty leads to ineffective reactive stance	**0.78**	0.09	0.08	0.01	0.23	0.09
CH9: Resourcing, developing, & hosting in house	**0.74**	0.05	0.22	0.17	-0.04	*- 0.37*
CH5: Own infrastructure, shadow IT, and proliferation of exposure	**0.68**	-0.01	-0.01	0.06	0.27	0.09
CH10: Expanding breadth and complexity of cyber domain	**0.63**	-0.03	-0.12	-0.01	-0.1	*- 0.33*
CH3: Contextual nature of normal versus anomalous behavioral phenomenon	0.01	**0.98**	-0.05	-0.03	0.03	-0.18
CH4: Lack of labeled incidents to focus detection	-0.06	**0.43**	**0.43**	0.12	*-0.25*	0.27
CH7: Traditional rules-based methods result in too many alerts	0.27	**0.43**	0.28	**0.40**	-0.07	- 0.08
CH1: Data preparation (access, volume, integration, quality, transformation, selection)	-0.11	0.01	**0.99**	-0.08	0.05	- 0.08
CH2: Unrealistic expectations proliferated by marketing hype	0.05	0.01	-0.09	**0.96**	0.05	-0.11
CH8: Program ownership, decision making, and processes	0.24	-0.03	0.01	0.03	**0.97**	0.02
CH11: Policy, privacy, regulatory, and fines	-0.07	-0.09	-0.04	-0.08	0.01	**0.55**

Proportion of variance, scree tests, and interoperability were all assessed to determine the optimal number of factors. Various factor solutions were trialed, and results were compared like-to-like balancing diagnostic measures with conceptual interpretability. Per methodological guidance, interpretability was a key objective, which included finding a balance between the relative number of variables composing each factor (Bartholomew et al. 2008; Harlow 2005; O'Rourke and Hatcher 2013; Osbourne and Banjanovic 2016).

Varimax (orthogonal) rotation was applied to clarify variable loading in order to maximize the variance in each factor loading. Challenge factor loadings were suppressed at below 0.4. A notation of negative loadings was made for inverse contributions less than −0.2 (negative numbers marked in bold italics in Table 3.22), but negative loadings were not adopted to frame interpretation as the focus was on clusters of explicitly surfaced themes.

For challenge (CH) responses, exploratory maximum likelihood (ML) factor analysis with SMC prior communality and varimax rotation produced an elbow in the scree plot at seven factors, implying a six factors solution at just below an eigenvalue of 1.0, capturing 83% of cumulative eigenvalue variation. Seventy-one percent of the aggregate variance was captured cumulatively within the six factors. All variables were accounted for and produced interpretable factors. Two variables had similar loadings and were adopted into dual factors as interpretability was aided (CH4 and 7).

Whole model and variable component significance tests were assessed to verify statistical validity. Model completeness evidenced significance with the hypothesis of no common factors (H_0) being strongly rejected at a P-value of <0.0001 and the hypothesis of more than six factors needed (H_1) being strongly rejected at a P-value of 0.41. Positive scores at 0.4 or above associated with the six resulting factors are shaded light grey in Table 3.22 (i.e., factor CH F1 includes CH6, 9, 5, and 10).

Per Kim and Muller, "the main objective of the extraction step in exploratory factor analysis is to determine the minimum number of common factors that would satisfactorily produce the correlations among the observed variables" (1978b). Six factors achieved both quantitative and conceptually interpretable measures of compactness, interpretability being a key threshold given the objective of identifying latent thematic context.

Having derived a statistically significant and conceptually promising set of six factors, the positively correlating significant challenge codes were mapped to the related factors, thematic interpretations were posited, and summary titles were interpolated, resulting in a summary captured in Table 3.23. Triangulated

Table 3.23 Factor interpretations—challenges (CH Fx)

Factor title	Component challenge (CH) themes	Latent thematic interpretation
CH F1 Expansive complexity	**CH6**: Uncertainty leads to ineffective reactive stance **CH9**: Resourcing, developing, and hosting in house **CH5**: Own infrastructure, shadow IT, and proliferation of exposure **CH10**: Expanding breadth and complexity of the cyber domain	Elements associated with this factor relate to challenges of uncertainty and understanding concerning what to secure given environments of increasing complexity, proliferating exposure, and expansive technology
CH F2 Tracking and context	**CH3**: Contextual nature of normal versus anomalous behavioral phenomenon *CH4: Lack of labeled incidents to focus detection* *CH7: Traditional rule-based methods result in too many alerts*	This factor addresses a complex of elements related to the challenge of pursuing data science treatments given a lack of context concerning phenomenon and events
CH F3 Data management	*CH4: Lack of labeled incidents to focus detection* **CH1**: Data preparation (access, volume, integration, quality, transformation, selection)	Associated with data management more specifically, this factor raises the dominant challenge of implementing end-to-end data management processes
CH F4 Expectations versus limitations	*CH7: Traditional rule-based methods result in too many alerts* **CH2**: Unrealistic expectations proliferated by marketing hype	This challenge theme raises a "between a rock and a hard place" quandary: practitioners express the challenge of struggling with overhyped expectations related to data science yet face declining efficacy with traditional methods
CH F5 Unclear ownership	**CH8**: Program ownership, decision-making, and processes	This factor raises the challenge of a lack of organizational clarity concerning who and what to implement due to confusion related to roles and responsibilities
CH F6 Data policies	**CH11**: Policy, privacy, regulatory, and fines	This factor observes the growing challenge of adopting data analytics methods in an environment where data compliance, legal, and regulatory strictures are in flux

sensitizing concepts from literature analysis, interview memos, interview codebook notes, and text analytics themes were referenced to aid in thematic interpretation.

Note that CH themes 4 and 7 co-occur in factors 2 and 3 and 2 and 4, respectively (as noted in *italics*). As the loading measures were nearly identical and retaining dual membership aided in nuanced interpretability, the results were retained. Otherwise, there was a clean and clear separability of theme responses between factor clusters. This spoke to the presence of clear, independent statistical clusters across challenge themes surfaced across interview responses.

Factor analysis was next undertaken on the range of best practice (BP) responses, replicating the same process, methods, and configurations. Exploratory maximum likelihood (ML) exploratory factor analysis with SMC prior communality and varimax rotation produced an elbow in the scree plot at seven factors, with six factors implied at an eigenvalue of 1.58, which captured 53% cumulative eigenvalue variation. Forty-one percent of aggregate variance was captured cumulatively in six factors. Factor loadings were suppressed at below 0.2, a lower threshold stemming from greater variability associated with the larger number of coded themes. A notation of negative loadings (negative numbers marked in bold italics in Table 3.24) was made for inverse contributions less than -0.25, noting a marked lack of response to the theme within the related factor.

All variables were accounted for and produced interpretable factors. Although less cumulative variation was captured due to the larger number of variables, the model evidenced completeness and interpretability. Three variables had similar factors loadings and were retained in dual factors to aid interpretability (BP15, 25, 11 in factors 1 and 6, 3 and 5, and 1 and 6, respectively). The hypothesis of no common factors (H_0) was strongly rejected at a P-value of <0.0001, and the hypothesis for more factors needed (H_0) was rejected at a P-value of 0.424. Again, themes associated with each columnar factor are denoted by light grey-shaded positive numbers (i.e., factor F1 includes BP10, 15, 4, 20, 19, 17, 6, 8, 11).

Having extracted significant and individuated factors from best practice responses, per the previous procedure, the six resulting factor themes were labeled and interpreted based upon an interpretation of the associated thematic responses. Again, sensitizing concepts from literature analysis and text analytics, along with interview codes and memos, were triangulated to support interpretations. Results are summarized in Table 3.25.

The goal of exploratory factor analysis being to identify latent themes, a key threshold for validity is theoretical and conceptual interpretability. Factor patterns extrapolated must pass a "common sense" litmus test as specific to the conceptual domain under scrutiny. In this case, themes surfaced and interpreted through factor clusters triangulated well with themes surfaced in adjacent mixed methods, namely, literature analysis, interview memos, response frequency analysis, and text analytics. Beyond quantitative diagnostics indicating factor validity, there was a strong concordance between macro-themes surfaced in factor analysis with those identified through adjacent methods.

Table 3.24 Best practice (BP) theme factor loadings

Factor Loadings: Best Practices (BP)	F1	F2	F3	F4	F5	F6
BP10: Track and label incidents and exploits	**0.70**	-0.14	-0.07	-0.01	0.04	0.12
BP15: Distinct exploration and detection architectures	**0.63**	0.16	0.06	-0.09	0.09	**0.32**
BP4: Scientific method as a process	**0.56**	0.10	0.13	0.19	-0.04	-0.07
BP20: Survey academic methods and techniques	**0.48**	0.05	-0.06	-0.04	-0.04	0.06
BP19: Human-in-the-loop reinforcement	**0.37**	*-0.38*	0.14	0.09	*-0.28*	-0.12
BP17: Deriving probabilistic and risk models	**0.37**	-0.01	-0.01	0.04	0.15	-0.08
BP6: Vulnerability, anomaly & decision automation to optimize operations	**0.32**	-0.18	-0.04	0.01	-0.12	-0.05
BP8: Model validation and transparency	**0.28**	-0.05	-0.04	0.02	0.17	-0.17
BP2: Building process focused cross-functional team	0.18	**0.96**	0.04	-0.03	0.03	-0.19
BP3: Cross-training team in data science, cyber, engineering	0.10	**0.73**	-0.17	0.16	0.11	-0.11
BP5: Instill core cyber domain knowledge	-0.06	**0.63**	-0.04	*-0.30*	0.09	0.12
BP16: Participate in data sharing consortiums and initiatives	-0.24	**0.26**	*-0.36*	0.04	0.08	-0.02
BP18: Upper management buy in and support	-0.21	**0.23**	-0.14	0.08	0.04	-0.03
BP7: Data normalization, frameworks & ontologies	-0.12	**0.23**	0.04	0.01	-0.03	0.19
BP22: Segment risk programmatically and outsource components	-0.20	-0.02	**0.95**	-0.18	0.06	-0.15
BP24: Preventative threat intelligence	-0.05	-0.01	**0.49**	0.17	-0.10	0.01
BP25: Hosting and pushing detection to endpoints	0.22	*-0.30*	**0.48**	0.19	**0.30**	-0.24
BP21: Cyber risk as general enterprise risk & reward	0.02	0.03	**0.20**	-0.01	0.03	*-0.31*
BP9: Data-driven paradigm shift away from rules & signatures	-0.01	0.03	0.17	**0.98**	0.02	0.06
BP1: Structured data preparation, discovery, engineering process	0.05	0.07	-0.10	**0.41**	-0.12	0.23
BP13: Understand own infrastructure & environment	0.05	-0.07	0.03	**0.23**	-0.14	-0.13
BP14: Cloud and container-based tools and data storage	-0.03	0.12	-0.03	0.08	**0.84**	0.06
BP12: Address AI hype and unrealistic expectations directly	0.09	0.09	0.02	*-0.28*	**0.36**	-0.14
BP23: Adding machine learning to SIEM	0.03	0.02	-0.01	-0.13	**0.25**	0.02
BP11: Cyclical unsupervised and supervised machine learning	**0.58**	0.03	0.14	-0.19	0.08	**0.75**
BP26: Honeypots to track and observe adversaries	0.01	-0.07	-0.02	0.16	0.03	**0.31**

Table 3.25 Factor interpretations—best practices (BP Fx)

Factor title	Component best practice (BP) themes	Latent thematic interpretation
BP F1 Scientific process	**BP10**: Track and label incidents and exploits **BP15**: Distinct exploration and detection architectures **BP4**: Scientific method as a process **BP20**: Survey academic methods and techniques **BP19**: Human-in-the-loop reinforcement **BP17**: Deriving probabilistic and risk models **BP6**: Vulnerability, anomaly and decision automation to optimize operations **BP8**: Model validation and transparency **BP11**: Cyclical unsupervised and supervised machine learning	Elements in this factor advocate the implementation of scientific methods as a set of processes. Themes of diagnostics, quantification, validation, verification, and human oversight are resident
BP F2 Cross-domain collaboration	**BP2**: Building process focused cross-functional team **BP3**: Cross-training team in data science, cyber, engineering **BP5**: Instill core cyber domain knowledge **BP16**: Participate in data-sharing consortiums and initiatives **BP18**: Upper management buy-in and support **BP7**: Data normalization, frameworks and ontologies	Organizational aspects are highlighted in this factor, emphasizing the need to connect cybersecurity and data science domain knowledge in cross-functional teams and processes. The need for standard processes and organizational alignment on goals and models are emphasized
BP F3 Risk management focus	**BP22**: Segment risk programmatically and outsource components **BP24**: Preventative threat intelligence **BP25**: Hosting and pushing detection to endpoints **BP21**: Cyber risk as general enterprise risk and reward	Risk themes are emphasized in this factor. Advocacy reflects a programmatic approach to assessing security risk, which includes soliciting external intelligence

(continued)

Table 3.25 (continued)

Factor title	Component best practice (BP) themes	Latent thematic interpretation
BP F4 Data-driven/data management	**BP9**: Data-driven paradigm shift away from rules and signatures **BP1**: Structured data preparation, discovery, engineering process **BP13**: Understand own infrastructure and environment	The dominant theme of data management as a process is surfaced here. This includes the importance of understanding context in terms of the environment to be safeguarded
BP F5 Focused tools	**BP25**: Hosting and pushing detection to endpoints **BP14**: Cloud and container-based tools and data storage **BP12**: Address AI hype and unrealistic expectations directly **BP23**: Adding machine learning to SIEM	Technology aspects are emphasized in this factor. Endpoint-based detection and implementing cloud and container-based CSDS solutions are advocated. Enhancing SIEMs with machine learning is advocated, as is addressing "AI hype"
BP F6 Structured discovery process	**BP15**: Distinct exploration and detection architectures **BP11**: Cyclical unsupervised and supervised machine learning **BP26**: Honeypots to track and observe adversaries	The importance of approaching cybersecurity discovery and detection problems separately is emphasized in this factor. This implies architectural and process segmentation with discrete cyclical overlaps. The need for threat intelligence is raised through advocacy on employing honeypots

3.3.5 Factor Substantiation

Exploratory factor analysis surfaced indications of latent macro-thematic patterns across CSDS coded interview responses. Interpretation of the factors derived was aided by triangulating themes raised in mixed method analysis, including literature analysis, interview memoing, text analytics, and frequency analysis of interview responses.

As the application of factor analysis to data derived from qualitative interviews may be considered novel for some organizational researchers, it was judged worthy to further substantiate the approach and to address possible methodological concerns.

The process of extrapolating structured data from the interviews applied best practices associated with the following: interviewing (Bernard and Ryan 2010; Richards and Morse 2007), codebook ordering (Glaser 1978, 1998; Goffman 1961; Polkinghorne 2015; van den Hoonaard 1996), and interview transcript coding (Bryman and Burgess 1994; Burnard 1991; Corbin and Strauss 2015; Saldana 2016).

A primary methodological question for organizational researchers may relate to the progenitorship of the quantitative dataset. Whereas a survey directly results in quantitative data, in this case structured data was extrapolated by the researcher from interview transcripts, cross-referenced with sensitizing concepts raised in adjacent mixed method research.

Following the initial codebook development phase, the researcher assessed each interview individually to determine if the interviewee raised a particular coded theme or not. This resulted in a binary dataset indicating the themes raised by each interviewee out of the total set of possible themes raised across all interviews. The resulting dataset is not dissimilar to data resulting from a survey. Through codebook extrapolation, the interviewer assessed and interpreted the range of responses possible across interviews. This interpretation was guided by sensitizing concepts derived from literature analysis and memoing.

After extrapolating the master codebook, each interview transcript was assessed according to whether the range of coded themes were raised or not. The researcher thus plays an active role in extracting and interpreting structured responses from interview transcripts. In comparison to a survey, there is greater interpretive initiative from the researcher. However, surveys can be said to embed implicit assumptions of interpretive meaning in the survey tool itself.

Once interview results were interpreted and quantified as a binary response dataset, factor analysis was applied to extrapolate latent themes across all interviews. Factor analysis has a long tradition of deriving latent variables in social science research (Kim and Mueller 1978a, b; O'Rourke and Hatcher 2013; Osbourne and Banjanovic 2016). However, the approach is less familiar in management and organizational research, which frequently pursues hypothetical-deductive model substantiation rather than pattern extrapolation.

As no target variable or causal model is assumed or asserted in factor analysis, there may be a question concerning what is being substantiated. Factor analysis aims to statistically examine the possible existence of multivariate latent patterns among a matrix of data. Across a set of interview or survey responses, this would indicate possible patterns across all responses to a survey or interview. This would indicate a tendency for a particular set of responses to co-occur, signifying a possible latent relationship.

Factor analysis diagnostic tests substantiate the probability of nonrandom multivariate patterns among a matrix of data. Significance testing through a chi-squared probability test indicates the level of confidence associated with a null hypothesis (H_0) of "no common factors" given the specifications and results associated with a particular factor solution.

The chi-square test indicates the probability of rejecting or accepting the null hypothesis (H_0) in a given multifactor, whole-matrix factor solution at a given threshold. For instance, achieving a 1% (0.01) or less threshold would indicate a 1% chance or less of finding evidence of an equivalent or stronger whole-matrix solution given a hypothetical random set of data. Substantiation of a whole matrix factor solution through rejection of the null hypothesis (H_0) at a set confidence level indicates statistical grounds for accepting that the patterns detected are at an $x\%$ confidence level of being resident against an equivalent hypothetical chance dataset.

Concerning the validity of factors identified through factor analysis, a possible critique from a skeptic might be the degree to which a randomly generated dataset could also produce significant factors. This is a question as to whether factor analysis might be capable of generating random "phantom patterns." This critique posits that a matrix of collected data should assert integrity beyond an equivalent set of random data.

In order to demonstrate the integrity of the factor analysis chi-square diagnostic test and to dispel the notion that random data might demonstrate interpretable results, ten random binary matrices were generated of equivalent size to the challenge and best practice interview results related to this inquiry, an 11-by-50 matrix for challenges (CHs) and a 26-by-50 matrix for best practices (BPs).

The datasets were run through an equivalent factor analysis process. Identical factor analysis was undertaken on the random datasets: exploratory maximum likelihood (ML) with SMC prior communality and varimax rotation with six factors.

One of the key results, observable in Table 3.26, was that all random datasets failed to reject the null hypothesis (H_0) of "no common factors" in chi-square diagnostic tests at a threshold below 0.10 (10%), which would be considered a generous threshold. This was an indication that the random datasets did not demonstrate sufficient structural pattern integrity within the rotated factor matrices generated to assert the existence of common factors. Chi-square tests of "no common factors" (H_0) on the equivalently sized random datasets failed to reject at a 0.10 threshold.

In contradistinction, the chi-square tests of the actual challenge and best practice interview data factors both achieved a P-value of <0.0001, achieving a confidence threshold of less than 0.01% in rejecting the null hypothesis (H_0) of "no common factors." In other words, there was appreciable statistical support for factor patterns achieved in the rotated factor matrix beyond random chance: only one in more than 10,000 random datasets would achieve a similar threshold.

Another observation in the randomization exercise was that exploratory factor loading of the resulting datasets was irresolute. Absolute loading of the randomized data required suppression at 0.2 or less, and the resulting factors were an even mix of positive and negative measures at a wide range of values. In distinction, the interview datasets demonstrate clean separability, a bias toward positive loadings, and a rough distribution around common loading values (i.e., the loading values were in a similar

Table 3.26 Randomized chi-square factor tests

Random	CH chi-square H_0	BP chi-square H_0
1	0.2818	0.7491
2	0.6752	0.2610
3	0.5245	0.9003
4	0.3076	0.1209
5	0.9464	0.1712
6	0.9426	0.3294
7	0.5579	0.5560
8	0.2912	0.7463
9	0.1968	*0.1670*
10	*0.1411*	0.6914
AVE	0.4865	0.4693

range, not too high or too low). Also, the actual interview factor solutions were interpretable when compared to macro-themes derived through mixed methods.

There is always a small chance of a random equivalent dataset appearing that would produce equivalently strong patterns. The lower the chi-square null hypothesis rejection threshold, the lower the chance of such a random instance occurring. However, it is important to emphasize that a factor solution ultimately is validated through a conceptual interpretation specific to the domain of inquiry. In addition to meeting a threshold of evidencing nonrandom patterns, factor solution results must additionally achieve a "common sense" interpretive assessment. Irresoluteness of the exploratory factor loadings in random datasets makes "common sense" interpretations difficult. Then, to the degree that random data also produces random indications in the context of the domain, a final "common sense" threshold of interpretability is extremely unlikely to be met.

The conclusion is that the derived coded interview dataset demonstrated resident patterns with statistically demonstrable nonrandom characteristics. Further, the exercise undertaken here shows that random data does not easily evidence demonstrable and interpretable factors. These premises are inherent to the application of factor analysis, but the demonstration of validity through a counterfactual demonstration on random data is useful to display the mechanism more robustly.

Once accepting the applicability and mechanism of factor analysis, there were remaining methodological issues specific to the application of factor analysis to this particular dataset that required additional substantiation. Factor analysis for both challenge and best practice responses met key benchmarks substantiating patterns extracted, namely, validity indicated by diagnostic measures, clear separation in factor loadings, conceptual conciseness, and theoretical interpretability.

However, an underlying methodological issue remained. Namely, factor analysis has not traditionally been advocated for dichotomous data except in exceptional cases (Kim and Mueller 1978a). This indicated an effort was necessary to substantiate exceptionality.

The underlying coded interview dataset subjected to factor analysis was binary (dichotomous), indicating whether an interviewee raised a codebook theme (1) or not (0). Although the thematic patterns surfaced in factor analysis triangulated closely with adjacent mixed method results, it was necessary to address concerns related to the suitability of the application of factor analysis. In order to substantiate the methodological approach, further effort to establish integrity was undertaken.

One of the exceptions permitted in applying factor analysis to dichotomous data concerns when the research goal is to search for clustering patterns (Kim et al. 1977). In this inquiry, the goal in applying factor analysis was to identify clusters of responses across themes raised. As the research goal was to search for clustered patterns in the dichotomous data, this exception was met.

A second exception concerns the cast and context of the dichotomous data. One of the typical assumptions in factor analysis is that measures are on a continuous scale, namely, interval or ratio data. Some allowances can be made when data conceptually represents a compressed Likert-type ordinal scale (i.e., agree versus disagree on a distributed scale). In the context of interpretation of the interview data, nonresponse to a theme was framed as equivalent to a "not manifestly interested" response, albeit by omission. In this sense, a case can be made that the dichotomous data conceptually represents a highly compressed Likert-type scale: 1 = interested in theme; 0 = not interested in theme.

While a critique might be offered that the absence of evidence is not evidence of absence, this relates to the format and context of data collection. In the case of the semi-structured interview format, interviewees were given free-range to frame the topics they viewed as being most important within the context of the questions. The 30-min format appeared to be sufficient in most cases for an exhaustive cataloging. As such, not raising a theme was more than an omission; it was viewed as an assertion that a particular theme was not viewed as being significant to the interviewee. However, understanding would be improved by follow-on research, for instance, more extensive interviews or a survey listing all coded themes and requesting an explicit "importance rating."

Finally, in order to rigorously substantiate findings given the exceptions invoked, supplementary exploratory factor analysis expressly configured for dichotomous data was undertaken based on guidance from Baglin (2014) and Lorenzo-Seva and Ferrando (2012). Based on guidance for treating dichotomous data, polychoric (tetrachoric) correlation matrix analysis was attempted utilizing the Factor software package (Lorenzo-Seva and Ferrando 2019a, b). However, the sample size was not large enough to yield clear results using this method.

Based on additional guidance from the same sources, a recommended proxy for treating dichotomous data is Pearson correlation matrix analysis (Lorenzo-Seva and Ferrando 2019a, b). In addition to addressing smaller sample sizes, Pearson correlation matrix analysis was also appropriate as a number of variables indicated negative skewness and kurtosis in diagnostic evaluation (Muthén and Kaplan 1992).

Pearson correlation matrix analysis was performed utilizing the Factor software tool, configured to apply the following: sweet smoothing (Lorenzo-Seva and Ferrando 2019a, b), unweighted least squares (ULS), robust factor analysis with mean and variance adjusted chi-square statistics (Asparouhov and Muthen 2010), and raw varimax for rotation to maximize factor simplicity.

In the case of challenge factors, Pearson correlation matrix analysis, as a recommended proxy for tetrachoric analysis due to the small sample size, produced a five-factor solution. The factor loading solution was compressed due to a two-variable per factor limit imposed by the method. Results gave similar loadings to previous exploratory factor analysis results, with the inevitable exception that the Pearson five-factor result led to some variables being distributed across the smaller set of factors. Results matched the original six-factor loadings closely except those noted in Table 3.27 in ***bold italic***.

This was viewed as reinforcing the factors extracted in the original six-factor maximum likelihood (ML) solution. Variants surfaced in Pearson factors were mainly duplicates, thus not cleanly separated into distinct factors. As the original six-factor ML solution evidenced clearer separability in terms of the individuation of variables, these results were retained.

Table 3.27 Comparative Pearson correlation matrix analysis—challenges

Challenge (CH)	Factor1	Factor2	Factor3	Factor4	Factor5	Factor6
1. Original ML	CH6, 9, 5, 10	CH3, 4, 7	CH4, 1	CH2, 7	CH8	CH11
2. Pearson	CH6, 9, 5, 10	CH3, 4, 7, *2*	CH4, 1	CH2, 7, *4, 3*	CH8, *5, 6*	n/a

Table 3.28 Comparative Pearson correlation matrix analysis—best practices

Best practice (BP)	Factor1	Factor2	Factor3	Factor4	Factor5	Factor6
1. Original ML	BP10, 15, 4, 20, 19, 17, 6, 8, 11	BP2, 3, 5, 16, 18, 7	BP22, 24, 25, 21	BP9, 1, 13	BP25, 14, 12, 23	BP15, 11, 26
2. Pearson	BP10, 15, 4, 20, 19, 17, 6, *X*, 11	BP2, 3, 5, 16, 18, *X*	BP22, 24, 25, *X*	BP9, 1, 13, *4, 24, 25*	BP25, 14, 12, 23, *8*	BP*4, 17, 18, 21, 25*

In the case of factor substantiation for best practices, a similar cross-check was undertaken with some variations resulting from the larger number of variables (26 best practice themes as opposed to 11 challenge themes). Pearson correlation matrix analysis was applied, configured for the following: sweet smoothing, unweighted least squares (ULS), no robust factor analysis (the larger range of variables resulted in an indeterminant solution under robust circumstances), and raw varimax for rotation to maximize factor simplicity with salient loading values specified as 0.2 or larger. This resulted in a six-factor solution which on the whole validated major patterns identified in the original ML factor analysis, with additions or omissions noted in ***bold italic***, per Table 3.28.

In the case of best practice factors, alignment between the original ML and Pearson analysis was considered to be fairly accurate for Factors 1, 2, 3, and 5, with one variable being different in each factor. Factor 4 maintained resemblance except for adding duplicated variables that met a rotation threshold in multiple factors. Similarly, Factor 6 contained a number of duplicated variables that appeared in multiple factors. As factors ideally cleanly rotate variables into unique factors, it was judged that overall integrity between the two methods was validated when duplicates were more strictly allocated one-per-factor.

In summary, despite caveats concerning the application of dichotomous data in factor analysis, it is asserted that the method and results are valid given that four conditions were satisfied:

1. Research goal is to search for cluster patterns (Kim et al. 1977).
2. Data qualifies as a modified or compressed Likert-type scale (Muthén and Kaplan 1992).

3. Cross-validated with Pearson correlation matrix analysis, a method specifically appropriate to dichotomous data (Baglin 2014; Lorenzo-Seva and Ferrando 2019a, b).
4. Conceptual clarity and interpretability of results substantiated through triangulated comparative methods (i.e., literature analysis, interview memos, frequency analysis, and text analytics).

It is asserted that these criteria were met and that the threshold for qualifying as an exception for the factor analysis treatment of dichotomous data was satisfied.

A final methodological caveat relates to the combined effect of dichotomous data and the small sample size. Although factor analysis was trialed on a combined challenge and best practice response dataset, this did not result in valid factor solutions. Factor analysis only produced valid solutions when examining challenge and best practice responses separately. The small sample size was clearly an issue in this respect. Combined, the two datasets encompassed 37 variables (26 best practice themes and 11 challenge themes). Guidance on factor analysis varies, but a 2 to 1 ratio of sample set to variables is considered to be at the outmost extreme limit, with most researchers advocating much larger ratios. Treating the two datasets separately supported factor solutions, whereas common patterns were too diffuse to detect in the combined dataset due to sample size.

It is possible that a larger sample size would support the identification of clear cross-category factors from a unified dataset. Indeed, text analytics applied to the combined challenge and best practice transcripts led to indications of key shared themes. Future research efforts should endeavor either to expand the number of interviews or to conduct a broader survey research effort to gain more samples. However, given the two valid factor solutions extrapolated, there was a remaining opportunity to examine potential connections between the two factor solution matrices. This effort was undertaken subsequently and led to results worthy of reporting.

3.3.6 Factor-to-Factor Fitting

Extrapolating challenge and best practice factors surfaced macro-themes resident across interview responses. Factor analysis reduced 11 challenge and 26 best practice themes to 6 macro-themes for each category, improving gross interpretability. Having extracted significant and explainable factors, it was next of interest to assess whether there were relationships resident *between* challenge and best practice factors. Applying correlation analysis and least squares fitting on the rotated factor matrices, significant challenge-to-best practice factor relationships were identified. This indicated latent practitioner views across interviews concerning the relationship between particular challenge and best practice macro-themes.

Per the conceptual model raised at the outset of this phase in Fig. 3.1, a reciprocal relationship was posited between CSDS challenges and best practices. This relationship was initially surfaced as a sensitizing concept raised in literature analysis. The connection was later observed as a phenomenon in interviews, indications that themes raised had dualistic challenge and best practice aspects. Outside the conceptual model, there are grounds to assert this as a general tendency in interviews. Interviewees tend to impress normative judgments on positive topics (e.g., the absence of X is unwanted versus more of X is deemed desirable). In the case of CSDS practitioner interviews, it was desirable to attempt to reduce and frame the latent unifying concepts resident between challenges and best practices.

The propensity to frame topics through value judgments was noted in interview memoing: respondents tended to frame views on CSDS practices with either a negative or positive valiance. Interview notes indicated that practitioners framed a theme as a challenge when they experienced a desirable outcome as missing and as a best practice when they were advocating a desirable prescription. It was hypothesized that this tendency to connect between challenges and best practices would be resident across coded responses, signifying macro-themes raised by respondents.

Indications of such relationships were identified earlier. In Sect. 3.3.3, nominal logistic regression model fitting identified significant relationships between several original challenge themes and a range of best practice themes. Attempts were made to extend this analysis by conducting correlation analysis across all challenge and best practice themes combined. Furthermore, an attempt was made to conduct factor analysis on a dataset combining all challenge and best practice responses. However, neither attempt yielded interpretable results. The small sample size combined with a large relative set of variables, 37 in total, complicated attempts to find interpretable relationships between combined challenge and best practice theme data.

Identifying clear statistical relationships between the two categories was likely difficult due to the range and complexity of combined challenge and best practice themes. The granular themes identified required consolidation to ease interpretability. Factor analysis, being a dimensionality reduction technique related to principal component analysis (PCA), was able to reduce challenge and best practice themes to compressed representations. Further, as a primary aspect of factor analysis is the assumption that latent variables are being extrapolated, the assertion is that the factors identified are latent macro-themes resident across interview responses.

Having extracted, substantiated, and interpreted latent themes through factor analysis, it was of interest to determine the degree to which specific challenge and best practice factor themes statistically co-occurred across responses. Factor-to-factor statistical co-occurrence would indicate a tendency for particular challenge macro-themes (as identified gaps) to surface along with particular best practices macro-themes (as gap-prescriptions). Significant relationships between factors could be interpreted as latent advocacy for addressing aggregated challenges with particular best practice treatments. Such observations would constitute evidence of *gap-prescriptions* advocated across CSDS practitioner interview responses.

Factor analysis reduction resulted in two factor matrices representing a compressed record of the original interview themes raised by all interviewees. It was determined to examine possible statistical relationships between the two matrices. While a unique approach, methodologically the procedure is roughly equivalent to examining the statistical relationship between two categorically distinct but related sets of PCA-derived variables.

PCA, having a close kinship to factor analysis as a related multivariate dimensionality reduction technique, is regularly used to reduce larger sets of variables by deriving statistical aggregates. Although there is smoothing-related information loss in such procedures, residual statistical measures are retained in the compressed representation. In principle, provided informational loss is not catastrophic, reduced components or factors should retain the ability to indicate cross-aggregate relationships for conceptually related variables. In this same respect, factor-to-factor correlation or fitting should identify statistically significant traces between conceptually connected factors, should such traces exist. Given that such relationships were detected earlier through logistic fitting, it was hypothesized that factor-to-factor fitting would identify similar connections, albeit mediated through compressed, latent factors.

Linear regression and correlation analysis were thus applied to examine factor-to-factor co-occurrence. Based on factor loadings, rotated factor scores were produced unique to each interview respondent derived from their relative contributions to the aggregated factor matrix. Thus, for each respondent, 12 rotated factors were derived unique to each respondent, constituting 6 challenge factors and 6 best practice factors. In total, two 6-by-50 rotated factor score matrices were produced, one for challenge factors and one for best practice factors. Each resulting vector of factor loadings was unique to each respondent based on their original responses, depending upon their relative contribution to the whole factor loading model.

It is worthy to note that factors are the result of whole-matrix analysis, meaning that factors are refined combinations of responses that incorporate all interacting variables in the matrix, including the absence of responses (negative propensity). Thus, for each respondent, unique factor rotations indicated a propensity toward agreement or nonagreement (through omission) with each theme relative to the aggregated loaded factors across all responses.

Following factor transformation, loadings per respondent per factor were captured as continuous variables and could be positive or negative, depending on the relationship to whole factor loading. As two 6-by-50 matrices captured the compressed factor loadings for challenges and best practices for all respondents, it was thus possible to assess correlations between the factor matrices. Measures of significance and correlation would indicate that particular challenge factor macro-themes tended to co-occur with particular best practice factor macro-themes. Fitting between the two factor loading matrices supported an assessment of statistical factor concordance between challenge and best practice factors across all responses.

Based upon the two rotated factor matrices, it was of interest to assess to what degree individual challenge factors might fit to the range of best practice factors

Table 3.29 Least squares fitting challenge-factors-to-best-practice-factors (Y-to-X)

Challenge factor	Significant	P-value	$R2$	P-val Significant best practice factors
CH F1: Expansive complexity	Moderate (at 0.15)	0.15	0.1865	0.044 BP F2: Cross-domain collaboration *(0.039) (BP F1: Scientific process)*
CH F2: Tracking/context	Strong (at 0.05)	0.0475	0.2468	0.005 BP F1: Scientific process *(0.138) (BP F3: Risk focus)* *(0.09) (BP F6: Structured discovery)*
CH F3: Data management	Weak (at 0.35)	0.3342	0.1415	0.054 BP F4: Data management 0.19 BP F2: Cross-domain collaboration
CH F5: Unclear ownership	Strong (at 0.05)	0.0539	0.2409	0.145 BP F2: Cross domain collaboration *(0.129 BP F1: Scientific process)* *(0.012) (BP F6: Structured discovery)*

(#.###) = negative correlation—indicating a "lack" of statistical response in theme

across all responses in a regression model. Least squares fitting was conducted to assess whether there were indications of significant relationships between challenge factors and best practice factors. Least squares was applied, as opposed to logistic regression, as factors were extracted and represented as an array of continuous rotated scores. Rotated factors were numerically cast as interval data on a quantified scale and thus were no longer discretely binary-nominal variables.

Results, summarized in Table 3.29, indicated significance in four particular challenge factors matching to unique combinations of best practice factors. This mirrored earlier findings that challenges generally correlated with more complex sets of best practices, the former hypothetically requiring a broader range of the latter for accommodation. This is to say that a single challenge generally implies a set of advocated best practice treatments, and not vice versa. Negative correlations, indicating a pronounced lack of tendency to report the best practice (*noted in parenthesized italics*), were also documented to aid interpretability.

Of note, "CH F3: Data management" produced a model with weak significance. However, subsequent correlation analysis showed a strong statistical relationship, and the related themes were judged as being conceptually strong.

Performing the same set of tests in reverse, fitting each best practice in sequence to all challenge factors, significant results were less apparent. Per Table 3.30, only one best practice factor demonstrated notable significance in factor-to-factor least squares regression, that of the best practice of implementing a scientific process as fitted with

Table 3.30 Least squares fitting best practice-factors-to-challenge-factors (Y-to-X)

Best practice factor	Significant	P-value	$R2$	P-val significant challenge factors
BP F1: Scientific process	Strong (at 0.05)	0.0157	0.2943	0.006 CH F2: Tracking/context *(0.043 CH F1: Expansive complexity)*

(#.###) = negative correlation—factor loadings indicate a statistical lack of response in theme

the challenge of tracking incidents and understanding context in the environment. Although reciprocally observable as a fit between CH F2 and BP F1 in the previous model profiled in Table 3.29, the result is notable as "CH F1: Expansive complexity" was observed as displaying inverse significance. This observes that those advocating scientific method related themes as treatments are *less likely* to surface themes associated with the challenge of expansive complexity, per Table 3.30.

The lack of significance displayed in fitting best practice factors to challenge factors is likely due to the greater range and complexity of best practice factors, as well as the context of conceptual framing in interview responses. Regression analysis identifies significant relationships whereby the presence of a dependent factor (Y) co-occurs statistically with a set of independent factors (X). In the models where dependent challenge factors (Y) match to independent best practice factors (X), a single challenge factor is explainable as the result of holding a particular set of best practice advocacy positions. That individual challenge themes are significant in relation to a set of best practice advocacies is explainable given the directional implication of challenges being addressed by an array of best practices.

As recorded in interview memos, there was an observable tendency to conceptually frame multiple treatments (best practices as solutions) to address singular challenges, whereas the reverse was less the case (a set of challenges being framed to a single advocated best practice). In this respect, a challenge theme is central and is addressed by an array of matching, conceptually atomic best practice treatments. It is less apparently the case, conceptually and quantitatively in the scope of the results, that a single best practice retroactively addresses an array of challenges. In interview feedback, best practices appear to be more conceptually atomic, whereas challenges appear to be associated with broad conceptual problems, which require a range of best practice solutions.

Having substantiated significant factor-to-factor relationships, cross-correlation assessment between challenge and best practice factors loadings was performed as a supplementary technique to substantiate assertions of relationships between the two sets of factors. As expected, results from correlation analysis directly reinforced earlier observations of significance in linear regression model fitting. Results are summarized in Table 3.31.

Having identified significant challenge-factor-to-best-practice-factor linear relationships reinforced by correlation analysis, a master factor-to-factor gap-prescription model was derived to represent results visually. Given the abstraction of factors, which are derived latent variables, the interpretation of factor-to-factor

Table 3.31 Factor-to-factor multivariate correlation

	BP F1 Scientific process	BP F2 Cross-domain mgmt	BP F3 Risk focus	BP F4 Data mgmt focus	BP F5 Focused engineer goals	BP F6 Segment discovery & detection
CH F1: Expansive complexity	- *0.29*	**0.27**	0.07	-0.03	0.11	-0.11
CH F2: Tracking & context	**0.37**	0.02	- *0.21*	0.09	-0.07	- *0.18*
CH F3: Data management	0.11	**0.16**	0.01	**0.28**	0.13	0.11
CH F4: Expectations vs. limits	-0.11	0.05	- *0.19*	0.02	0.07	-0.02
CH F5: Unclear ownership	- *0.24*	**0.20**	0.01	-0.12	-0.01	- *0.39*
CH F6: Data policies	-0.01	- *0.16*	0.04	- *0.22*	**0.13**	-0.11

correlation is more subtle than response-to-response coded theme matching. Challenge and best practice response codes associated with each factor were thus included as an aid to interpretability.

Two models were derived, a full model tracing the relationship between all responses and factors, including significance and correlation measures (including negative correlations) (Fig. 3.5), and a reduced model focusing only on those factors which evidenced notable positive statistical significance (Fig. 3.6). The latter figure summarizes the focused results.

As with factor solutions, conceptual interpretability of the factor-to-factor results should be considered a central threshold for judging the merit of results achieved. A "common-sense" assessment of the degree to which particular challenges align conceptually with particular best practices is necessary. This is addressed in a structured manner in the final research phase, *Phase III*, following in Chapter 4.

Concerning whether structured results could be achieved through random data alone, extending the earlier analysis of random factor solutions, the two best performing random factor solutions derived and evidenced in Table 3.26 were utilized to assess factor-to-factor fit: challenges (CH) random dataset 10 (chi-square H_0 0.1411 chi-square) against best practices (BP) random dataset 9 (chi-square H_0 0.1670). The results from fitting the two highest performing random factors were not conclusive. Per Table 3.32, three challenge factors matched to single best practice factors with moderate-to-weak significance (at, variously, 0.15, 0.25, and 0.35 thresholds). In comparison, per Table 3.29, stronger P-values (<0.1) and correlation measures were achieved in factor-to-factor fitting of the interview data factors.

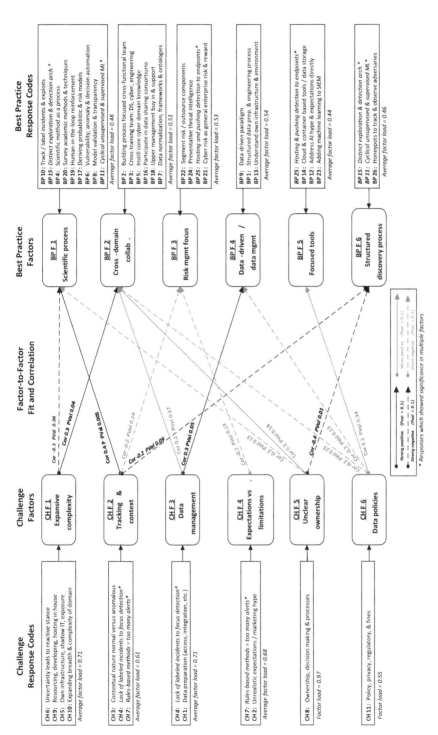

Fig. 3.5 Full factor-to-factor diagnostic gap-prescription model

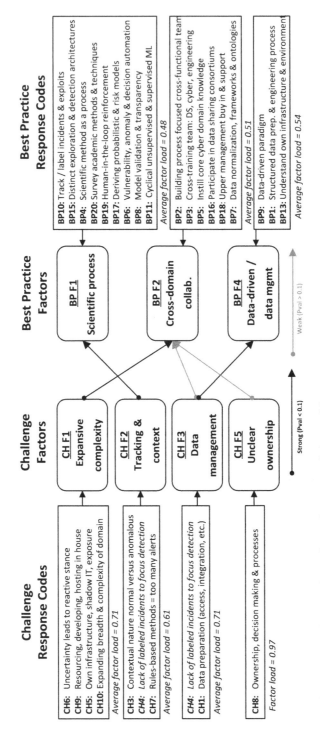

Fig. 3.6 Significant factor-to-factor diagnostic gap-prescription model

Table 3.32 Comparative factor-to-factor fitting derived from random data

Challenge factor	Significant	P-value	$R2$	P-value significant BP factors
CH F1	Weak (at 0.35)	0.3175	0.1449	0.049 BP F2
CH F3	Moderate (at 0.15)	0.1488	0.1900	0.0298 BP F1
CH F6	Moderate-weak (at 0.25)	0.2465	0.1609	*(0.0578) BP F3*

(#.###) = negative correlation—indicating a "lack" of statistical response in theme

3.3.7 Gap-Prescription Extrapolation

Having extracted latent factors associated with challenges and best practice interview responses, four challenge factors indicated significant relationships to three best practice factors, per Fig. 3.6. For these factors, there was a statistical propensity across all interviews to raise the associated underlying challenge and best practice themes together. As challenges and best practices are indicative of perceived gaps and advocated treatments, respectively, it is asserted that the extrapolated constitute latent diagnostic gap-prescriptions. Table 3.33 summarizes the resulting gap-prescriptions statistically extrapolated from the entire set of CSDS practitioner interview responses.

Recalling the CSDS conceptual model derived to frame the interview inquiry at the outset of this phase (see Fig. 3.1), technical trends were viewed as framing cybersecurity challenges and/or data science best practices. Further, challenges and best practices were framed as being reciprocal: best practices could be framed as a challenge and vice versa. Practitioners might frame a challenge as an inhibitory agent when a necessary function was viewed as lacking. Similarly, a functioning requirement was viewed as an enabler when implemented as a best practice. The conceptual model thus anticipated interviewees raising themes as issues requiring attention (challenges) and/or as facilitating enablers (best practices).

The systematic process of coding themes, aggregating themes as factors, and substantiating significant connections between factors framed key guidance across the practitioner interview responses. However, to derive interpretable guidance, clarifying insights are helpful. Following again from the original conceptual model, CSDS best practices are not only prescribed treatments but also ideal states in operational practice. It is thus appropriate to view best practices as the perceived ideal end-goal of programmatic CSDS implementations.

It is insightful to represent the significant best practice factors identified as being the result of the array of combined challenge and best practice themes that tie to the derived factors. To summarize these findings, a graphical interpretation is useful. Given the implied relationship leading to the key best practice factors, a "cause-and-effect" diagram modeling approach (otherwise known as a "fishbone" or "Ishikawa" diagram) was utilized (Ishikawa 1990). Positioned in these frameworks, the best practice factor outcomes are linked to the aggregated, statistically correlating challenge and best practice themes underpinning the associated factors. This allows for a detailed understanding of all elements, which together centrally contribute to each best practice factor (BP Fx) as ideal end outcomes.

Table 3.33 Key extrapolated CSDS gap-prescriptions

Challenge factors: diagnosed gaps	Best practice factors: prescribed treatments
CH F1: Expansive complexity	**BP F2**: Cross-domain collaboration
CH F2: Tracking and context	**BP F1**: Scientific process
CH F3: Data management	**BP F4**: Data-driven/data management
	BP F2: Cross-domain collaboration
CH F5: Unclear ownership	**BP F2**: Cross-domain collaboration

Each diagram below captures the collection of challenge (CHx) and best practice (BPx) themes associated with one of the three significant best practice factors (BP Fx) identified. The diagrams summarize the range of themes contributing to and delivering the ideal best practice. Best practice factors (BP Fx) are in this sense presented as desired gaps-outcomes: ideal, although absent, outcomes defined by a combined range of associated challenges to be overcome and best practice prescribed treatments. The three gap-prescriptions resulting from mixed method analysis are detailed in the following sequence:

Gap-Prescription 1: Data Management (BP F4)—Best practice factor, "BP F4: Data management," represented in Fig. 3.7, has a relatively simple set of linked contributing challenge and best practice themes. In addition to the themes associated with the root best practice factor, the challenge factor "CH F3: Data management" is associated. This is an expected relationship, as conceptually the challenges of data management are addressed by data management best practices. This finding observes the conceptual assertion of reciprocity between challenges and best practices, as well as the integrity of the factor-to-factor fitting approach.

The associated themes charted to address the BP F4 gap imply the need for an organizational effort to embrace a paradigm involving the centrality of data analytics over traditional rule-based approaches in cybersecurity. It is also advocated that a concerted effort be made to comprehensively map and track phenomena particular to the environment being safeguarded. There is advocacy for a process-driven approach to data management, uniting methodological goals with engineered solutions.

Gap-Prescription 2: Scientific Processes (BP F1)—In examining the root themes associated with the second gap-factor, "BP F1: Scientific processes," per Fig. 3.8, the goal of framing and operationalizing scientific processes is highlighted. In factor-to-factor correlation and fitting, the challenge factor "CH FC2: Tracking/context" matched with this best practice. Thematically, we can interpret the collection of associated themes as advocacy for a variety of processes to support scientific inquiry. Notably, scientific processes are multiple and diverse, with the advocacy supporting a variety of treatments including operational optimization, distinct exploratory and predictive methods, evidence gathering and framing, contextual framing of models, approaches supporting risk quantification, and methods to validate and substantiate models.

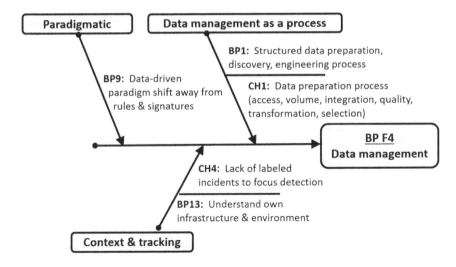

Fig. 3.7   BP F4—data management fishbone diagram

Of note, practitioner feedback identified scientific methods in the plural. This is key and notable as an oversimplified representation would be to advocate a singular approach to scientific inquiry. Here, it clear that CSDS practitioners recognize that there are multiple methods and processes supporting scientific inquiry and that these should be framed and supported both individually and in unison, to the degree they must interact. As this raises a range of methodological prescriptions, this topic will be examined in greater detail in the following phase (*Phase III*).

Gap-Prescription 3: **Cross-Domain Collaboration (BP F2)**—In the last cause-and-effect diagram derived, per Fig. 3.9, the ideal end-state factor is "BP F2: Cross-domain collaboration." When identified as a gap, a desirable-yet-absent-end-state, the array of associated challenges and best practices are represented to clarify requirements for delivery.

This factor represents an ideal outcome, whereby CSDS is operationalized in an organization through optimal training and resource coordination. The derived factor emphasizes the need to connect cybersecurity and data science domain knowledge in cross-functional teams and processes. In the cause-and-effect diagram, all statistically associated challenge and best practice themes are aggregated, per the central best practice source factor and the challenge factors "CH F1: Expansive complexity," "CH F3: Data management," and "CH F5: Unclear ownership."

Five summary categories are positioned to associate similar challenge and best practice themes together. The gathering of systematic evidence is noted, which can be viewed as a necessary step toward undergirding and substantiating credibility in the application of data science methods. Data management practices are also noted,

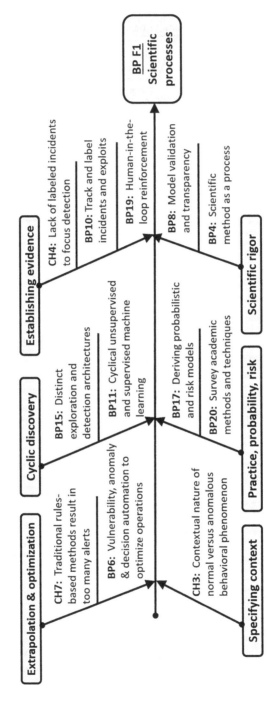

Fig. 3.8 BP F1—scientific processes fishbone diagram

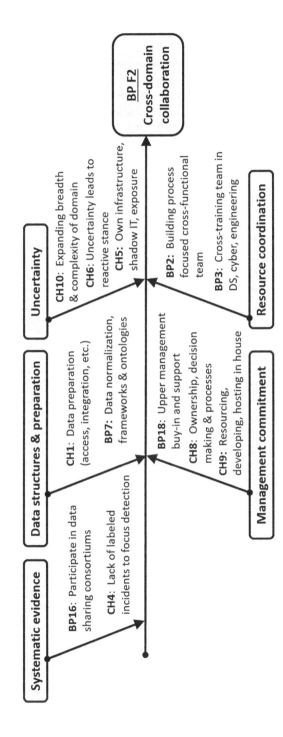

Fig. 3.9 BP F2—cross-domain collaboration

as lapses in data quality will logically erode the efficacy and credibility of data science methods.

Themes related to uncertainty are also raised, which position complexity and lack of clarity in organizational programs as impeding CSDS interdisciplinary collaboration. The final two categories involve management exhortations to clarify ownership and decision-making associated with a CSDS programmatic implementation. This includes concerted efforts to build cross-functional processes and to cross-train multidomain CSDS teams such that shared understandings are facilitated.

3.4 Summary of Extrapolated Insights

3.4.1 Mixed Method Diagnostic Gap Analysis

This phase combined qualitative and quantitative methods to surface and summarize themes encountered in interview research. For the qualitative component, diagnostic opinion research was undertaken in the form of interviews with 50 CSDS practitioners. This was followed by a quantitative component seeking patterns among interview response data. Analysis served to clarify and frame key themes to inform diagnostic gap analysis. Whereas literature analysis surfaced similar themes, this component triangulated a refined understanding of the relative importance of and relationships between these themes based on practitioner perspectives.

The qualitative interview component addressed a research lacuna concerning practitioner perceptions of the emerging CSDS professional domain. This provided a supplement to literature analysis to explore key themes related to CSDS professionalization in greater depth. The complex of themes that emerged from interview research surfaced a range of organizational, process, and technological topics. Structured analysis led to the derivation of an ordered codebook summarizing central themes raised. Frequency analysis and text analytics emphasized the relative prevalence of particular themes. Interview memos provided supplementary insights.

In the subsequent quantitative component, a range of multivariate statistical pattern analysis methods were applied to coded interview data to refine a more nuanced understanding of themes surfaced. This supported diagnostic gap analysis related to challenges and best practices in the CSDS domain. Logistic regression demonstrated the presence of significant connections between challenge and best practice coded themes. Based on this indication, factor analysis was applied to extrapolate latent themes resident across coded interview results. Finally, factor-to-factor linear fitting and correlation analysis resulted in a set of key CSDS gap-prescriptions. Results framed shared practitioner macro-themes advocating approaches to promote CSDS domain professionalization.

Table 3.34 Summary of CSDS gap macro-themes surfaced across mixed method analysis

		Key CSDS domain gap themes			
		Data management	Risk management	Scientific methods	Organizational factors
Method	1. Literature analysis	✓	✓	✓	✓
	2. Interview memoing	✓		✓	✓
	3. Text analytics	✓	✓	✓	✓
	4. Theme frequency	✓		✓	✓
	5. Factor-to-factor	✓		✓	✓

Three central gap-prescription themes were extrapolated:

1. Data management
2. Scientific processes
3. Cross-domain collaboration

Interpretation and framing of patterns surfaced through quantitative analysis were supported by themes raised in previous qualitative findings, bridging literature analysis, interview memoing, text analytics, and thematic frequency analysis. The key CSDS gap-prescription macro-themes raised and shared across the mixed methods are summarized in Table 3.34.

Triangulating mixed methods across qualitative and quantitative components provided variant approaches to substantiate central themes surfaced. The newness and complexity of the domain advocated the application of mixed methods aimed at exploring and extrapolating hidden patterns.

The application of quantitative methods (logistic regression, factor analysis, and factor-to-factor fitting) led to greater refinement and specificity concerning the array of themes associated with the key macro-themes. Whereas interview research surfaced 11 challenge and 26 best practice themes, the application of factor analysis and factor-to-factor fitting supported a rigorous approach to both refine macro-themes and to associate an array of subthemes with the macro-themes.

A core methodological aspect of factor analysis is its focus on surfacing latent themes. Researcher interpretations are necessary to frame factors. The array of mixed methods applied in this effort supported rigor in thematic interpretations. The methods applied combined inductive and deductive approaches to substantiate themes. Qualitative mixed methods supported the interpretation of factors, which themselves demonstrated statistical individuation and diagnostic integrity. Patterns were indicated in logistic regression analysis of challenge-to-best practice responses.

Factor-to-factor fitting substantiated challenge-to-best practice significance through correlation analysis and linear fitting. Key CSDS gap-prescriptions surfaced are thus the result of both qualitative and quantitative extrapolation.

Whereas literature analysis surfaced a similar range of key themes, combined interview and diagnostic gap analysis research conducted here has iterated and refined an understanding of the nature of, relative importance of, and relationships between key CSDS themes. To aid in interpretability of the gap-prescriptions surfaced, the three macro-themes extrapolated in quantitative analysis were framed along with the array of associated underlying best practice and challenge themes. Results were surfaced in fishbone cause-and-effect diagrams; the details are replicated in tabular form to summarize findings (Table 3.35).

In extrapolating practitioner views of CSDS gap-prescriptions, a foundation for subsequent design science problem-solving is established. Design problem-solving is a natural extension to the diagnostic gap analysis foundation established. Deriving practitioner views of perceived gaps in the CSDS domain serves to frame focused design requirements.

Design problem-solving necessitates the clarification of structured requirements (Verschuren and Doorewaard 2010). The following phase will surface requirements for each of the practitioner gap-prescriptions raised and advocate design solutions to address the requirements. This will address the CSDS gap-prescription macro-themes of data management, scientific methods, and cross-domain resource coordination. Results from research undertaken here thus frame problem-solving design requirements to be addressed comprehensively in the following concluding phase.

3.4.2 Research Relevance

Interview research undertaken here has informed diagnostic gap analysis regarding challenges facing the CSDS field. Mixed methods served to surface and frame themes across interviews with CSDS practitioners. The relative importance of and relationships between these themes were analyzed. Through concerted refinement, applying both qualitative and quantitative methods, a substantiated set of focused gap-prescriptions were extrapolated. In framing guidance to foster professionalization of the CSDS field, this research has established a foundation for design-oriented problem-solving.

As CSDS practice expands, body of theory gaps will become self-limiting as practitioners increasingly encounter conflicting understandings concerning proper practice. Literature analysis previously explored CSDS body of theory gaps, while interview research conducted here has iterated understanding by extrapolating and framing gap-prescriptions from the perspective of active practitioners. Interview findings have surfaced shared understandings of CSDS practice and thus promote CSDS body of theory maturation.

Table 3.35 Summary of CSDS gap analysis macro-themes surfaced

Key gap	Category	Practitioner themes (interview codes)
BP F4 Data management	Paradigmatic	**BP9**: Data-driven paradigm shift away from rules and signatures
	Data management as a process	**BP1**: Structured data preparation, discovery, engineering process **CH1**: Data preparation process (access, volume, integration, quality, transformation, selection)
	Context and tracking	**CH4**: Lack of labeled incidents to focus detection **BP13**: Understand own infrastructure and environment
BP F1 Scientific processes	Scientific rigor	**BP4**: Scientific method as a process **BP8**: Model validation and transparency
	Specifying context	**CH3**: Contextual nature of normal versus anomalous behavioral phenomenon
	Establishing evidence	**CH4**: Lack of labeled incidents to focus detection **BP10**: Track and label incidents and exploits **BP19**: Human-in-the-loop reinforcement
	Extrapolation and optimization	**CH7**: Traditional rule-based methods result in too many alerts **BP6**: Vulnerability, anomaly and decision automation to optimize operations
	Cyclic discovery	**BP15**: Distinct exploration and detection architectures **BP11**: Cyclical unsupervised and supervised machine learning
	Practice, probability, risk	**BP20**: Survey academic methods and techniques **BP17**: Deriving probabilistic and risk models
BP F2 Cross-domain collaboration	Systematic evidence	**BP16**: Participate in data-sharing consortiums **CH4**: Lack of labeled incidents to focus detection
	Data structures and preparation	**BP7**: Data normalization, frameworks and ontologies **CH1**: Data preparation (access, integration, etc.)
	Uncertainty	**CH10**: Expanding breadth and complexity of domain **CH6**: Uncertainty leads to reactive stance **CH5**: Own infrastructure, shadow IT, exposure
	Management commitment	**BP18**: Upper management buy-in and support **CH8**: Ownership, decision-making, and processes **CH9**: Resourcing, developing, hosting in house
	Resource coordination	**BP2**: Building process focused cross-functional team **BP3**: Cross-training team in DS, cyber, engineering

Together the three gap-prescriptions surfaced imply that delivering a CSDS program must address a complex of interconnected themes. The range of challenges and required treatments span organizational, process, and technical prescriptions. This aligns findings with the MIS tripartite model advocating multi-contextual approaches to information systems solution designs.

Findings indicate a clear awareness on the part of CSDS practitioners of the need for greater rigor in scientific inquiry. Referring to the process of professionalization and highlighting the need to properly frame scientific practice within the professions, Greenwood notes:

> The importance of theory precipitates a form of activity normally not encountered in a nonprofessional occupation, viz., theory construction via systematic research. To generate valid theory that will provide a solid base for professional techniques requires the application of the scientific method to the service-related problems of the profession. Continued employment of the scientific method is nurtured by and in turn reinforces the element of rationality. (1957)

This observation reflects central advocacy raised during interviews that CSDS practice necessitates greater scientific rigor. The results of diagnostic gap analysis have highlighted this as a central issue. Further, quantitative analysis has improved understanding by charting and associating an array of linked subthemes. This has clarified a set of discrete requirements necessary for bolstering scientific rigor in CSDS practice. The systematic "working out" of details concerning gaps facing the CSDS domain is asserted as being a valuable research contribution to researchers, practitioners, and managers interested in furthering the CSDS profession.

A focused research contribution concerns establishing a fruitful foundation for follow-on research. One such research thread is addressed directly in the subsequent phase (*Phase III*), that of treating the CSDS gap-prescriptions surfaced here as design science requirements. However, other threads are implied by the results suggested here. A promising next step from the qualitative interview component would be to apply learnings from interview themes derived to design and conduct subsequent survey research. Findings are suitable for framing and designing a survey instrument. Survey research would raise the opportunity to gather additional results at scale.

As caveated and discussed, the small sample size resulting from interview research constrained the application of factor analysis. Cross-validation through mixed methods was undertaken, and a novel approach was applied through factor-to-factor fitting. The integrity of findings was cross-validated and bolstered through several comparative qualitative and quantitative approaches.

Ideally, future research would focus on obtaining a larger sample. It is suggested that future research efforts be undertaken to either expand the number of interviews or to conduct a survey research effort at scale. A beneficial outcome of the mixed method effort undertaken here is that it has produced a wealth of guidance appropriate for designing a robust survey tool. The results obtained in mixed method thematic analysis provide a solid foundation for follow-on survey research.

3.4.3 Managerial Relevance

Results raised and summarized will be of interest to managers, institutional stake-holders, and strategic planners seeking to implement CSDS programs. Results frame an understanding of key themes impeding CSDS professional maturation in the context of MIS research. The centrality of organizational factors in particular was surfaced clearly. Results highlighted the importance of management support, clarity in organizational ownership, and a structured approach to cross-training and collaborative teaming. A primary goal identified concerned the central importance of clarifying and strengthening organizational processes for the rigorous application of the scientific method within CSDS practice.

In extrapolating gaps and guidance in the CSDS field, advice is provided to managers seeking to pursue programmatic implementations. Summarizing guidance extracted from the key fishbone cause-and-effect diagrams (Table 3.35) according to the classical MIS tripartite framework of organization, process, and technology:

1. *Organizational*: Ensure that management is engaged and knowledgeable concerning the inherent complexities associated with CSDS. Organizations should introspect concerning how the breadth and complexity of the CSDS domain complicates and obfuscates ownership, decision-making, team collaboration, and cross-domain knowledge. The following themes were advocated by practitioners as key CSDS organizational enablers:

   (a) Management commitment
   (b) Data-driven paradigm shift
   (c) Uncertainty reduction/risk management focus
   (d) Operational optimization orientation
   (e) Multidisciplinary team orchestration
   (f) Cross-domain training

2. *Process*: A drive toward a process-driven approach to enable both data management and scientific methods is advocated. Data management and applied scientific methodological processes are recognized as having distinct overlaps in addressing the questions of *what to track*, *what data to collect*, *what is normal*, and *how to analyze*. These are topics that require concerted scientific inquiry to frame, extrapolate, and substantiate. Scientific processes were surfaced as being multifarious: problem framing, discovery, feature selection, categorization, validation, detection, prediction, and optimization. The following key themes were advocated as CSDS process enablers:

   (a) Data management and structuring
   (b) Mapping and modeling environmental security context
   (c) Systematic evidence gathering, framing, and labeling
   (d) Disassociating normal versus abnormal phenomenon

(e) Risk quantification

(f) Data and model validation

(g) Linking distinct exploratory and predictive methods

3. *Technology*: Although explicit technology topics were raised in interview feedback (i.e., cloud, honeypots, endpoint detection, SIEMs, data engineering, unsupervised vs. supervised ML), it was notable that their direct appearance was muted in surfaced gap-prescription themes. This does not imply technology is unimportant to CSDS best practices but rather that practitioners seemed to indicate that technology is an enabler and that organizational framing and process alignment goals take precedence. This suggests technological solutions are viewed as the outcome of organizational direction and process goals, rather than a primary driver. Given the preponderance of engineering-driven approaches in the broader security field, the muted focus on technology was both surprising and laudable.

In summary, guidance advocated the need to integrate organizational and process-driven treatments, especially to align human and technical resources through organizational processes. Key enabling processes framed centered around data management and applied scientific inquiry. Results suggested that resource coordination, data management, and scientific methods must together conjoin to drive CSDS as a functional organizational process. This presented CSDS as necessitating multiple, overlapping processes, particularly cyclical data discovery and scientific inquiry.

The need for cross-domain coordination and collaboration to integrate data science methodological approaches with cybersecurity operational gaps was highlighted. CSDS presents as a recent development, having emerged as a professional title in the last 3 years. The domain can be classified as being *syncretic*. Participating professionals have entered either from a cybersecurity or data science background and build cross-competencies in the junction of the two domains. This highlights the need for clearer, more explicit cross-disciplinary bridges.

CSDS ideally enables a symbiotic collaboration between the parent domains to orient data science practice to address focused cybersecurity prescriptions. The implicit *iterative* interaction between data processing and scientific inquiry was revealed by practitioners to be essential to delivery. This can be summarized conceptually as a model in which cybersecurity and data science gaps combine to frame CSDS treatments (Fig. 3.10).

Fig. 3.10 CSDS gap analysis model

3.4.4 Research Questions Addressed

See Table 3.36.

Table 3.36 Research questions addressed—opinion and gap analysis

Management problem 2	**What prescriptive treatments address categorical CSDS gaps?**
	• A range of organizational, process, and technical best practices are advocated by CSDS practitioners to address focused challenges surfaced in interview research
	• Key CSDS gap-prescription areas surfaced for treatments:
	  – Data management
	  – Scientific processes
	  – Cross-domain collaboration

Research objective 2	***Diagnose* gaps impeding professionalization based on qualitative research**
	• An analysis of themes surfaced in CSDS interview results identified three key gap-prescription areas: data management, scientific processes, and cross-domain collaboration
	  – Interpretation and coding of interviews resulted in a set of challenge and best practice topics
	  – Factor analysis of best practices and challenges extrapolated central themes
	  – Correlation of factor themes surfaced key macro-gap themes associated with best practice and challenge prescriptions

Research question 5	**What gaps can be diagnosed in the emerging CSDS field from practitioner perspectives?** Key challenge themes surfaced in factor analysis of interview feedback: • Expansive complexity • Tracking and context • Data management • Expectations versus limitations • Unclear ownership • Data policies Key CSDS gap-prescription areas requiring focused treatments: • Data management • Scientific processes • Cross-domain collaboration

Research question 6	**What treatments are prescribed to address gaps based on practitioner input?**
	• Based on correlation of challenge and best factor themes extrapolated from interview feedback:
	  – Improve rigor of scientific processes
	  – Improve approaches to data management
	  – Improve cross-domain organizational collaboration
	• Details of treatments are suggested by the array of subthemes aligned to the key gap-prescription areas

Chapter 4
Phase III: CSDS Gap-Prescriptions— Design Science Problem-Solving

4.1 Research Objectives

As a capstone to this research inquiry, the final phase frames and advocates design-derived gap-prescriptions. CSDS has thus far been systematically explored through triangulated diagnostic methods as an emerging practitioner discipline. In the preceding phases, practice-oriented diagnostic research has been undertaken, encompassing background analysis (*Phase I*), opinion research (*Phase II*), and gap analysis (*Phase II*). Per guidance from Verschuren and Doorewaard (2010), a design approach is a natural accompaniment to conclude diagnostic analysis in problem-solving research.

Focused guidance extrapolated from gap analysis findings are further explored here to establish a foundation for *design science problem-solving*. Design problem-solving necessitates the clarification of structured requirements based upon robust definitions of a focused problem (Gregor and Hevner 2013; Hevner et al. 2004; Wieringa 2014). This phase undertakes detailed diagnoses of the three gaps surfaced and substantiated in the previous interview inquiry and derives problem-solving requirements leading to advocated design prescriptions.

The three key CSDS gaps framed for design prescriptions relate to data management, scientific methods, and cross-domain collaboration. The overarching MIS disciplinary focus roughly frames these gaps within technology, process, and organizational contexts, respectively.

A design science problem-solving process is applied, leading to requirements and recommended prescriptions. Standing challenges in the cybersecurity profession and methodological debates in the field of data science are explored and worked out in order to provide practical CSDS guidance. Reflecting the partial problem-solving scope of this research effort, designs are framed as nascent designs (Gregor and Hevner 2013), conceptual principles advocated for future implementation.

Results should be of interest to managers and practitioners pursuing programmatic CSDS implementations. As well, the design prescriptions outlined should

S. Mongeau, A. Hajdasinski, *Cybersecurity Data Science*,
https://doi.org/10.1007/978-3-030-74896-8_4

Table 4.1 Summary of research and results—gap-prescription designs

Chapter	Phase	Method	Result
4. CSDS Gap-Prescription Designs	III. Design problem-solving	Design science	Design requirements & prescriptions

Table 4.2 Summary of design research questions

Management problem 3	What guidance can be offered to orchestrate the implementation of CSDS programs?
Research objective 3	*Prescribe* design treatments based on gap analysis and extrapolation from literature
Research question 7	How should CSDS methodological challenges be addressed?
Research question 8	Which data science methodological treatments are prescribed for the CSDS domain?
Research question 9	What processes are advocated to implement prescriptions?
Research question 10	How can CSDS prescriptions be implemented as organizational processes?

serve as guidance for researchers interested in pursuing follow-on case, design, and/or action research implementations. An overarching goal is to iterate CSDS body of theory in the service of the development of the CSDS profession. Results should thus also be of interest to institutional stakeholders seeking to promote the development of the CSDS field.

The larger goal in the context of the research effort is to advance CSDS body of theory by systematically examining key *as-is* gaps in the domain and to espouse recommended *to-be* treatment prescriptions. Table 4.1 summarizes the research approach and outcomes achieved in this concluding phase. Focused research objectives and questions addressed are listed in Table 4.2.

4.2 Framing CSDS Design Prescriptions

4.2.1 CSDS Gaps in Context

Background literature analysis highlighted central challenges to advancing the maturity of the CSDS profession. This has manifested in difficulties codifying a CSDS body of theory necessary for professional maturation. A summary of key challenges raised:

- Cybersecurity as a CSDS parent domain is increasingly challenged by rapidly evolving threats and expanding vulnerabilities due to the proliferation and growing complexity of digital infrastructure and devices.

- Data science as the accompanying CSDS parent domain often lacks codified theoretical and methodological consensus due to a lack of resolute disciplinary boundaries.
- CSDS as a hybrid domain presently lacks a codified body of theory to guide the systematic diagnosis of security problems leading to design prescriptions.

The MIS tripartite model of technology, process, and organization has provided a guiding framework for structured inquiry into the CSDS domain. Based upon interview research conducted in the previous phase, practitioner perspectives surfaced three key gaps framing requirements for CSDS professional development. The key gaps identified: data management, scientific processes, and cross-domain collaboration. Mapped to the MIS tripartite model, these gaps can be roughly aligned to the areas of technology, process, and organization, respectively (Fig. 4.1). This phase will systematically examine each of these gaps sequentially, resulting in suggested design treatments to address surfaced shortcomings.

Technology—Data Management: A fundamental perceived challenge in the domain relates to data management: gathering, integrating, cleansing, transforming, and extrapolating key measures from the fragmented, voluminous, and fast-streaming sources that underlie modern digital infrastructure. Notably, this objective was framed by practitioners in interviews as transcending pure engineering. Technology planning necessitates the application of a design process to select, refine, and validate datasets and data-driven models prior to operationalization.

Process—Scientific Processes: Concerning the most complex and far-reaching gap, scientific processes within the CSDS domain, there is a perception amongst practitioners, reflected in literature, that the field must develop structured processes to facilitate the application of scientific methods. This requires an understanding of which scientific treatments need to be applied to particular use cases, available methods being multiple and dependent on context. A corresponding challenge is

Fig. 4.1 CSDS gaps in the tripartite MIS model

that CSDS practitioners work in high-pressure, time-driven tactical environments. A key opinion expressed across interview feedback was that advancement of the domain requires a mandate and the wherewithal to undertake staged research—some combination of scientific hypothesis framing, experimentation, testing, and validation. Practitioners often cautioned that the resources and time to undertake core research was typically absent in operational environments.

Organization—Cross-Domain Collaboration: The third gap identified by CSDS practitioners, cross-domain collaboration, invokes organizational challenges associated with the sheer breadth of the combined cybersecurity and data science domains. The complexity framed by the hybridization of CSDS implies that even seasoned professionals can only hope to gain expertise in specific areas combined with a general understanding of others. This necessitates organizational approaches to ensure cross-domain collaboration across hybridized teams of professionals. As well, it implies the need for overlapping organizational process frameworks to ensure the integration of collaborative expertise.

In summary, despite the great promise of applied data science to cybersecurity, focused challenges confront CSDS practitioners and managers. Gaps surfaced in interview research reinforce themes gleaned from CSDS literature analysis. A comparative review of key CSDS texts identified four topical gaps in practitioner literature: risk quantification, scientific methods, data management, and organizational management.

A key concept identified in literature and interview research indicated a reciprocal relationship between CSDS challenges and best practices. Security challenges resulting from innovation frame new defense goals and vice versa. In interview responses, challenges raised by interviewees implied gaps as an absence of a best practice. Best practices, conversely, advocated approaches to address gaps. This reciprocal relationship between CSDS challenges (gaps) and best practices (prescriptions) was represented per Fig. 4.2.

In CSDS practice, the reciprocal relationship between CSDS parent domains frames a design science process approach to problem-solving: diagnosing gaps, framing problem-solving requirements, and prescribing data science treatments. This approach provides a foundation for driving design science prescriptions vis-à-vis an extrapolation of structured requirements from diagnosed gaps. Gaps frame problems which, when diagnosed in a detailed manner, frame requirements. Design prescriptions are thence suggested to address the surfaced requirements.

However, gaps in CSDS body of theory impede clarity in framing CSDS prescriptions. As raised in literature analysis, the relative immaturity of the data science domain forestalls clear methodological treatments to cybersecurity gaps. Advancement of the CSDS domain thus necessitates bridging this shortcoming by exploring, framing, and recommending data science design approaches to resolve gaps, per the symbiosis framed in Fig. 4.3. In this phase, CSDS gaps will be explored to suggest practical design approaches to resolve this quandary.

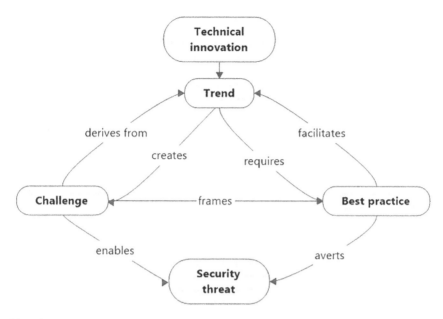

Fig. 4.2 CSDS challenge-best practice reciprocity conceptual model

Fig. 4.3 CSDS symbiosis between cybersecurity and data science body of theory (BoT)

4.2.2 Cybersecurity Challenges Redux

Following the exposure of the Internet to commercial usage in the early 1990s (Hafner 1998; Leiner et al. 1999), the exponential growth of networked computer and communication technologies opened a vast new realm. The emergent nature of the complex technologies and networks which have sprung up within scarcely three decades has occurred at such a pace that it has been a challenge to track the array of new devices and applications now populating the cybersphere. Cybersecurity, at base, is charged with stewardship of this domain, assuring that "information systems can operate as intended in a hostile environment" (Shostack 2012).

The cybersecurity domain is in a relatively mature state of professionalization: sanctioned by society, supported by a professional community, encompassing a well-expressed body of knowledge, and hosting a range of focused training offerings, standards bodies, and professional certifications. A brief review of key cybersecurity goals and challenges is appropriate to frame objectives for prescriptions. In background analysis, a four-quadrant model for security assurance was extrapolated

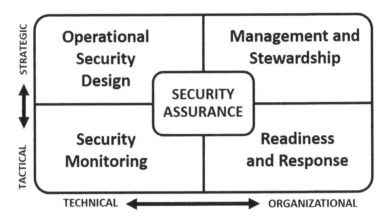

Fig. 4.4 High-level security assurance functional model

to summarize key organizational goals (Fig. 2.7, as simplified here in Fig. 4.4). Whereas CSDS potentially offers treatments across all these areas, security monitoring is the most immediate target for the application of data-driven methods (Greengard 2016; Verma et al. 2015; Zinatullin 2019).

However, as a function, cybersecurity monitoring faces a perfect storm of challenges (Fig. 4.5):

- Growing sophistication of threats circumventing traditional monitoring measures (Kirchhoff et al. 2015; Morgan 2017a; Poppensieker and Riemenschnitter 2018; Security Brief Magazine 2016)
- Expanding indicators and anomalies leading to a spiraling surfeit of false alerts (Cisco Systems Inc. 2017; Greengard 2016; IDC 2014; Lohr 2011; SANS Institute 2016)
- Cyber data and alert overload from the continual expansion and growing complexity of digital infrastructure, systems, and devices (Ponemon Institute 2017)
- Challenges in integrating systems and data to achieve situational awareness (Leuprecht et al. 2016)
- Overworked and undertrained cybersecurity resources struggling under expanding vulnerabilities and threats (Garande 2019; Giles 2018b; Greig 2019; Oltsik 2019b; Palmer 2019; SANS Institute 2015; Shostack 2012)
- Increasingly geographically distributed, diffuse, and porous organizational networks making traditional perimeter-based defenses increasingly ineffective (Lehto 2015; Leuprecht et al. 2016; Wolff 2014), including potentially longer-term shifts towards mass teleworking as a result of COVID-19 (Burr and Endicott 2020; Lichfield 2020)

The result for many organizations has been a growing siege mentality, a sense that the network may already have been compromised (Connors and Endsley 2014; Shostack 2012). Exacerbating factors include limitations on the ability to restrict access to digital infrastructure, limited scope to shape organizational policy, the

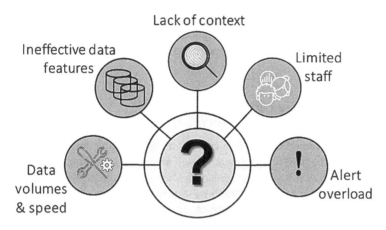

Fig. 4.5 Key cybersecurity challenges

rapidity of technical proliferation and change, and the expanding breadth of the cybersphere.

While CSDS presents as a promising palliative, the nature of challenges facing the cybersecurity domain are complex and overlapping, requiring systematic multicontextual responses. For example, cybersecurity "big data" volume and velocity present data engineering challenges necessitating processing at great volumes and speed. However, data engineering solutions do not inherently address challenges surrounding data quality, integrity, and veracity. A primary exploratory effort is necessary to determine which data to process and how to transform, route, and integrate this data to drive operational detection. Forestalling such efforts are a general lack of theoretical foundations in the security realm to ground and orient such inquiry (Schneider 2012).

The selection and framing of data in turn are subject to the needs of contextual model development and operationalization. The data requirements of models must be anticipated, tested, and validated. More than an engineering challenge of processing, transforming, and routing data, a systematic effort must be staged to establish the scientific efficacy of the data to be processed. This occurs in the context of assessing the conceptual integrity of the suitability of datasets to particular models. Assuring this chain of contextual dependencies requires development efforts involving iterative design testing and revision, a cyclical process superseding traditional linear "waterfall" systems development.

Framing a chicken-and-egg quandary, implementation and utilization of big data in an operational context cannot be properly validated without an exploratory trial-and-error process. Data framing must be staged to drive model development, validation, and operational testing. In this context, CSDS-focused data engineering is a step in an iterative cycle of exploration, experimentation, validation, and iteration, not a one-time engineering solution.

Multifaceted and overlapping challenges such as this are characteristic of CSDS gaps: proposed solutions are contingent upon satisfying an iterative array of underlying technical, process, and organizational requirements. The singular shared element is the demand for scientific rigor in methodological approaches to assure the validity of applied methods and the relevance of operational outcomes.

4.2.3   Data Science Challenges Redux

CSDS has been framed in literature analysis as an emerging domain applying prescriptions from the field of data science, itself an emerging and maturing discipline. An idealized CSDS perspective views data science as implementing applied methods in cybersecurity operational settings to achieve practical outcomes. However, background analysis revealed a dearth of clear theoretical underpinnings to anchor the application of data science methods, which raises questions of methodological integrity.

Literature analysis raised the challenge of codifying CSDS methodological prescriptions given the hybrid grab-bag of methods which data science harbors. This complication was reinforced in CSDS practitioner interviews, surfacing as a pronounced lack of concordance regarding which specific analytical methods were most suitable for application in the cybersecurity domain. Practitioners cited eight distinct external domains as providing methodological guidance to CSDS, ranging from fraud analytics to bioinformatics. Further, there was discord resident regarding the degree to which statistical versus machine learning methods should be emphasized in CSDS practice.

CSDS inherits dysfunctionality through its data science parent in being methodologically promiscuous and thus adverse to committing to individuating standards. A central challenge facing data science as a professional domain is the expanding range of methods and disciplines to which it lays ostensible claim.

As a hybrid interdisciplinary field with commercial and practitioner origins, data science continually assimilates new analytical techniques into a voluminous and growing "toolbox." The downside of this phenomenon is that it becomes difficult to distinguish what is, and what is not, data science in terms of codified methodological practice. Whereas data science seemingly adopts and assimilates *any* type of analytical method, grounds for professional practice standards blur. Data science, as a parent of CSDS, thus inhabits a paradoxical status of being both a blessing and a curse.

Disciplinary haziness has not inhibited efforts to pursue data science engineering solutions, namely, the rush to apply data science methods through technical tools in commercial settings. This has not otherwise improved efforts to establish rigor in data science practice. To the degree data science has become popularized by commercial interests, noise from the marketplace has muddled attempts to clarify disciplinary boundaries and methodological standards (Oltsik 2019b; Ross 2019; Smith 2018; Violino 2019b).

The commercial hype surrounding data science, big data, machine learning, and AI has led to a rising number of concerted critiques. Examples include *The AI*

Delusion (Smith 2018) and *Weapons of Math Destruction* (O'Neil 2016). A commercial focus on gross automation through machine learning, with the implicit implication of disintermediating the overhead of human labor, has been a reoccurring theme that has distracted efforts to establish rigorous professional standards.

Many commercial interpretations of machine learning subtly demote or outright refute the notion of the practitioner-as-researcher as a rational agent playing an irreducibly necessary role in framing, validating, interpreting, and assuring scientific integrity. A perplexing myth, first popularized in the 2008 Wired article "The End of Theory: The Data Deluge Makes the Scientific Method Obsolete" (Anderson 2008) and later promulgated through the big data movement, implies the much-disputed assertion that scientific discovery can be computationally automated (Pigliucci 2009). Related misinformation has been promulgated, such as the notion that bigger datasets are ipso facto "better" and that they transcend issues such as measurement error (van Smeden et al. 2019).

The rapidity with which data science has emerged in both business and academia has left a range of central methodological concerns unaddressed. The rush to deploy data engineering solutions for commercial interests often pushes fundamental issues concerning data-generating processes (DGPs), data quality, selection bias, model validity, predictive integrity, and interpretation to the wayside in the drive to operationalize predictive machine learning solutions. The phenomenon of big data has ignited an unresolved practitioner debate between statisticians and machine learning advocates.

A related growing critique concerns challenges integrating data science-based decision automation into organizations. A growing number of firms have eagerly deployed automated machine learning driven decision mechanisms into operation with questionable or absent efforts to validate and explain the theoretical assumptions underlying the algorithms. This has led to pointed ethical critiques that question the ease with which automated algorithms can make incorrect or morally suspect decisions (O'Neil 2013, 2016).

Data science, to the degree it promulgates over-enthusiasm concerning the abilities of machine learning, perpetrates a myth of decision automation and transhuman technical efficacy. This face of data science overemphasizes *technê*-based engineering to the diminution of *logos* driven *epistêmê*, which results in a poverty of theoretical rigor. Gaps emerge between data science engineering-focused practice and a relative lack of generalizable theoretical rigor. Emphasizing data science as an engineering project destabilizes foundations for framing and codifying professional best practices that inculcate scientific rigor. Body of theory grounding, as a result, lags behind engineering implementations.

Background analysis framed this schism as a symptom of overemphasizing engineering-focused predictive efficacy (*technê*-focused) and regularly ignoring theoretical and methodological rigor (*logos*-facilitated *epistêmê*). A key conclusion was that disciplinary debates associated with the data science movement stand unresolved and ironically inhibit the advancement of efforts towards practical operationalization to the degree that standards for validating implementations remain unclear.

These debates will not be resolved solely through the demonstration of engineering prowess.

In summary, standing in the path of substantiating CSDS as a proper scientific undertaking, literature analysis asserted that immature body of theory in the data science domain remains an inherent challenge to asserting claims of applied scientific rigor. This frames the distinction between *technê*- and *logos*-mediated *epistêmê* in applied technology. Whereas *technê*—skill, art, or craft—focuses on pragmatic results, *logos*, the expression of rational, structured thought, supports the development of *epistêmê*, understanding, knowledge, and scientific theory.

Core methodological debates from the field of statistics, particularly the applied role of computational methods and tools to scientific inquiry and theory formation, remain stumbling blocks in CSDS professionalization. Whereas professionalization in CSDS is challenged by fundamental questions concerning the methods of data science, it is asserted that clarifying a way forward necessitates a step back to re-examine fundamental methodological issues. To reframe data science in terms of disciplinary rigor, one must return to deeper academic disciplinary foundations that have emerged stepwise over the course of several decades. Academia is by nature slow, methodical, and conservative in its conduct and consideration.

As academic institutions, despite protestations, often have commercial interests, or at the very least a motivation to address popular interests, there has been a spate of new degree programs in analytics and data science that have emerged in the past several years. The rapidity with which the analytics discipline has grown has propelled the establishment of a range of data science academic degree and professional certificate programs, as well as a cottage industry of one-off courses, technical training offerings, MOOCs, and so-called boot camps.

Covering a broad conceptual territory, each data science training offering makes efforts to establish views on the disciplinary boundaries associated with the domain. However, data science, as a rapidly emerging practitioner discipline, has expanded in advance of academic-level methodological rigor. Methodological issues and debates resident in data science thus lie in wait for efforts to apply data science methods in a focused domain.

The uncomfortable reality is that data science, in a traditional sense, currently has very weak grounds for asserting knowledge individuation, a symptom of which is its insubstantial claim to formal academic disciplinary status. To be sure, degree programs are operating, faculty are titled and focused, and the topic is lectured upon. Yet there is a lack of a theoretical coherency resulting from diffusion amongst the broad and growing interdisciplinary domains, which data science, as a magpie discipline, agglomerates.

Data science variously spans, amongst others, computer science, AI, machine learning, business intelligence, statistics, econometrics, operations research, and the management of information systems (including the sub-domains of decision support systems (DSS), expert systems, management support systems, and executive information systems). To the degree that these domains host unresolved critiques of their neighbors, data science struggles with inherent internal disciplinary quandaries.

However, as raised in literature analysis, there is some promise for CSDS through the route of domain specialization. A central conclusion of literature analysis was that, given a lack of clear theoretical underpinnings in the data science domain, CSDS must pursue theoretical individuation and differentiation through cybersecurity-specific use cases. This otherwise frames and motivates design science approaches to applied CSDS treatments as a path to body of theory maturation.

Specialized fields have individuated via the field of analytics by framing focused use cases and domain challenges as a path to codify the targeted application of analytical methods. By focusing on domain-specific examples, theoretical and practical analytical treatments emerge (Kiron et al. 2014). Examples include operations management, financial analysis, and marketing/consumer analytics. By focusing on domain-specific problems, a foundation for working out focused methodological best practices and, significantly, general theory, is established. It is in this sense that CSDS is framed: conscious of the gaps in data science theoretical foundations but viewing this as an opportunity for individuation through specialization.

The path of specialization implies that structured efforts to define and explicate focused domain problems-as-use cases are worked out. By systematically defining focused challenges as analytical problems, a diagnosis is undertaken resulting in requirements. These requirements then frame and suggest potential prescriptive solution designs that can be methodically trialed, iterated, refined, and adopted. This otherwise invokes and advocates the discipline of design science as a promising and practical approach to iterating CSDS body of theory.

4.2.4 Towards Design Science Prescriptions

In the preceding phases, CSDS as an emerging profession was explored through triangulated diagnostic research, encompassing background analysis (*Phase I*), opinion research (*Phase II*), and gap analysis (*Phase II*). Extrapolating from integrated practitioner and academic literature, *Phase I* conducted background analysis, providing a conceptual foundation for CSDS as an emerging field.

Phase II undertook interview-based opinion research, framing a set of key challenges and best practices resident in emerging practice. Based on extrapolated findings, gap analysis was conducted utilizing quantitative methods. Together, opinion and gap analysis research suggested a set of requirements for iterating CSDS body of theory to address surfaced gaps.

As an iteration of this inquiry, *Phase III* frames design science prescriptions to guide CSDS practitioners and programmatic implementations. Design science is a methodology focused on iteratively building artifacts with practical use in implementation to address identified gaps (Gregor and Hevner 2013; Hevner et al. 2004; Vaishnavi and Kuechler 2015; Wieringa 2014). Gap analysis frames problem-solving requirements, which are utilized to extract and advocate design prescriptions. This applies an iterative approach aimed at building knowledge to frame and improve design artifacts, as evoked by Wieringa (2014), per Fig. 4.6.

Fig. 4.6 Design science problem-solving activities

Treatment implementation

Implementation evaluation / Problem Investigation
- Stakeholders? Goals?
- Conceptual problem framework?

Treatment validation
- Artifact X Context produces Effects?
- Trade-offs for different artifacts?
- Sensitivity for different contexts?
- Effects satisfy Requirements?

Treatment design
- Specify requirements
- Requirements contribute to Goals?
- Available treatments?
- Design new ones!

Fig. 4.7 The design engineering cycle

Design problem-solving entails the clarification of structured requirements. Verschuren and Doorewaard outline four types of requirements necessary for design-oriented problem-solving: functional, contextual, user, and structural (2010). Extrapolating from surfaced CSDS domain gaps, this phase explores functional, contextual, and user requirements with a view to framing structural requirements. Structural requirements specify characteristics required of design prescriptions, which are subsequently developed, framed, and advocated. Wieringa frames the centrality and application of requirements within the design engineering cycle (2014), per Fig. 4.7.

As framed by Drench et al., "artifacts are designed and created to effect some change in a system, solving problems and allowing for a better performance of the system as a whole" (Dresch et al. 2015). Designs are embodied as artifacts, which can be physical, technical or engineering-based, or systematic conceptual constructs such as procedures or policies to be implemented. This presents a pragmatic approach to frame and address gaps in cybersecurity practice. This view is mirrored by Landwehr in his article "Cybersecurity: From Engineering to Science" (Landwehr 2012).

Design science admits and embraces the organizational importance and context of artifacts-as-solutions, ensuring that designs meet practical and contextual goals. Wieringa frames the process of formalizing design science artifacts as interacting with the needs of organizational stakeholders to achieve a goal (2014), per Fig. 4.8.

As has been raised throughout this inquiry in the context of the MIS tripartite model, cybersecurity can be interpreted in multiple, overlapping frames: organizational, process-oriented, and technological. To the degree that design science encompasses artifact design in terms of organizational goals and to the degree

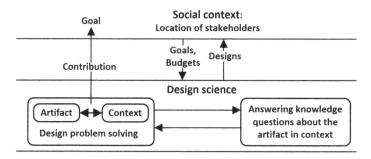

Fig. 4.8 Interaction of design artifacts with context contributing to stakeholder goals

artifact specifications can have both process-focused and technical aspects, design science addresses the full range of MIS contexts.

As has been reinforced in the review of cybersecurity and data science challenges, there are a range of overlapping, mutually dependent process gaps in the CSDS domain. This is particularly the case concerning the mutual dependence between scientific methods/rigor and data engineering processes. Design advocacy in this analysis thus focuses on process design and specifies where designed processes must interact.

In a design science perspective, processes are viewed as artifacts that must be configured and implemented to address organizational context and goals, which, writ broadly in the CSDS domain, is a dynamic understanding of cybersecurity integrity, a knowledge-focused problem. Both cybersecurity and data science gaps identify problems of inculcating methodological scientific rigor in organizational processes.

By focusing on overlapping process-based design prescriptions, there is a recognition that knowledge (i.e., situational cyber awareness) is the result of interdependent mechanisms-as-process-artifacts: technical, methodological, and organizational. Beyond *technê*-focused engineering solutions, this establishes a foundation for continuous *logos*-facilitated *epistêmê* theory development and validation in CSDS organizational practice.

In terms of the outcomes of design science efforts, March and Smith specify four outputs:

1. *Constructs*: conceptual vocabulary arising from the problem/solution domain
2. *Models*: propositions expressing the relationships among constructs
3. *Methods*: a set of steps used to perform a task to realize a solution
4. *Instantiations*: operationalization of constructs, models and methods
March and Smith (1995)

Within the scope of this inquiry, a foundation has been established for problem domain constructs through the review of problem-solving requirements. What follows will iterate by developing solution domain constructs, amplifying these with a series of models and advocated methods. Implementation is framed as a follow-on step for practitioners and/or researchers to operationalize.

Designs extrapolated are framed as prescribed requirements but are not carried through to implementation. Design science is thus applied to the point of advocating follow-on implementations. The intention is to provide a foundation for subsequent design and action interventions in research and practice. Recommendations are suitable for practitioners pursuing CSDS programmatic implementation, as well as for researchers undertaking subsequent design implementation, case study, and/or action research.

4.3 Design Prescription I: Data Management

4.3.1 Problem-Solving Requirements

A key challenge raised across both literature analysis and interview research concerns the need in CSDS practice for improved approaches to data management and processing. There was a clear need expressed for more rigorous approaches for processing security data prior to the application of analytical methods. This was framed as a challenge given the characteristically fragmented, voluminous, and fast-streaming data sources that underlie modern security environments.

Data preparation was the most frequently cited theme in interviews with CSDS practitioners, 84% ($n = 42/50$) of respondents having framed this topic as both a challenge and best practice. A frequently cited sentiment amongst practitioners was that data preparation should be a part of a structured process and that it should be promoted as a core aspect of improving CSDS practice. However, there was a frequent perception of organizational challenges in gaining visibility, time, and resources to properly address data preparation needs.

Practitioners indicated a need to structure and rationalize the array of data treatments available, encompassing selecting, gathering, integrating, cleansing, transforming, and extrapolating key measures. Further, it was emphasized that better approaches to integrating data preparation processes and procedures to address analytics context were needed. It was advocated that data preparation should be more clearly integrated into analytical model development and scientific inquiry processes. The challenge and best practice interview themes associated with data management surfaced in gap analysis are revisited here, per Table 4.3.

Table 4.3 Problem-solving requirements: data management

Key gap	Category	Practitioner themes
BP F4: Data management	Paradigmatic	**BP9**: Data-driven paradigm shift away from rules and signatures
	Data management as a process	**BP1**: Structured data preparation, discovery, engineering process **CH1**: Data preparation process (access, volume, integration, quality, transformation, selection)
	Context & tracking	**CH4**: Lack of labeled incidents to focus detection **BP13**: Understand own infrastructure and environment

What follows is an extrapolation of design prescriptions for addressing CSDS data management gaps framed for the benefit of program managers, practitioners, and researchers seeking to pursue CSDS implementations.

4.3.2 Data Preparation as a Process

Gartner defines data preparation as "an iterative and agile discovery process for exploring, combining, cleaning and transforming raw data into curated datasets for data integration, data science, data discovery and analytics/business intelligence (BI) use cases" (Zaidi 2019). Data preparation was identified as the single largest challenge in CSDS practice, with advocacy exhorting greater rigor in organizational implementation.

A clear point of advocacy from practitioners was that data management should be approached as a well-defined, structured process. A lack of organizational management focus and attention was often cited as an underlying challenge. CSDS practitioners raised a reoccurring observation, reinforced in domain literature, that CSDS data management should be viewed as more than a rote mechanical engineering process, although this was how many practitioners felt it was perceived in the wider organization.

Cybersecurity data is, by nature, voluminous and fast-moving. It also tends to be unstructured, locked in log files or other sources, where some effort is required to extract variables and entities suitable for analytical treatment. Sources also tend to be distributed and fragmented, requiring efforts to bring together and integrate data in order to assemble a composite picture. Events, when tied to entities for tracking, occur sporadically and irregularly. Clear signals can be challenging to extrapolate from large security datasets as there is a great deal of variation and noise.

Additionally, the nature and quality of data extracted from security sources is uncertain. There are typically primary quality issues that require efforts to clean and treat. Administrative metadata sources, which depend upon human curation to track key entities (i.e., user and device tracking repositories), are often out-of-date or inaccurate. There are also doubts concerning the veracity of data: skilled adversaries will typically seek to cover their tracks by modifying log files and erase records of their passage and actions. In aggregate, this creates a combination of data quality issues and concerns which need to be addressed before analytical treatments can be performed.

Data management is central to CSDS practice as preparatory data treatments have a direct impact on the subsequent quality and reliability of analytical model results. When operationalized, data preparation leads to analyst productivity and faster insights (Zaidi 2019). Data processing embeds knowledge and assumptions concerning the context of and relationships between key entities in a particular domain. Decisions related to the preparation and treatment of data also embed assumptions concerning the phenomenon under examination. Additionally, distinct analytical approaches require specific data types, formats, and treatments. A robust

CSDS data preparation approach must therefore undertake data processing as a structured development process, bearing in mind the goals associated with the intended analytical methods to be applied.

A rigorous CSDS data design process involves feedback mechanisms and validation checks, facilitating alignment between data and phenomenon under assessment. The data processing in preparation for analytics involves multiple steps, and each step may lead iteratively backwards as new findings suggesting alternate interpretations of, and therefore treatments to, data. To the degree CSDS strives towards scientific rigor, assumptions embedded in the processing of data are ideally exposed and assessed through diagnostics and experimental testing.

Efforts to structure and prepare cybersecurity data for analytics includes treatments such as extraction, parsing, derivation, transformation, calculation, normalization, integration, and cleansing. Given the wide variety of sources and formats for cybersecurity data, the amount of effort necessary to process data varies widely but is generally agreed to be nontrivial by nature (Domingos 2012; Kent 2016).

Data preparation and quality assurance are consistently reported to consume the largest portion of data scientist attention and effort. According to a survey of 179 data scientists, "over half identified *addressing issues related to data quality* as the biggest bottleneck in successful AI projects. Big data is so often improperly formatted, lacking metadata, or 'dirty,' meaning incomplete, incorrect, or inconsistent, that data scientists typically spend 80 percent of their time on cleaning and preparing data to make it usable, leaving them with just 20 percent of their time to focus on actually using data for analysis" (New 2019).

Regarding CSDS, some interviewees cited figures of 90% or 95% of their time being spent on data quality issues. Some even suggested that problems associated with data quality were so deep that it inherently restricted and hobbled their primary attempts to conduct analytics. Although the methods and processes associated with data preparation are well-understood in aggregate, they are difficult to automate and are inherently time-consuming given a strong dependence on human judgment.

Processing data in preparation for analytical methods requires consideration of how decisions regarding the treatment of data embeds context. Ensuring quality in the data preparation process involves addressing a range of considerations:

- *Epistemological*: e.g., How was the data generated and what phenomenon does it represent? What is the underlying data-generating process (DGA)? Is the data source trustworthy and consistent? Does data framed and collected adequately represent the targeted phenomenon within the frame of the intended inquiry? Does data volume present spurious correlations or phantom patterns? Are there seasonal or exogenous effects outside the scope of the data which should be considered?
- *Structural*: e.g., What format and structure should the data be stored and represented within? How should data elements be linked, joined, or merged as conceptual entities? What explicit and/or latent categories are resident in the data? Should treatments be implemented to improve the contextual representation,

such as transformations to accommodate parametric distortions or accommodations for sparseness and outliers?

- *Technical*: e.g., Can the sources issuing data be verified for generative accuracy? Are there missing elements or variables or periodic lapses in the record suggesting technical production breakdowns? How are challenges of speed and volume addressed in terms of data aggregation and summarization while controlling for the potential loss of coherency?

The strong link between data quality and the reliability of analytical outcomes is logical and well-accepted by both practitioners and researchers (Hazen et al. 2014, 2017). It is also highlighted as a reoccurring challenge practitioners face in operational environments, especially to the degree there is a pressure to deliver rapid results. To address operational challenges in a structured way, standards and guidelines for assessing data quality are increasingly available. Guidelines provide a mechanism for assuring data quality prior to analytical treatments, applying statistical or broadly analytical means to facilitate quality assurance.

An example of standards for data quality from the domain of statistics is the EUROSTAT "Handbook on Data Quality Assessment Methods and Tools" (EUROSTAT 2007). This set of standards frames institutional recommendations for assuring data quality. Key aspects include ensuring alignment with organizational factors, management systems, and leadership in particular. Noteworthy is that statistical validation procedures are advocated in data preprocessing and postprocessing to assure machine learning interpretability and explainability. A critique surfaced in interview research observed that many organizations implement machine learning with the mistaken assumption that efforts to substantiate data quality rigor through diagnostic approaches are not necessary.

By emphasizing organizational and institutional context alongside technical and methodological treatments, data quality is framed as the result of a process that requires continual organizational oversight and transparency (Kreuter and Peng 2014). This stands in distinction to the tendency to frame the management and production of data as a purely technical engineering challenge. Formal data preparation processes are advocated as a path to overcoming a reported tendency to ignore or de-prioritize data preparation efforts in CSDS organizational settings.

4.3.3 Specifying Data Structures

Consideration of how data will be represented, structured, and stored is of central importance to data preparation efforts. Cybersecurity data frequently is presented in the form of raw, unstructured text. Practitioner research identifies log files, which are typically unstructured or semi-structured, as the most frequently utilized form of cybersecurity data (Security Brief Magazine 2016).

To extract structured data from unstructured sources, decisions must be made concerning central concepts, for instance, key entities, elements, and variables. In the CSDS domain, these decisions are guided by cybersecurity domain knowledge, combining general principles, e.g., technical knowledge relating to components, codes, entities, indicators, etc.; grounding in the environment under scrutiny, e.g., knowledge of local technical architecture and the ways interactions between users and devices are tracked in log files and other repositories; and context related to the intended goals of analysis, e.g., BI reporting, anomaly generation, or predictive modeling, etc. These factors provide guidance regarding which measures to extract as related to an understanding of what is being measured and how the measurements should be best represented.

A CSDS measurement may be a direct quantification (e.g., number of bytes sent) or a derivation involving a summarization and/or transformation (e.g., the ratio of internal-to-external bytes sent per hour). Measurements are further specified by attributes, properties, and characteristics, suggesting quantitative or qualitative representation. Decisions are thence made concerning allocating data types, for instance, framing variables as continuous (interval or ratio) or categorical (nominal or ordinal). Categorical variables may subsequently be cast and stored as numerical or textual data types.

Until relatively recently, many data structuring decisions would need to be worked out and specified up-front, as traditional relational database management (RDBM) or data warehouse storage mechanisms demand that data be formatted and represented appropriately prior to loading into structured storage. Databases operationally require data to be "cleaned" (i.e., treated, formatted, structured, and normalized) prior to ingestion. This invokes the principle of schema-on-write, the notion that data must be pre-formatted to meet structured definitions prior to writing to storage. Relational storage and retrieval mechanisms depend upon pre-structured data models and, subsequently, support the use of structured query language (SQL) for retrieval.

Normalization necessitates that data also conform to reductive relational structures and tabular entities, prior to being ingested and loaded into storage. In this approach, reductive structural relationships between formally differentiated entities are strictly specified via relational schemas in a database as links between tables. For example, an authentication event might be pre-specified for storage in an authentication event table as a combination of a time stamp, validated credential indicator, authentication protocol, and an IP address bound to a registered userid. Authentication protocol may in turn be linked by a code to a foreign table with further specifications on each protocol, which in turn could be linked to a table cataloging authentication repositories. These reductive specifications impose strict overhead on the pre-preparation of data and force particular conceptual restrictions upon subsequent access to and utilization of the data. Great clarity is gained through the normalization and relational enforcement process, but it results in a loss of general flexibility. Schema-on-write restrictions associated with databases and data warehouses impose substantial overhead in terms of time-consuming data preparation efforts (Chen et al. 2012; Jacobs and Rudis 2014; Strawn 2016).

In the last decade, big data associated technologies have proliferated due to efficient and inexpensive mechanisms for storing large volumes of data, Apache

Hadoop and the associated ecosystem of standards and tools being the prime progenitor (Chen et al. 2012; Ekbia et al. 2015; Gerhardt et al. 2012; Sivarajah et al. 2017). A major advent of the rapid rise of cloud-based computing and big data storage repositories has been the increasing popularity of schema-less data storage, allowing raw, unstructured data to be stockpiled without the overhead of preprocessing to conform to a pre-defined schema. Big data associated storage and retrieval mechanisms support unstructured data to be stored in bulk and structured upon retrieval, thus schema-on-read.

However, while the advent of cloud-based repositories greatly speeds and increases the efficiency of storing data in bulk, it does not obviate the need to structure data upon extraction, prior to use. While some data formats do not require essential structure for primary use, i.e., unstructured text, audio, and video, most other types require a structured extraction approach prior to utilization, i.e., schema-on-read translation. This is especially the case for preparing data for use in analytics models, as most analytical models require data to be formatted, cast, and presented in particular ways prior to statistical or algorithmic processing. Most analytics models, both during development and in implementation, imply requirements for a structured representation of data prior to ingestion and treatment.

A central observation and critique of the commercialization of big data approaches is that they risk inaccurately implying that efforts to "clean" and structure data, which often involve substantial amounts of effort and overhead, are unnecessary. The reality is that prior to systematic analytics, unstructured data must also be structured and prepared in some fashion, albeit at the time of read rather than write (Chen and Zhang 2014; Ekbia et al. 2015; Liu et al. 2016; Loshin 2013). In this sense, there is no "free lunch" concerning the need to invest effort in overlaying semantic structures on-top-of raw data. Structuring data representations is a uniquely human and epistemological activity, drawing upon deduction and associated theoretical concepts concerning "how the world works and why."

There have been advances in automated methods for algorithmically analyzing unstructured data, for instance, via deep learning, a machine learning technique utilizing nested neural networks, but most of the promising results have involved data types with inherent complex latent structure, such as images, video, and audio. There are examples of the implementation of deep learning on structured datasets that demonstrate this approach in a cybersecurity context (Berman et al. 2019; Tuor et al. 2017).

However, outside these special cases, there are few reliable or credible indications that automated algorithmic methods such as deep learning can outperform, like-to-like, in analyzing completely unstructured data, versus data which has gone through human-moderated curation imposing structure, quality, and context. The central assertion is that there is no reliable automated mechanism to infer semantic structure from any and all datasets, and naïve analytical algorithms lack sufficient semantic context to extract insights from unstructured datasets (Domingos 2012).

Overcoming the unfounded (and irresponsible) implication that all types of unstructured data can be effectively analyzed by automated algorithms, practitioners are faced with the need to specify schemas, structured formats for the storage of and/or reading of primary data, whether on write, in the case of relational database storage, or on read, in the case of extracting data from cloud or Hadoop-based big data repositories.

Significant to CSDS and efforts to reify approaches to CSDS data management, schemas can provide crucial context by imposing and encoding conceptual assertions regarding the nature of digital phenomena and the particular technical domains under examination. A typical CSDS schema imbeds structural understandings of engineered IT systems and networks upon which digital traffic flows, and further is grounded by contextual understandings regarding patterns of connection, exchange, access, and trust in and amongst network carriage, through to and amongst users and devices.

To the degree that social and behavioral factors underly digital phenomena and to the degree that technical complexity itself increasingly imposes an obfuscating veil over linear and deterministic definitions of phenomena, it is appropriate to frame schemas as theoretical semantic propositions concerning digital phenomena. The associated danger is that data scientists ignore the implicit assumptions embedded in this encoding and decoding process. Essentially, the schema-based representation of structure in a dataset is a semantic, and thus knowledge-based, assertion regarding the nature of and relationship between key entities in a particular domain.

Schemas can vary in type from very highly structured and specified (i.e., fifth normal form relational structures) to more loosely specified (e.g., key-indexed objects, tabular datasets, or arrays) or ad hoc representations (e.g., free text blobs with a flexible parser overlaid to extract an on-demand JSON representation). Semi-structured and alternately structured data is increasingly facilitated through NOSQL (not-only-SQL) storage and retrieval mechanisms, such as graph, triple-store, key-value, columnar, document, or object-based databases. These are alternatives to the rigor of relational databases and often address special purposes or fulfill particular operational needs such as speed of retrieval, high-frequency input/output rates, or rich semantic encoding.

The special topic of semantic engineering and semantic frameworks was raised and discussed previously in literature analysis as associated with a number of US government-sponsored efforts to rationalize and standardize aggregated schemas for encoding and representing cybersecurity knowledge. Two approaches to systematizing schema-based knowledge representations were raised: (1) taxonomies, tree-based hierarchical semantic descriptions of a domain, and (2) ontologies, formally encoded specifications of conceptual relationships (subject-verb-object tuples or triples, such as "IP address-specifies-network device" in the cyber context).

Ontologies are of special interest to CSDS as they specify knowledge-based representations of domains that can be used to support experimental efforts, for instance, to guide efforts to test statistical aspects of relationships or to validate hypotheses. Ontologies also offer flexibility, as an ontology can be used to extract taxonomic representations on demand. Finally, ontologies are promising in-of-themselves in terms of offering a mechanism towards facilitating systems interoperability, automation, and machine-inference. Ontologies raise opportunities for semi-automating select aspects of rote scientific inquiry such as hypothesis framing and testing (albeit requiring human stewardship to curate and present evidence, frame underlying conceptual models, specify semantic connections, and interpret the causal and theoretical implications of results, etc.).

4.3.4 Exploratory Data Analysis (EDA)

In the SAS course "Feature Engineering and Data Preparation for Analytics" by J. Laramore, several focused challenges associated with data preparation in the context of analytics are raised:

- *Massive datasets*: the reduced cost of data storage has vastly outstripped the cost of rigorous analytics
- *Transaction and event data*: transformation, summarization, and contextualization are required to situate discrete events in models
- *Non-numeric data:* categorical data can pose special challenges in modeling
- *Outliers and exceptions:* exceptional, extreme, and missing values
- *Stationarity:* models may degrade as conditions change from the time a model was built

Laramore (2017)

Addressing these challenges requires staged efforts to explore, diagnose, and frame potential treatments to surfaced challenges. This suggests a structured approach to data preparation.

John Tukey, a luminary in the field of statistics and data analysis, framed exploratory data analysis (EDA) as the process of investigating and treating data in order to improve analysis (Tukey 1962, 1977). His prescient advocacy for data analysis and statistical computing has served as a core foundation for what has developed into the analytics movement and data science today (Donoho 2015).

The scope of EDA encompasses a staged process of data exploration and discovery intended to facilitate conceptual reframing, new efforts at data collection, novel combinations of data, broader experimental approaches, and/or alternate hypotheses. Methods include visualization, outlier analysis, multivariate analysis, causal analysis, econometrics (linear regression-based methods), and nonparametric techniques.

The underlying assumption of the exploratory approach in EDA is that the more one knows about the data, and, in turn, the measurement and data-generating (DGA) processes, the more effectively data can be used to develop, test, and refine theoretical propositions (Hartwig and Dearing 1979). There are a range of multivariate analysis techniques within the scope of EDA which promote deeper understandings of data, provide insights into exception handling, frame and explain outliers, isolate anomalies, and prepare data for the implementation of particular analytical techniques.

Through applying EDA, one develops a deeper understanding of the domain and phenomenon under scrutiny. This includes dynamics related to individual variables and amongst variables. In a CSDS context, this means gaining insights, for instance, into probability distributions for key variables, identifying outliers and determining how to treat them, identifying correlations amongst key variables, identifying potential composite measures and transformations, and determining variable significance in multivariate datasets.

Undertaking EDA prior to applying analytics to a dataset allows for opportunities to frame and treat the data in order to improve analytical procedures and analytical results (Verma et al. 2015). This supports the selection and calibration of algorithms for optimal efficacy in that one embeds conceptual understandings in datasets during the data preparation phase, leading to greater theoretical integrity and practical efficacy based on an understanding of the core phenomenon under scrutiny.

Two applied examples of central interest to CDSD of how EDA associated approaches can improve data preparation and engineering include the handling of time-series and big data. Concerning temporal factors in cybersecurity phenomena, achieving accurate real-time prediction of incidents is a central goal of continuous monitoring. Ideally, cybersecurity professionals would like to be alerted concerning adversarial events in progress, as the events are happening. However, confusions can occur regarding how to develop real-time prediction models. Legacy cybersecurity approaches to real-time detection have utilized rule- and signature-based approaches, isolated indicators that have broadly led to an unmanageable surfeit of alerts in security operations environments.

Refined real-time models require development, testing, and validation based upon retrospective time-series or longitudinal data. That is, isolated indicators in real-time, cross-sectional data samples are typically insufficient to reduce and focus alert efficacy, to reduce false positives. In order to refine alerts in terms of relevancy, the researcher ideally examines longitudinal data, a sequence of linked events over time which lead up to incidents.

Ideally, this includes forensic data capturing actual or invoked (i.e., red teaming or penetration (PEN) testing-based) recorded incidents as they occurred through a process—a series of discrete steps over time. This requires the analysis of time-series data, ideally contextualized in order to represent a structured process as a chain of events. By mapping a series of linked events over time, an analyst is able to trace the hypothetical causal order and sequence of events. From here, key incident-related multivariate indicators can be identified as staged signals occurring in some combination or order. Utilizing combined contextual signals to frame alerts allows for much greater refinement and differentiation than those provided by isolated point indicators.

The ability to track processes in digital environments is complicated by the computational overhead associated with processing, storing, and analyzing immense volumes of events in motion. There is a balance between goals in time-series process tracking and operational capacity to track and store the events efficiently. This necessitates approaches towards abstraction: compressing and summarizing sets of variables associated with events and behavior such that they contain traces of causal signals, yet do not overwhelm the capacity to store and process data. A full-scale

Fig. 4.9 Theoretical
optimality in temporal
event abstraction

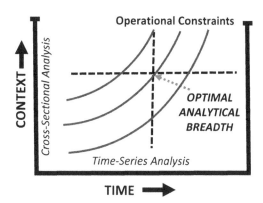

map of the world is unwieldy, maps being useful to the degree they are portable abstractions. This frames a theoretical balance between richness in storing and tracking events over time, the breadth of data summarized, and computational capacity, as represented in Fig. 4.9.

Methods associated with multivariate data analysis, particularly econometric and multivariate diagnostic validation techniques, can help identify features and mechanisms for optimizing event-based tracking. Such mechanisms ideally help to achieve a balance between computational efficiency and operational efficacy.

Temporal data analysis ties to the focused domain of process mining (van der Aalst 2016), as supported by focused techniques such as probabilistic causal analysis and graph analytics (Aggarwal and Wang 2010). Process mining and supporting techniques drive efforts to frame, develop, and test real-time alerting models. The application of process mining techniques often leads to the extrapolation of entirely new measures, such as graph-derived metrics (i.e., centrality, density, modularity, reach). Examples of graph-related security monitoring implementations in particular are becoming more popular in CSDS related research (Akoglu et al. 2015; Fourer 2010; Harley et al. 2018). An in-depth discussion of such methods is out-of-scope, apart from emphasizing that the underlying analytical methods provide new insights into temporal causality, leading to improved algorithmic detection mechanisms.

An investment in time-series process analysis, although involving nontrivial overhead, has outsized benefits to the efficacy of subsequent machine learning and algorithmic models. Properly investigating retrospective patterns of behavior frames approaches to summarize events over time, which can ground, inform, and refine the ability to detect anomalies with a resulting reduction in false-positive rates. Such efforts ideally leverage known examples of incidents to refine and guide representations.

The second context whereby staged econometric approaches are efficacious is directly related to the first, namely, inherent challenges presented by big data. In terms of framing, the term big data is largely an umbrella concept rather than a discrete technical specification, which can quickly lead to confusion. Big data does not refer to a singular aspect of data as datasets may be "big" in a range of respects. In literature, practitioners have proposed a series of V-titled designations associated with big data, which in various representations characterize "bigness" as:

1. *Volume:* the amount of data
2. *Variety:* the range of data types
3. *Velocity:* speed (Gandomi and Haider 2014; Kelleher and Tierney 2018; Mahmood and Afzal 2013)
4. *Veracity*: quality and trust in data (Chen and Zhang 2014; Jin et al. 2015; Morabito 2015; Voulgaris 2017)
5. *Variability:* the degree to which context shifts
6. *Visualization:* ability to summarize visually
7. *Value:* the ability to evidence value (McNulty 2014)

Methodologically, in the context of data processing, all such characteristics bear relevance. However, the most immediately relevant to CSDS practitioners relate to the primary three big data characteristics: volume, variety, and velocity. From a statistical analysis perspective, these attributes characterize a dataset as being, variously: *tall*, many instances (volume); *wide*, containing many measures or variables (variety); and *long*, many measures recorded over time (velocity). Fig 4.10 positions these three aspects of big data in context.

While the availability of big data is frequently framed as a great benefit, and indeed opens many possibilities to improve security monitoring, the presence of big data also poses challenges (Mahmood and Afzal 2013). Big data, in terms of both opportunities and challenges, is related to CSDS in three key overlapping respects:

1. The proliferation of big data has increased cyber risks as more, and more valuable data is stored and exposed.

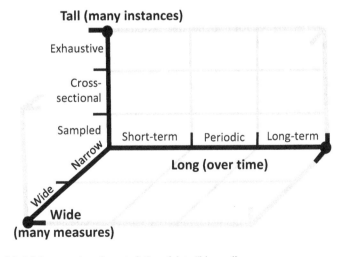

Fig. 4.10 Model for assessing characteristics of data "bigness"

2. Cybersecurity data sources, data about data, and the IT infrastructures managing data are typically "big" by nature (Curry et al. 2013).
3. Big data presents both methodological opportunities and operational challenges to data science practitioners (Liu et al. 2016).

Big data challenges can be categorized as being associated with data, process, and management factors (Sivarajah et al. 2017; Zicari 2014). Data quality related to big data is a key concern, especially as greater volumes of data are now accessible. The rate of acquisition and storage threatens the speed with which quality can be properly assessed. The rapid growth of cluster and cloud computing means that organizations can inexpensively store large volumes of unprocessed and often unstructured data with the hope of later parsing and extracting meaning.

Induction versus deduction is framed as a concern when addressed in terms of emerging big data approaches to inference. Researchers are presented with challenges related to both measurement and inference. Measurement concerns verifying that the data accurately represents the phenomenon under examination. Inference addresses an understanding of the data sample and related measures (Kreuter and Peng 2014). A key misperception is that big data somehow overcomes questions concerning the data-generating process that the datasets presented are all-encompassing and address all concerns related to data origin, composition, and robustness.

Data science often risks attaching itself to a misleading assertion associated with big data: an epistemological claim that simply collecting enough data will naturally result in scientific insights. This amounts to a declamation of the primacy of theoretically exhaustive induction, whereby if enough naïve measurements are recorded concerning some phenomenon, insights will spring forth from the raw data. Data science risks positioning machine learning as a magical mechanism whereby computers will reveal the hidden nature of any phenomenon. There have been claims that correlation with a sufficiently large enough dataset itself precludes the need to apply methods associated with classical scientific inquiry (Hey 2010).

However, outside a few select demonstrations involving substantial human curation, training, and guidance, there is no credible evidence that unaided computer algorithms combined with a large set of data can generate and test scientific theory independently. The claim of pure inductive correlation from big data has led to embarrassing missteps in practice, as cited in "The Parable of Google Flu Trends" (Lazer et al. 2014). In "A Few Useful Things to Know About Machine Learning," Domingos raises and addresses a number of the methodological missteps that can occur and advocates treatments (2012).

As raised by McNeely and Hahm regarding predictive machine learning utilizing big data, "correlations take precedence and may lead to potentially dangerous decisions, especially as data are not neutral but gain meaning only in context. They require recognition of the subjective and context-dependent nature of knowledge... the socio-technical limitations of big data rest on questions of context and meaning" (2014). This notion is backed by an appreciation of general consensus in the philosophy of science, which holds that pure induction alone is inadequate to achieving scientific proof.

A fundamental critique leveled against relying on induction is that there is always some degree of uncertainty concerning the next case. This gives rise to Karl Popper's example of the black swan: European explorers assumed all swans were white until encountering the Australian black swan. During the twentieth century, the scientific positivist movement was critiqued on this basis, with Popper declaiming the poverty of induction and framing falsifiability, rather than proof, to be the central demarcation for scientific integrity in hypothetical-theoretical propositions (Popper 1963, 1992). Outside largely sidelined positivist holdouts, deductive theory, guided by extrapolated context and meaning, is generally regarded as being an inherent foundation for building and testing explanatory scientific models.

Large datasets typically by nature contain patterns that are random artifacts of the dataset itself, that is, the result of chance associations. The challenge is to identify chance or spurious patterns and to disassociate them from patterns with replicable, predictive power. This relates to the problem of generalizability versus accuracy. A generalizable model applies to a variety of cases and typically has demonstrable predictive power. Accuracy identifies highly similar cases but is susceptible to overfitting, a condition wherein a predictive model is uniquely tied to the dataset used to derive the predictive model.

Data scientists recognize that effective models must justify and situate themselves between overgeneralization and overfitting with reference to the underlying phenomenon under scrutiny (Domingos 2012). Analyzing big data thus requires focused attention to measurement, framing, and statistical techniques. The explanatory validation of underlying phenomenon leading to theoretical constructs cannot be abandoned. Inferential statistical methods, in particular, are important in substantiating descriptive models (Grimmer 2015).

Statistician T. Hastie, along with co-authors, has written a series of well-regarded books on the application of inferential statistics, including methodological guidance on addressing various combinations of data "bigness" (Efron and Hastie 2016; Hastie et al. 2009; James et al. 2013). Hastie and co-authors cover methods for regularizing data, selecting algorithms, and assessing and interpreting models. Issues of particular interest to big data are addressed, for instance, high-dimensionality and preventing model overfitting. Hastie specifically addresses special considerations associated with tall and wide data in lecture notes (Hastie 2016). These themes are expanded upon in works such as "Feature Selection for High-Dimensional Data" (Bolón-Canedo et al. 2015).

Challenges addressed by inferential and multivariate statistics relate to the aforementioned issue of achieving a balance between representation and completeness, the role of sampling, ascertaining significance amongst multivariate datasets, and addressing the problem of spurious correlations that emerge naturally from tall and wide data. This last topic should be of particular concern to CSDS practitioners in regard to the misguided practice of utilizing anomaly detection models to train predictive machine learning when labeled incidents are lacking.

It is methodologically suspect to codify unvalidated anomalies detected in big datasets into predictive models without a primary effort to substantiate that the observed anomaly is an actual security incident (a "true-positive"). Operationalizing anomaly detection in predictive models should be accompanied by efforts at root cause analysis and statistical substantiation, ideally involving expert review and, if possible, reference to known examples of logged incidents. It is also advocated that anomaly-associated data be thoroughly analyzed through a combination of EDA and feature engineering prior to modeling.

4.3.5 Feature Engineering

The rapid growth in the popularity of machine learning has seen a corresponding growth in the importance of feature engineering. Feature engineering is the process of applying treatments to datasets to frame measures that optimize the performance of predictive models. Feature engineering and associated treatments can be considered to involve knowledge discovery as it provides the analyst deeper insights into the phenomenon under examination through insights gained from data processing (Nelson 2018; Voulgaris 2017). There is thus an inherent principle of *design* inherent in feature engineering during which staged knowledge of the underlying phenomenon is tested, validated, and embedded via iterative decisions that become encoded in the structured representation of the dataset (Domingos 2012).

The process of feature engineering is an increasingly integral aspect of data science initiatives, especially to the degree that raw data at hand is not initially suitable for algorithmic treatment. Represented as a multi-staged process, various steps cited by practitioners include *data cleansing*, sometimes known as wrangling or munging, which includes addressing quality (e.g., outliers, missing data or observations, etc.); *feature selection*, identifying an impactful sub-set of variables; *feature extraction*, condensing representative meta-measures, often machine-facilitated; *feature creation*, extrapolating new measures and representations; and *feature calibration*, refinement of measures resulting from diagnostic testing.

Feature engineering, as seen in an idealized representation per Fig. 4.11, has a roughly linear end-to-end flow. However, as observed by the recursions indicated in the figure, in practice, there are often iterative retrenchments, whereby a previous stage is revisited, leading to modified representations of the dataset as it is revisited

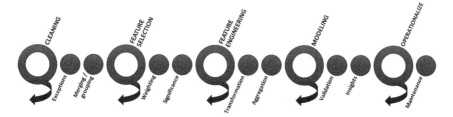

Fig. 4.11 Idealized feature engineering process

and reprocessed. In the context of quality assurance, this observes that data processing involves circular checks to assure completeness and rigor prior to presentation to an analytical algorithm.

Feature engineering typically involves the application of statistical diagnostic and validation methods, particularly econometric, linear regression-associated techniques. As an example, statistical tests will often be applied to determine variable significance in datasets intended for a predictive model, informing feature selection and refinement (Guyon and Elisseeff 2003). Diagnostic statistical procedures thus drive the specification, individuation, combination, and/or abstraction of key variables in a dataset. The process embeds knowledge gained from diagnostic and multivariate treatments into the dataset, improving representation to the benefit of subsequent analytical procedures and models.

A knowledge of the predictive analytical target method or methods which will be applied and assessed should be closely considered when pursuing feature engineering. For example, linear regression-associated models (e.g., multivariate regression) requires attention to and validation of the assumptions inherent to the method. Predictor variables need to be substantiated as formally independent and noncollinear (predictors should not demonstrate a strong correlation between one another). Target variables for prediction (used for model training) need to display constant variance (homoscedasticity) and nonrandomness (weak exogeneity). Linearity should be evidenced between the predictors and target variable.

Violations of linear assumptions compromise performance or invalidate linear approaches. However, other linear approaches have looser assumptions. For example, logistic regression does not assume normally distributed targets as they are categorical by type. This otherwise emphasizes that a clear understanding of the assumptions of and data required by the intended algorithmic treatment is essential to rigor in feature engineering.

A number of machine learning models incorporate linear aspects and approaches, albeit in many cases providing for greater adaptability and flexibility in boundary framing (e.g., boundaries may adapt to and "bend around" dynamic predictors). The larger advocacy is that attempts to understand the parametric nature of predictor and target variables should not be entirely abandoned even when applying inherently nonlinear algorithms.

Even as nonlinear machine learning techniques are increasingly popular, often to the express degree that they reduce the overhead associated with feature engineering, it may be the case that checking and improving representations to generate greater linear or parametric fidelity may also improve predictive efficacy. In the least, parametric exploration improves understanding of the fundamental underlying phenomenon. For the analyst, there is a danger of using claims of nonlinear efficacy, implementing "fancy" or fashionable techniques such as deep learning, to abandon any attempts at data exploration or efforts towards improving coherence, focus, and clarity in the underlying dataset.

Treatments, such as binning variables or applying logarithmic or exponential transformations, may be instrumental in turning a poorly performing raw dataset into a highly efficacious sample. This may also help reduce the risk of overfitting to random patterns in large datasets or of losing coherency due to overgeneralization. Again, "The Parable of Google Flu" is a discrete case which is highly instructive in this regard (Lazer et al. 2014).

Attention to parametric aspects of data and seeking methods to improve signal coherency is especially important in treating cybersecurity data. Cybersecurity data, network traffic data in particular, is typically sparse though voluminous by nature. Summarized network traffic variables (e.g., number of connections made, total bytes sent, total ports used) typically evidence degree, logarithmic, or exponential distributions, reflecting power law, small world, and Pareto principle dynamics broadly observable in many types of nonrandom scale-free networks.

In summary, a potentially overlooked implication of feature engineering concerns the epistemological implications of exploring and treating data. Through feature engineering, the analyst is confronted by the inherently abstract, representative nature of data. The analyst embeds theoretical assumptions as well as potential errors in the process of selecting, gathering, recording, and integrating data. Outside this comes a consciousness that data itself is the result of a process of measurement and that the measurement mechanisms themselves may embed and possibly hide underlying technical, methodological, or epistemological errors or misunderstandings.

Statisticians have a central notion of the data-generating process (DGP) and keep the possibility of measurement, sampling, and other errors in mind when treating and analyzing data (Fig. 4.12). In this context, the outcome of feature engineering and model testing may well result in reframing the fundamental inquiry, leading to new approaches towards measurement or data collection, either at a technical level or in efforts to collect entirely new sources of data.

For the CSDS practitioner, this distinction is manifested as the need to question the relationship between a dataset, as an abstract representation, and the underlying phenomenon under observation. For instance, a representation of a list of servers and workstations from a configuration management database (CMDB) is data representing organizational assets, otherwise lacking specific security-related context. As is often the case with CMDB data, it is likely out-of-date, and the labels implied may be tied to functional purposes distinct from security purposes (e.g., machines

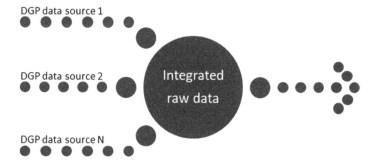

Fig. 4.12 Integrating data-generating processes

may be labeled and tracked with an asset tag for accounting and maintenance purposes, rather than to identify a functional designation relevant in a security context). Carefully considering data-as-an-artifact in terms of origination, production, context, and quality is essential to preventing potential misalignments with the intended goals and methods of analytical modeling.

4.3.6   Analytics Process Models

In aggregate, data processing has several distinct but overlapping frames of reference. The analyst must balance the relative importance of these frames in terms of the intended goals and outcomes of the processing exercise, which may be variously associated with:

- *Treating:* preparing data for mechanical storage and retrieval to meet operational goals (as associated with technical capabilities)
- *Describing:* preparing data for presentation or visualization, supporting contextual representation that may reveal hidden insights to experts and analysts (as associated with statistical testing)
- *Diagnosing:* explaining and validating understandings of phenomena (as associated with statistical testing)
- *Predicting:* presenting data to an algorithmic model (i.e., supervised machine learning) in order to optimize pragmatic predictive goals while balancing between generalizable integrity and overfitting (as associated with feature engineering)
- *Operationalizing:* driving and fulfilling a broader knowledge-driven process, such as an investigation, which provides context in terms of the origins, interrelationships, meaning, and goals associated with data (as associated with organizational understanding)
- *Scientific context:* providing insights which inform testable theoretical hypotheses leading to follow-on experimentation, categorical differentiation, falsification, model iteration, derived theoretical propositions, or reframed assumptions which suggest new sources of data (as associated with hypothetical and theoretical understanding)

These goals and modalities are not mutually exclusive, and ideally practitioners mix approaches and methods. For instance, applying descriptive and diagnostic approaches to improve representations of data may centrally improve predictive efficacy. Integrating EDA statistical and feature engineering procedures improves and assures data in preparation for analytical implementation and should be considered a standard best practice. Further, insights gained from the implementation of machine learning models may lead iteratively backwards, implying revisions to and modification of the data (Verma and Marchette 2020).

There are many opportunities whereby combined predictive and explanatory data management approaches can be self-reinforcing and improve rigor, both in theory building and in refining predictive models. In "To Explain or Predict?", Shmueli details a range of ways in which predictive approaches can enrich and inform the scientific process, including theory building and substantiation. In Shmueli's enumeration, efforts to build and refine predictive models are helpful to explanation as:

1. New causal mechanisms and explanations may be revealed
2. New measures are potentially surfaced
3. Hidden patterns may be uncovered
4. Potential disconnects between theory and practice can be assessed
5. Metrics for comparing competing theories are framed
6. Benchmarks for predictive accuracy are established

Shmueli (2010)

Model development, implementation, testing, and validation are part of a broad, attenuated data-processing value chain that embeds knowledge and understanding in a dataset. Machine learning model deployment, as such, is not an isolated activity: it has inherent context to a wider data management and preparation process. Developing and deploying models without examining and treating the data risks operationalizing errors, ranging from overfitting randomness and spurious correlations, to operationalizing erroneous assumptions, to analyzing the wrong data or pursuing an improperly framed analytical question.

Framing data processing and management as a part of an extended process incorporating model development and implementation understandably introduces complexity. It is understandable that potential confusions crop-up in operational environments when it is proposed that algorithmic model logic is inherently part of a broad chain of process dependencies, and that model development may require iterative retrenchments across the full range of data treatments. This, in particular, creates challenges for staging data engineering development efforts to operationalize data processing pipes. If revelations gained during model development alter the premises and procedures surrounding data handling, data engineering implementations require subsequent modification. In larger operational settings, a facile stance may understandably emerge whereby data scientists simply "make do" with the data that is handed to them through distant and disconnected data engineering processes.

To the degree that interdependencies in data management vis-à-vis model development requires integrated responses, process-based approaches to data preparation in the context of machine learning model development, and subsequent model selection, testing, implementation, and validation, have emerged. The explosion of interest in applying machine learning algorithms, particularly predictive machine learning, has led to the development and specification of *analytics process models* to establish standard organizational guidelines.

Of benefit to CSDS is that there is structured guidance on the topic of integrated machine learning model development and management issuing from the analytics domain. The popularization of analytics and data science has led to the development and socialization of a range of codified process-based best practices to support and facilitate integrated data preparation and model development. Data science practitioners are encouraged to apply a process-based approach to guide analytics inquiries. Distinct from linear software development systems development life cycle (SDLC) approaches, typical of technical development initiatives, process-based approaches to data preparation are framed as involving iterative learning, and progress may lead back to earlier process stages.

Several analytics process models have been proposed over the last two decades, some with more or less focus on particular analytics methods. Data scientists typically follow some general representation of these processes to treat and extrapolate insights from data in preparation for modeling. The most commonly cited analytics process models are summarized in Table 4.4.

Table 4.4 Summary of analytics process models

Framework	Focus	Summary	Reference(s)
ASUM-DM: Analytics Solutions Unified Method for Data Mining/Predictive Analytics	Data mining	A refined and extended version of CRISP-DM framed by IBM which specifies detailed project management approaches	Haffar (2015)
CRISP-DM: CRoss Industry Standard Process for Data Mining	Data mining	A popular and long-standing methodology to guide the implementation of data mining projects (framed by ESPRIT as an EU EC initiative)	Chapman et al. (2000), Shearer (2000), Wirth and Hipp (2000)
FMDS: Foundational Methodology for Data Science	Data science	A multi-stage methodology to guide data science initiatives framed by IBM	Rollins (2015)
INFORMS Data Analytics Process	Data analytics	INFORMS professional society representation of key steps in the analytics process	INFORMS (2013)
KDD: Knowledge Discovery in Databases	Data mining	A methodology concerned with extracting models and patterns of interest from large databases	Fayyad et al. (1996), Fayyad and Stolorz (1997)
PPDAC: Problem, Plan, Data, Analysis, and Conclusions	Statistics	A cyclical problem-solving framework for applying the scientific method to an analytical or research question	MacKay and Oldford (2000), Spiegelhalter (2019)
SAS Analytics Life Cycle	Data analytics	An integrated, iterative process from SAS Institute to guide the development and implementation of data analytics efforts	SAS Institute (2018)
SEMMA: Sample, Explore, Modify, Model, and Assess	Data mining	A sequential set of steps to guide the implementation of data mining initiatives framed by SAS Institute	Azevedo and Santos (2008)
TDSP: Team Data Science Process	Data science	An agile, iterative methodology and set of tools to facilitate implementation of data science initiatives from a consortium including Microsoft	Microsoft (2017)

Key differences amongst the analytics process models are tied to intended outcomes as related variously to data mining, data analytics, machine learning, and/or data science. Several researchers have compared and critiqued various of the frameworks, deriving best practices and approaches from findings (Azevedo and Santos 2008; Foroughi and Luksch 2018; Rohanizadeh and Moghadam 2009). While there are many versions and interpretations, generally the processes, in aggregate, seek to frame a systematic approach to the analysis of data in an attempt to establish rigor and to assure quality in analytical results. Some version of the following sequential steps are typically reflected:

- Business problem framing
- Analytics problem framing
- Data exploration, processing and engineering
- Model development and selection
- Model validation and deployment
- Assessment of and interpretation of results
- Operationalization and maintenance
- Iterative inquiry potentially framing new problems and/or data

The SAS Analytics Life Cycle model is a well-socialized exemplar, framing data analytics as an iterative process that integrates data preparation with analytical modeling and deployment, per Fig. 4.13 (SAS Institute 2019b).

A key point of advocacy resident across the various models is that analytics requires staged verification and validation checks at multiple points across the iterative life cycle. Data exploration processes are aimed at surfacing patterns and verifying assumptions, while statistical diagnostic procedures are intended to validate treatments. When assumptions are proved erroneous or treatments invalidated, the results suggest revisions, some of which may reframe earlier stages in the life cycle. This otherwise serves to bolster the observation that data preparation for analytics typically involves cyclical processes, particularly in the junction between framing and treating data in preparation for modeling, and subsequently in the process of composing, testing, and refining analytical models.

Analytics process models generally have a bias towards supporting machine learning model development and deployment specifically. However, to the degree that organizational implementations of a process approach become monomaniacally focused on predictive machine learning model deployment, there are risks of ignoring verification and validation efforts. Abandoning efforts to explore assumptions and validate treatments risks ignoring underlying phenomena (Sivarajah et al. 2017).

Fig. 4.13 SAS Analytics Life Cycle

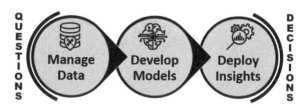

A well-documented misstep in the implementation of predictive machine learning models is to push large, unexamined, untreated datasets into a single model, for instance, a deep learning machine learning algorithm, and to fail to conduct diagnostic validation, operationalizing the raw results which issue forth (Chalvatis 2017). Such approaches risk ignoring hidden biases and sources of error, for instance, exogenous causal factors, spurious correlations resulting from transient noise, ignoring the effects of seasonality or other temporal influences, improperly treating outliers or anomalies, or failing to contextualize parametric idiosyncrasies such as logarithmic or exponential distributions (Lazer et al. 2014). In particular, operationalizing models operating off of purely correlative effects not only risks embedding misunderstandings (Lazer et al. 2014), but potentially operationalizes erroneous, and potentially unethical, decisions (O'Neil 2013, 2016; Smith 2018).

Once a rigorous analytics process model is framed and adopted in the organization, a firmer foundation for both data engineering and model operationalization is established. Having established rigor in the integrated treatment of data and analytical model development, further structured approaches to operationalize decision-making through engineering and technical development can be pursued. As an example, Fig. 4.14 positions data management and analytics model development in the larger context of infrastructure implementation and organizational decision-making (SAS Institute 2019a):

As the application of machine learning to cybersecurity increases, the need for process-driven best practices specific to CSDS will grow. The emergence of focused CSDS frameworks was discussed earlier in Sect. 2.5.5 with reference to the Carnegie Mellon SEI cyber intelligence framework (Fig. 2.20). This model positions machine-driven data analysis as a central facilitator in security assurance.

The Carnegie Mellon SEI also recently published focused guidance on the application of machine learning to cybersecurity in the report "Machine Learning in

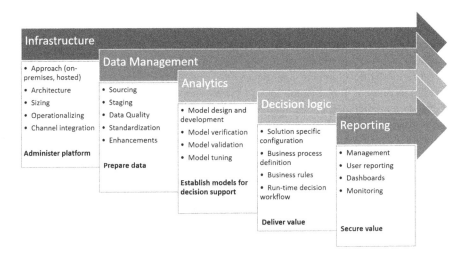

Fig. 4.14 Context for operationalizing analytics

Cybersecurity: A Guide" (Spring et al. 2019). The report advocates seven key questions to address in scoping the deployment of applied machine learning to cybersecurity challenges:

1. What is your topic of interest?
2. What information will help you address the topic of interest?
3. How do you anticipate that an ML tool will address the topic of interest?
4. How will you protect the ML system against attacks in an adversarial, cybersecurity environment?
5. How will you find and mitigate unintended outputs and effects?
6. Can you evaluate the ML tool adequately, accounting for errors?
7. What alternative tools have you considered? What are the advantages and disadvantages of each one?

Spring et al. (2019)

Such scoping questions prompt planners and practitioners to properly frame and validate CSDS machine learning approaches.

Addressing practitioner advocacy for more rigorous standards in CSDS model design and validation, a prescribed best practice is to ensure that questions such as those framed above are addressed within an analytics process (per Table 4.4). Ideally, CSDS will see the future emergence of a CSDS-specific analytics process model as consensus develops concerning domain-specific best practices.

4.3.7   Summary of Design Prescription I: Data Management

A central CSDS challenge raised by data management is that analytics model logic is distributed across a continuum of frames, from the data-generating process through to the representation of model results and subsequent interpretation. As such, careful attention needs to be paid as to how model assumptions become embedded in various phases of the continuum and how these assumptions influence and potentially bias upstream representations. Focused efforts to apply verification through diagnostic-driven exploration and validation are essential to assuring rigorous results.

The process of preparing and treating data, including exploratory data analysis, although laborious and time-consuming, is also a learning process. Domain knowledge and context are gained as the analyst prepares data. Decisions made concerning the relative importance of particular data elements, their transformation, and presentation of the final multivariate dataset for algorithmic modeling embeds crucial context that centrally influences analytical results. Decisions made during data preparation, which codify the dataset selected for algorithmic treatment, should thus be viewed as an integral aspect of the model itself. Equally, the analyst should always be open to returning to problem framing, data collection, and fundamental aspects of diagnostics and feature engineering during the modeling phase.

Table 4.5 Summary of data management design prescriptions

Framed design prescriptions: data management
• Data management is best framed as an integrated process with organizational, methodological, and technical aspects   – Data assurance is central to the organizational integrity of analytical inquiry results   – Data management context spans problem framing, data collection, data exploration, feature engineering, model development/validation, operationalization, and monitoring/maintenance   • Data management for analytics is an iterative development process with overlapping steps   – Analytics process models are useful to assuring rigor in organizational efforts   – A robust analytics process ensures rigor in conjoining data and models, establishing a foundation for operationalization, and grounding organizational decision-making   – The organizational implementation of an analytics process model ideally embeds diagnostic data exploration and feature engineering, themselves mutually reinforcing   – Semantic and theoretical assumptions concerning the phenomenon and domain under examination are embedded throughout the data preparation process   – Data preparation is an iterative and recursive exploratory development process—staged results may suggest revisions up and down the extended process, from collecting new sources of data and reframing the inquiry to modifying model parameters   • Exploratory data analysis (EDA) and feature engineering approaches should be applied in combination to inform and validate efforts to process data in preparation for analytical treatments   – Overfocus on predictive machine learning to the dearth of data exploration and diagnostic validation risks operationalizing erroneous assumptions and decisions   – Training predictive machine learning with naïve anomalies is methodologically weak   – Development of real-time detection models ideally involves statistical analysis of retrospective multivariate time-series data, analysis of confirmed events, process analytics, and reference to cybersecurity process frameworks (e.g., MITRE ATT&CK)   – Big data requires special considerations and treatments, particularly regarding potential spurious correlations (correlation is not causation)   – Feature engineering is a multi-staged design and discovery process aimed at improving datasets in preparation for analytical treatments   – Common feature engineering steps, which often overlap, include data cleansing, feature selection, feature extraction, feature creation, and feature calibration

Table 4.5 provides a concluding summary of key design propositions advocated for data management in CSDS practice.

4.4 Design Prescription II: Scientific Processes

4.4.1 Problem-Solving Requirements

As framed previously, CSDS professionalization is inhibited to the degree that unaddressed challenges are resident in the parent domains. While a great deal of practical research has been conducted and is underway in the field of cybersecurity, there is a lacuna concerning scientific foundations, particularly regarding the implementation of scientific methods to the analysis of cybersecurity phenomena resulting in general theory (Meushaw and Landwehr 2012).

Similarly, while the data science domain proposes a broad array of treatments, the field struggles with confusion resulting from methodological diffusion. As data science brings forward a voluminous toolbox of methods, there is a need to determine which tools are most appropriate to which tasks. As raised in the previous exploration of data management treatments in CSDS, pursuing methodological prescriptions without anchoring in scientific rigor is a fraught path, potentially leading to erroneous insights and decisions.

Of the three gaps surfaced in CSDS practitioner interviews, the need for scientific processes was the broadest. Associated themes overlap with the two other gaps identified, data management and cross-domain collaboration. To the degree challenges associated with scientific rigor are confronted in CSDS practice, associated goals related to data management and cross-domain collaboration are clarified. Scientific grounding for processes related to validating datasets, categorizing phenomena, framing hypotheses, experimental testing, and establishing theoretical propositions serve to both orient data management and to clarify goals for collaboration.

In interview research, CSDS practitioners evoked the need for process-based scientific methodological approaches in CSDS practice (68%, $n = 34/50$). The need for stricter standards surrounding scientific rigor was also raised as related to analytics model validation (62%, $n = 31/50$). During interviews, these themes were associated with reference to an array of related challenges. Practitioners cited challenges associated with collecting experimental evidence, a lack of rigor in hypothesis framing and testing, and an overarching lack of theoretical foundations in the security domain.

A general sentiment expressed by CSDS practitioners was that the cybersecurity domain generally lacks a strong scientific basis and theoretical foundation. This view was validated in literature (Craigen 2014; Longstaff 2012; Maxion et al. 2010; Meushaw and Landwehr 2012; Pavlovic 2012; Tardiff et al. 2016). An overt engineering orientation in the cybersecurity domain often results in deference to deterministic and engineering-associated understandings of security dynamics. This in turn leads to rules-based and mechanical, rather than probabilistic and theoretical, views of security phenomena.

The result is that cybersecurity practice is heavily influenced by engineering perspectives. Theoretical propositions regarding security phenomena are often rudimentary and are validated in observation rather than through comprehensive diagnostic exploration (e.g., frequent connections to "junk" URLs are observed to be an indicator of malware infections). The bulk of security research reflects this engineering bias and results in a focus on technical implementations. It is less common for security researchers to describe and apply rigorous scientific methods, to frame formally testable hypotheses, or to socialize repeatable research (Landwehr 2012).

An underlying related challenge is that both cybersecurity and CSDS practitioners frequently work in demanding high-pressure, time-driven environments.

Theoretical rigor involves formally framing and substantiating hypotheses, requiring carefully staged experimentation, cross-testing, diagnostics, and efforts to share repeatable findings. The relative priority for staging research efforts in demanding operational domains, often under the pall of a siege mentality, is typically low.

Prior to framing and advocating factors and approaches to operationalize scientific processes in CSDS practice, it is appropriate to motivate the topic by emphasizing the need for scientific perspectives on security. Pavlovic, in the article "On Bugs and Elephants: Mining for Science of Security," proffers that:

> Conjoining cyber, physical, and social spaces by networks gives rise to new security problems that combine computational, physical, and social aspects. They cross the boundaries of the disciplines where security was studied before, and require new modeling tools, and a new, united framework, with a solid scientific foundation, and empiric methods to deal with the natural and social processes on which security now depends. In many respects, a scientific foundation for the various approaches to security would have been beneficial even before; but now it became necessary. (Pavlovic 2012)

Accepting the premise that there is a pronounced need for scientific approaches to security, related questions concerning scope and desired outcomes are appropriate. Meushaw and Landwehr (2012) cite a set of grounding questions framed at a workshop on the "Science of Security" sponsored by the National Science Foundation (NSF):

- Is a science of cybersecurity possible?
- What might a science of cybersecurity look like?
- How can we reason about problems that seem impossibly hard?
- Is it possible to have scientific security metrics?
- What lessons can we learn from other disciplines?

To add to this enumeration with specific reference to CSDS, a central question to pose is: *if CSDS constitutes a scientific discipline, what type of science is it?* This provocative, if oversimplistic, question does force a perspective demanding clarification. Namely, what is the systematic mechanism through which CSDS diagnoses security challenges *as scientific questions* and prescribes data science treatments *as scientific methods*?

The breadth of the security domain, as has been reinforced several times in this inquiry, presents an array of challenges that frame multitudinous scientific questions. Equally, the breadth of methods on offer through the field of data science supports a potentially overwhelming range of methodological treatments. As a hybridization, CSDS suffers from a surfeit of choice.

The challenge for framing a scientific approach to CSDS is to systematically navigate the junction between the two parent domains. CSDS, as a solution provider, occupies a mediating status as a gap-diagnosis mechanism. This can be posed in formal logical form as an equation which represents the uniting of focused cyber-security challenges with complementary data science treatments, per Fig. 4.15.

$$C = \bigcup (O_1, O_2, O_3, \ldots O_n)$$

$$D = \bigcup (M_1, M_2, M_3, \ldots M_n)$$

$$P = \bigcap (C, D) = \begin{bmatrix} O_1, M_1 & \cdots & O_x, M_n \\ \vdots & \ddots & \vdots \\ O_n, M_x & \cdots & O_n, M_n \end{bmatrix}$$

Where:

$C = $ *cybersecurity gap* (*challenge*)
$O_n = $ *cybersecurity objectives*
$D = $ *data science treatment* (*best practice*)
$M_n = $ *data science methods*
$P = $ *cybersecurity data science* (*CSDS*) *prescription*

Such that diagnostic gap analysis takes the form of:
$$P = C \cap D$$

Fig. 4.15 CSDS gap-diagnosis logical equation

In this model, CSDS prescriptions reside in efforts to work-out approaches to domain-specific problems: to explicate security problems as formal requirements and to prescribe focused data science treatments. This frames the application of a design science approach to developing problem-solving artifacts. Specified scientific treatments to address CSDS requirements can, in this respect, be framed as design artifacts.

Once the notion of a singular prescription for embedding scientific rigor in CSDS practice is dispelled, it is possible to structure and discuss a range of treatments. Science itself is not singular and admits to a range of approaches and methods. It is appropriate to speak of CSDS scientific processes in the plural, recognizing that there are many ways to introduce and pursue scientific rigor.

There are distinct approaches and methods appropriate to, variously, categorizing new phenomena, gathering and validating evidence, framing hypotheses, testing, experimentation, and building theoretical propositions (which in turn serve to guide new hypotheses and experimental forays). As CSDS proposes to apply itself to a breadth of cybersecurity phenomena in multiple capacities, it must make efforts to catalog and clarify a range of discrete scientific approaches and treatments.

Key questions related to the application of science and the scientific methods to the CSDS domain are implied:

1. What is the nature of science and the scientific method as related to the cyber-security domain?
2. How should systematic inquiry vary across and respond to the range of inter-disciplinary contexts in which cybersecurity resides?
3. How should the range of data science methodological offerings be rational-ized to address different categories of cybersecurity challenges in a struc-tured way?

Table 4.6 Problem-solving requirements: scientific processes

Key gap	Category	Practitioner themes
BP F1: Scientific processes	Scientific rigor	**BP4**: Scientific method as a process **BP8**: Model validation and transparency
	Specifying context	**CH3**: Contextual nature of normal versus anomalous behavioral phenomenon
	Establishing evidence	**CH4**: Lack of labeled incidents to focus detection **BP10**: Track and label incidents and exploits **BP19**: Human-in-the-loop reinforcement
	Extrapolation and optimization	**CH7**: Traditional rules-based methods result in too many alerts **BP6**: Vulnerability, anomaly, and decision automation to optimize operations
	Cyclic discovery	**BP15**: Distinct exploration and detection architectures **BP11**: Cyclical unsupervised and supervised machine learning
	Practice, probability, risk	**BP20**: Survey academic methods and techniques **BP17**: Deriving probabilistic and risk models

The following exploration will seek to work out pragmatic approaches to these questions by framing requirements and advocating accommodating design treatments. While there is not a singular "silver bullet" treatment to deliver rigor in CSDS practice, working out and describing a range of practical approaches provides a solid foundation. Categorizing a range of approaches offers much-needed context for practitioners and researchers seeking to rationalize, frame, and implement CSDS operational treatments.

To work out various process-based prescriptions to introduce scientific rigor to CSDS practice, it is helpful to diagnose the central gaps inhibiting CSDS scientific process development. For this purpose, themes associated with scientific processes raised in CSDS practitioner interviews are summarized in Table 4.6.

The themes surfaced in interview research raise a number of overlapping requirements. The overlapping nature of the topics suggests processes that interlink. The following extrapolation frames and advocates a set of process-based designs to address the range of requirements surfaced.

4.4.2   Operationalizing Scientific Discovery

Rapid technological change pressures security to reinvent approaches to monitoring, from a castle-and-moat paradigm to persistent and pervasive surveillance. The traditional goal of securing "hard" network perimeters and individual devices has all but faded in the context of pervasive computing, prolific interconnection, and distributed systems. This necessitates a transition from deterministic, rules-based methods to probabilistic and algorithmic models for identifying anomalous behavior.

Predictive machine learning is one such approach with great promise. However, a key aspect of predictive approaches is that they require training examples, which in the CSDS context means consumable representations of known attacks and incidents recorded in data. As raised in interviews (56%, $n = 28/50$) and validated in literature analysis, there are a pronounced lack of labeled representations of security incidents suitable for training predictive models.

The fallback typically involves attempts to surface focused anomalies through descriptive analytical techniques, which may be statistical (e.g., multivariate outlier detection) or unsupervised machine learning-based. Anomaly detection and outlier analysis techniques represent a focused set of methods of central interest to cyber-security. Anomaly detection and outliers are a well-covered topic in data science, with researchers detailing a range of sophisticated methods in literature (Aggarwal 2017; Aggarwal and Sathe 2017; Bhattacharyya and Kalita 2014; Mehrotra et al. 2018).

Anomalies are the result of efforts at statistical differentiation and can be quite sophisticated, for instance, supporting nonparametric or nested multi-stage algorithmic methods. Nonparametric algorithmic models are able to adapt dynamic boundaries, tracing nonlinear distinctions between statistically "normal" and "abnormal" multivariate measures. However, there is a natural limit to refining anomaly generation through such sophisticated processes. The methods are self-limiting to the degree that they are initially naïve—lacking explicit context concerning the nature of expected versus unexpected phenomena, as encoded in datasets, the model itself, and/or expert guidance.

Specific to the cybersecurity domain, anomaly detection models face an immense amount of noise issuing from streaming security data, as abstracted in Fig. 4.16 (Aggarwal 2017). Even sophisticated methods struggle to differentiate signal from noise given the ocean of highly variable, streaming events characteristic of the bulk of modern digital environments. Again, lacking overt human context, such systems are likely to result in the self-same result as rule and signature-based systems: alert overload.

Noise, resulting from growing underlying complexity, is increasingly common to digital environments. Security monitoring must confront this complexity through improved approaches to configure and calibrate anomaly detection mechanisms. As advocated by Bonabeau, "the more options you have to evaluate, the more data you have to weigh, and the more unprecedented the challenges you face, the less you should rely on instinct and the more on reason and analysis" (Bonabeau 2003).

Addressing complexity through reason and analysis can take a range of forms, encompassing improving data preparation processes, utilizing more sophisticated analytical methods and mechanisms, embedding greater understanding of domain context in data and models, extending and improving efforts to train discerning models, and embedding the capacity for continual learning into model implementations. Whereas initial anomaly detection models are often naïve, inculcating processes for iterative refinement can lead to progressive gain.

However, despite the rising adoption of more advanced anomaly detection models, methods, and tools, false alert overload persists as a commonly reported problem in operational security settings. Alert overload is a byproduct of an inability to reliably disassociate normal and abnormal behaviors in complex digital environments. The inability to specify context, to accurately segment malicious from

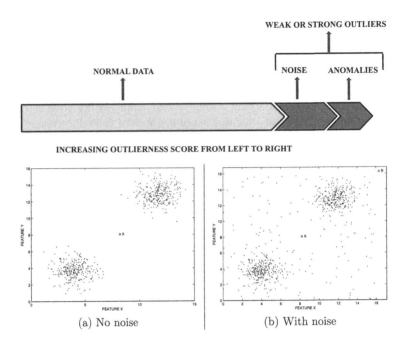

Fig. 4.16 Anomaly detection amongst noise

"acceptable-but-anomalous" behavior, indicates fundamental confusion regarding what is being tracked and why.

CSDS interviewees identified a lack of operational context to disassociate normal from anomalous behavioral phenomena as being a frequent challenge in practice (60%, $n = 30/50$). Context ideally is the result of domain knowledge, both general and localized, paired with descriptive and diagnostic methods. Context requires basic understandings of the local environment, baselines for expected user and device behavior, knowledge of key vulnerabilities and threats, and awareness of common transactional patterns.

As noted cybersecurity data scientist Joshua Neil notes concerning his work, "we mostly model what's normal and quantify deviations from normal... seeking evidence against normal" (Neil 2019). Through establishing a context for "normal," it is possible to establish "expected" statistical baselines to differentiate normal versus abnormal behavior and thus ultimately to reduce false alerts.

Achieving operational alert reduction through refined contextual understanding evokes three integrated requirements:

1. *Context structuring through data:* engineering and validating key measures which encode contextual understanding leading to demonstrable anomaly identification results

2. *Context extrapolation through models*: developing models demonstrably able to disassociate acceptable from abnormal phenomena with accuracy and precision
3. *Context learning through feedback:* context gained through staged feedback resulting in targeted anomalies which are in turn utilized to refine data and models iteratively

An integrated approach, per Fig. 4.17, connects these three approaches in a virtuous loop: categorization and specification (data), inquiry and insight (models), and experimentation and revision (operational reviews and refinement). In terms of inculcating rigor, this relies upon careful staging and validation during each phase. The first step, refining data through exploration and feature engineering, is particularly impactful in terms of efforts leading to subsequent higher-quality results (Bhattacharyya and Kalita 2014).

This process can be asserted to be loosely scientific in form, the key principle being that there is a cyclical effort to extrapolate insights, model and test treatments, and refine understandings based on staged results. The general form is of cyclical framing, experimentation, and model revision. Such a process, as with experimental approaches broadly, benefits greatly from structured investments in scoping, staging, calibration, statistical validation, etc.

If approached as a naïve process, for instance, with untreated data and untested models, initial results will generally be of poor quality. Improved results depend upon the determination to continue the exercise iteratively until staged improvements begin to emerge through adhering to the operational feedback loop. Human oversight, judgment, and statistically rigorous feedback are assumed essential in the process. Unaided automated processes, lacking deductive insights rooted in expert domain understanding, are not capable of complex inferential and symbolic conceptual reasoning.

Fig. 4.17 Integrated self-improving modeling process

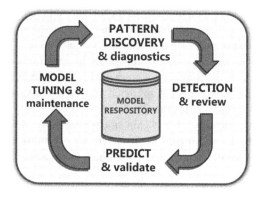

It is appropriate to consider the process of validating anomalies and improving the gain of the anomaly detection mechanism as a staged *calibration process.* Calibration in this context involves an empirical approach in which staged observations of anomalies issuing from the system are observed and examined. The results are determined to either be true-positives (correct, verified real incidents) or false positives (type 1 errors, false alarms). An inherent byproduct is the identification of true-negatives (correct, verified non-incidents) and false negatives (type 2 errors, incidents incorrectly identified as non-incidents). Tracked and quantified, the aggregation of these designations results in a confusion matrix as a benchmark for efficacy. Results are subsequently used both to refine the dataset and to refine the model (i.e., the algorithmic anomaly generation mechanism).

It is useful to describe and illustrate the high-level operation of a hypothetical "toy" self-improving anomaly generation system. Firstly, streaming security telemetry is fed into a process that periodically rolls up the behavior of particular entities (i.e., users or devices) into aggregated summary measures (e.g., byte traffic, connections made, scanning events, endpoint behavior, etc.). An anomaly generation mechanism is established, which in initial implementation might be quite simplistic (e.g., flag userids that evidence X standard deviations higher Y events than the general population per day).

To the degree that particular entities begin demonstrating statistically similar measures, they can be categorized for refined tracking (i.e., as statistical peer groups). Statistical groupings are used to refine and focus anomaly assessments. Monitoring continues, with several cycles of examination, introspection, investigation, and calibration. The core data feed, summary measures, and anomaly detection model are refined iteratively based on staged insights, per the representation in Fig. 4.18.

Together, naïve anomalies and associated baselines (summarized metrics establishing "normal" behavior) gained from systematic monitoring can be leveraged to refine anomaly detection models via contextual explanation, validation, and calibration. A key aspect of the process is that greater and greater context is embedded in the dataset and the anomaly detection model as learnings from insights and results are applied iteratively. This includes labeling and categorizing entities as appropriate to refined understandings of their inherent security relevance.

A challenge that afflicts naïve anomaly generation systems is an initial general lack of evidence and context. Namely, there is often a deficit of entities segmented into security-relevant categories in operational environments. In a typical cybersecurity environment, a great deal of complexity initially confronts the CSDS practitioner. One logical approach may be to consult with local subject-matter experts to understand the domain.

One may wish to gain an understanding of how entities are categorized native to the local domain, for instance, by obtaining configuration management database (CMDB) data to understand how entities are categorized in the local environment.

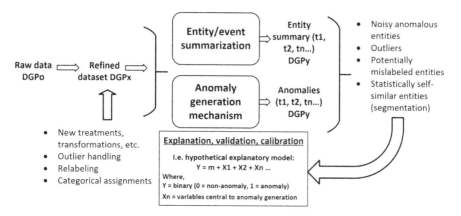

Fig. 4.18 Anomaly calibration step 1—validating and calibrating tracking

As well, expert input may contribute background and context useful to framing data, i.e., understanding the effects of network segmentation decisions and idiosyncrasies associated with the local definition of key entities.

However, experts and local resources may represent IT entities (users and devices) with reference to non-security-related criteria, for instance, in terms of organizational hierarchy (e.g., the department associated with a user or device). In discussion with CSDS practitioners in interviews, it was frequently noted that security monitoring systems are often configured to track users within organizational groupings (e.g., department, cost center, location). That is, security professionals frequently operate, often under explicit instruction, with entities categorized by organizational designations or labels (e.g., functional department, asset tracking). These designations rarely are demonstrably security-relevant. For instance, organizational structure typically has little demonstrable statistical relevance to security phenomena. The security relevance of categorical designations is theoretical and unvalidated until some type of statistical diagnostic procedure is undertaken to validate the significance of the designated labels.

Experience in the field has suggested that organizational designations typically have no statistical validity when related to security-related behavior. For instance, there is little confirmed evidence that an organizational department is strongly predictive of network behavioral patterns as represented in metrics such as external IP connections, bytes transacted, or scanning events per period of time. Quite the reverse, this practitioner has often observed that there is no statistical significance between network usage behavioral measures and organizational structural categories. Organizational categories that lack explicit security-relevant behavioral context seldom perform better than random classifiers.

As noted by Herbert Simon, "an early step towards understanding any set of phenomena is to learn what kinds of things there are in the set—to develop a taxonomy" (Simon 1996). To evidence scientific rigor in categorization, there must be a statistical basis for asserting security-focused individuation and differentiation.

Whereas techniques for behavioral segmentation are known, including staged multivariate inferential statistics and descriptive machine learning methods (e.g., unsupervised approaches), interpretability can at times pose a challenge. A related challenge is that initial attempts to segment behavioral designations confront a great deal of noise and unusual behavior on networks, some of which requires local knowledge and interpretation.

As an example, a user may have a highly unusual job role that requires that they perform activities that generate immense amounts of network traffic, at times many multiples beyond the mean of the average user. Network measures often have extreme long-tailed, exponential-logarithmic distributions. For instance, it is not uncommon when examining daily user byte traffic to see a small number of users that are 30 or 40 standard deviations from the population mean.

In operational implementation of anomaly detection systems, it is common and expected that initial anomalies identify noisy entities (e.g., unusually active devices or users) which are improperly segmented (e.g., users with an administrative or security role) or which are mislabeled (e.g., userids used by a machine or process, not a person). For instance, initial results might identify users with unusual functional roles in terms of high-volume network usage or accessing many devices, such as systems administrators or security testers. Meanwhile, measures which frame "normal" behavior and parameters are issued from the system and can be used to establish and refine statistical baselines.

As the structural substrate underlying much of the behavior being tracked involves networks, an observation can be made that the effort to rationalize entities is aligned with efforts to define and refine, or prune, highly interconnected network, or graph, models. Whereas according to the Erdős–Rényi model, outliers and exceptions can be considered a natural property of graphs (Bollobás 2001; Erdős and Rényi 1959; Gilbert 1959), there are two associated concerns. For one, we must clearly define the premises underlying the graph structure, namely, the theoretical definition of the entities and links.

For instance, in a graph of connecting userids (e.g., via network-based authentication and resulting IP-to-IP connections), it would be useful to prune a subgraph representation of only human userids which does not include machine userids, especially if our goal is to examine similarities and dissimilarities in aggregate human behavior. A machine userid may be active 24 h a day, whereas humans must rest and sleep. It is easier to identify anomalies in usage when a subgraph of similar entities is pruned.

Secondarily, once a graph is conceptually pruned, some effort to examine the natural outliers resident is useful to understand the dynamics and principles which give rise to outlier behaviors. For instance, observing that penetration testers and systems administrators, as subclasses, may evidence several magnitudes higher levels of connectivity and traffic than normal workers is instructive (e.g., in terms of bytes sent, connections made, ports used, etc.). Having pruned the graph conceptually and defined the principles whereby exceptional nodes appear, it then becomes much more straightforward to identify and alert entities that are misbehaving or are indeed interlopers in the graph, as the case may be.

In order to build and substantiate refined security categories, a primary effort to segment and interrogate statistical groupings is required. Initial efforts to cast security-relevant statistical groupings confront classical operational problems such as:

- *Mislabeled entities*: userids being used by systems or processes, thus nonhuman users.
- *Extreme outliers*: highly unusual entities, many of which are acceptable exceptions related to unusual roles or designations (e.g., a security PEN tester generating highly unusual traffic).
- *Unexplained anomalies*: investigate and consult with local experts to determine if there are reasons to remove the observations from the dataset, for instance, to identify as a security event or an approved unusual exception.
- *Policy violations*: it is not unusual that users bend the rules regarding local security policies, for instance, sharing a userid amongst a group or sharing access to a device or application amongst a team—the reasons may be entirely pragmatic, for instance, to facilitate specialized work. However, the violation of the underlying policy creates noise and confusion, for instance, userids that exhibit superhuman behaviors such as working 24 h a day (in the case a 24 × 7 team shares a userid across shifts to continue work).
- *Events in progress*: events and behaviors may potentially indicate events in progress—these need to be investigated to determine whether they are true- or false positives.

Segmenting unusual, noise-generating entities from homogenous populations is necessary to refine entities into coherent, statistically similar groupings. A typical process involves staged structuring with outliers being surfaced through temporal summarization. Temporal summarization allows the surfacing of aggregate behaviors and patterns for comparative evaluation.

In the example represented in Fig. 4.19, a CSDS practitioner aggregates a summary of network behavior over time. All data is aggregated, such that "normal" behavior and flagged anomalies are included. From here, a structuring process summarizes activities into time epochs (i.e., per minute, hour, or day). Entities designated as users, devices, or unknowns are initially segmented based on organizational records (e.g., organizational CMBD or asset records). However, there is no assumption these records are accurate or up-to-date. For instance, it is common that entities designated as an ostensible human user may in fact be a machine or process. Initial efforts should focus on identifying these mislabeled entities. Further summarization is undertaken which leads to the identification of highly unusual cases (e.g., statistical outliers, users with highly unusual roles, other potentially mislabeled entities).

When a data structuring process, such as in the diagramed example, is operationalized, results continually improve knowledge of local environmental dynamics. The primary goal is to improve mislabeled entities and to segment exceptions that create noise. CSDS practitioners should treat deterministic organizational labels with inherent doubt concerning relevance and quality and seek to statistically

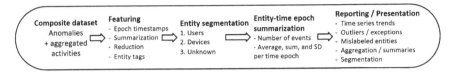

Fig. 4.19 Example process for surfacing summary behaviors for handling

validate, re-segment, and "clean up" wrongly designated entities. This reduces statistical behavioral variation amongst entities.

Having identified and categorized all mislabeled userids which are actually machines (or applications or userids that are shared by a team), it is then possible to treat userids with the assumption that a single userid is equivalent to a single human user. Greater conceptual integrity is assured, substantiating assumptions associated with the data-generating process. From there, a statistical baseline can be established, for instance, to assert that 95% of human userids are actively using the network generally no more than 40 h a week and that the mean of usage is between 5 and 15 h a week.

As the operationalized process continues to track usage patterns over time, it becomes possible to establish longitudinal (temporal) baselines for behavioral thresholds. This allows for the possibility to add temporal deltas (change-based) anomalies into the anomaly detection mechanism. For instance, an exception could be generated when a userid exceeds their own moving average for network traffic generation by X standard deviations. Developing a clearer picture of what is expected as "normal" in the environment, as associated with statically labeled entities and groups, allows for the iterative contextual refinement of anomaly thresholds, per Fig. 4.20.

As anomalies are generated and investigated, feedback concerning the results of investigations return as evidence substantiating or refuting the premises of the anomaly generation model. Initial results can be used to improve feature engineering and anomaly detection. As cyclical review and refinement continues, the monitoring process will begin building a library of labeled security incidents (true-positives) which are suitable for training predictive machine learning models. A historical dataset summarizing "normal" behaviors combined with flagged security incidents provides a basis for predictive machine learning model training and testing.

To summarize the "toy system," having operationalized exception remediation and having used the refined results to derive security-relevant categories, statistical baselines can be operationalized for anomaly detection. Categorical assertions (statistically relevant group labels) can then be added to anomaly detection models (i.e., as nominal features). Anomaly detection guided by refined categorical indicators which have been validated through both statistical and organizational means leads to baselines with much greater coherency (i.e., tighter and more reliable probability distributions). These categories can then be operationalized in continuous monitoring efforts, with anomalies and exceptions put forward for investigation. The results of investigation (true or false positives resulting from investigation efforts) can then

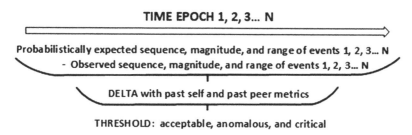

Fig. 4.20 Temporal delta-based anomaly detection

be used to further refine and orient semi-supervised and supervised machine learning mechanisms.

The concept of feeding continual results into a supervised machine learning model is embodied in the method of *online machine learning*. The main principle is that labels arrive over time to update the training dataset, leading to improvements in model accuracy. Two specific characteristics of security implementations should be noted that require active human monitoring and model maintenance. For one, changes in fundamental environmental dynamics, as represented by independent variables, may require periodic refitting and potential re-evaluation of the core model. A second aspect assumes expert human reviews of the cases put forward as labeled instances.

While there may be simple automated use cases associated with rules-based detection (e.g., detection of virus or malware signatures), wider use cases in the security domain typically involve more complex sequences involving human agency (e.g., staged incursion, escalation of privilege, exfiltration, etc.). This element of adversarial behavioral craft introduces a great deal of uncertainty and novelty, necessitating expert human reviews to make determinations. The adage "set a thief to catch a thief" is otherwise instructive. To the degree that model reviews and investigative triage can be streamlined, operational efficiencies emerge. Eventually, regularizing such monitoring opens potential opportunities for applying reinforcement learning—enabling goal-oriented agents to proctor and improve both model review (discovery) and investigative (detection) processes.

While online supervised machine learning and reinforcement learning approaches hold great promise, there is no assertion that current approaches are capable of realizing complete autonomy. There is a danger in assuming that model-based processes can lead to full automation and thereafter operate autonomously in perpetuity. In security monitoring, both existing and changing conditions need to be accommodated through periodic staged human reviews.

Phase changes in security fundamentals create the risk of catastrophic interference (forgetting) in trained models, whereby a model abandons former learning to adopt a new regime, which itself may be temporary. Changes in fundamentals, for instance, the introduction of a zero-day exploit, raises the potential for catastrophic interference when conditions shift to the degree that embedded model context is forgotten. Such risks can be addressed via incremental learning approaches.

However, given the complexity and shifting nature of the security domain, the best palliative involves a periodic revisiting of the parameterized model, possibly leading to a return to feature engineering. Such processes involve expert oversight and involvement, and it is asserted no completely reliable, fully automated mechanism can guarantee model outperformance over human-curated models in perpetuity given present approaches and techniques. This is a notable assertion as it runs counter to suggestions made by some commercial interests that full automation is feasible.

While active monitoring and feedback from investigations provides a clear path to extrapolating evidence, there are other routes. Staged incidents can be inflicted through red teaming or pen testing. The resulting labeled instances can be used to refine prediction, per the iterative training process represented in Fig. 4.21.

Notable in this idealized representation, humans play an active, central role, both in reviewing anomalies and outliers and in providing labeled examples to train and refine machine learning detection. In aggregate, such an approach can be considered to be a type of "cyborg categorization," in that humans and computational processes interact in an integrated, staged process. Humans provide behavioral and organizational context, whereas algorithms provide computational "heavy lifting" to focus and refine detection.

The ability to cross from descriptive and explanatory methods to predictive methods observes the inherent relationship between unsupervised and supervised machine learning (Fig. 4.22), and explanation and prediction more broadly. In environments of great complexity, statistical and descriptive techniques provide a stepwise foundation for building contextual knowledge, which when operationalized and refined provides a potential stage for supervised methods.

Subsequently, the anomaly extrapolation and review process may continue stepwise, but with a growing focus on refining predictive machine learning, per Fig. 4.23 instead of naïve and/or unsupervised mechanisms.

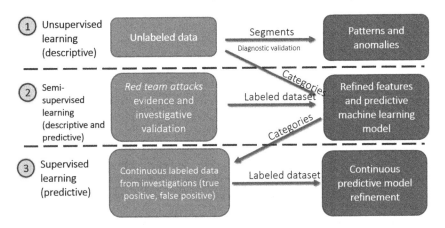

Fig. 4.21 Process for extrapolating staged evidence

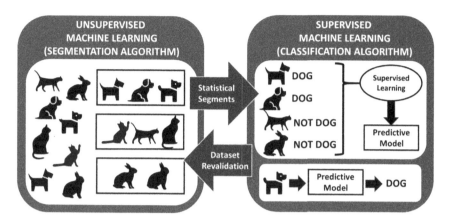

Fig. 4.22 Categorical reinforcement between unsupervised and supervised mechanisms

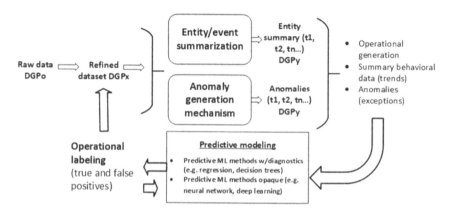

Fig. 4.23 Anomaly calibration step 2—operationalizing prediction

Predictive modeling is assumed to involve champion-challenger testing between a variety of methods. Namely, that a range of algorithmic modeling methods are trialed and compared like-to-like to identify the best competitively performing candidate. A detailed representation of the cyclical development process involving trialing multiple methods is represented in Fig. 4.24.

The benefit of an integrated, cyclical approach is its adaptive capacity. As conditions change and shift, the design process will adapt and shift to improve data representations and detection models. This addresses the requirement for continuous adaptation to changing conditions as adversarial methods evolve around countermeasures.

In approaching anomaly detection refinement as a process, it is appropriate to differentiate different types of anomalies in terms of the "gain" achieved as the system improves. This establishes benchmarks for classifying different types of

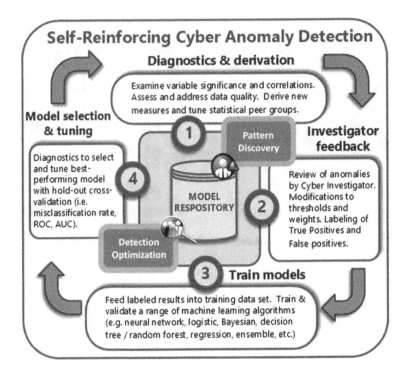

Fig. 4.24 Detailed self-improving modeling process

anomalies in terms of their level of refinement in an operational context. An example scheme to differentiate informational gain in operational anomalies which clarifies the additive effect of context:

1. *Rule trigger*: a naïve anomaly resulting from a univariate rule (e.g.,>10 failed log-in attempts)
2. *Threshold trigger*: an anomaly issuing from a set statistical threshold (e.g., >5× a set threshold)
3. *Flagged outlier*: an entity with a combination of many high univariate measures (e.g., 99th percentile across five measures)
4. *Anomaly*: an unusual observation identified via statistical comparison to the population (e.g., devices which connect to ten standard deviations more external IP addresses than the sample mean)
5. *Baseline anomaly*: a multivariate anomaly resulting from comparison to a range of expected measures across the target population
6. *Contextual anomaly*: an anomaly with an unusual statistical status based on comparison to self-similar entities (or to own behavior) assessed over time
7. *Alert*: an anomaly with a security-relevant designation based on similarity to a known violation or condition
8. *Incident*: anomaly confirmed by an investigation as indicating a security event

The central principle observed is that anomalies improve through contextual informational gain and that there are distinct levels of quality associated. Efforts to refine context improves the relevance of anomalies. An anomaly, in this sense, is a theoretical proposition concerning "unusualness" with respect to a particular domain. To the degree that a refined understanding of "normalness" in a domain is embedded in data preparation and modeling, the ostensible validity of the proposed anomaly as an "abnormality" is bolstered and refined.

A final observation significant in an operational context is that computational resource requirements are assumed to be distinct between exploratory discovery and operational detection phases of model development. The requirements for refining and fitting naïve anomaly detection models are assumed to align with big data repositories and heavy in-memory or computationally distributed exploratory efforts. Refined detection models, for instance, tested and validated predictive models, have a different set of needs as they are associated with highly focused, operationally engineered datasets.

Technical architectures for predictive detection have more discrete data storage and handling requirements but benefit from iterative exploratory discovery findings. In this respect, it is expected that technical architecture stages discovery efforts separate from detection mechanisms, per Fig. 4.25. While architecturally segmented, there is an assumption that the two communicate fluidly as new findings in discovery repositories suggest new predictive detection operational models.

4.4.3   Extrapolating Experimental Evidence

The exhortation to "measure what is measurable, and make measurable what is not so" is attributed to Galileo Galilei. In security settings, this means collecting data in context. As was explored in the discussion on data management, measurement requires context and scale. Whereas great volumes of security data are available, raw data must be situated and scaled to be made measurable. The associated irony

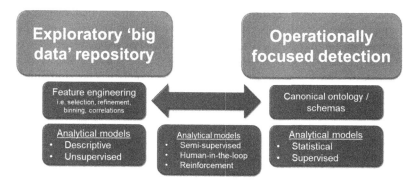

Fig. 4.25 Segmented data architectures

is that security environments often are afflicted by a "water, water everywhere, but not a drop to drink" paradox, whereby there is a surfeit of raw data, yet a lack of clear context concerning what is being, or should be, measured.

Treatments which facilitate the measurement of validated incidents are beneficial to CSDS practitioners as evidence. A central shortcoming raised in interviews and reinforced in research literature relates to the difficulty of staging security-related scientific inquiry due to a lack of evidence relating to confirmed incidents (Verma et al. 2015). Practitioners observed that in security operations settings, it is not common for incidents to be recorded in a detailed and structured form. Detected incidents, being relatively rare, more often than not, are not recorded in terms of capturing a comprehensive picture: the context of what occurred, effecting which users and systems, when, across what timeframe, etc.

A scientific effort involves framing a proposition as a hypothesis, ideally grounded by theory, clarifying the associated testable expectations or predictions of the hypothesis, subsequently subjecting the hypothesis to experimental testing aimed at validation, or otherwise sampling observations to produce evidence, and conducting statistical assessments of the explanatory and/or predictive integrity of results. As Pavlovic frames science specifically within the security domain:

> In general, science is the process of understanding the environment, adapting the system to it, changing the environment by the system, adapting to these changes, and so on. Science is thus an ongoing dialog of the system and the environment, separated and conjoined along the ever-changing boundaries. Dynamic security, on the other hand, is an ongoing battle between the ever-changing teams of attackers and defenders. Only scientific probing and analyses of this battle can tell who is who at any particular moment. In summary, if security engineering is a family of methods to keep the attackers out, security science is a family of methods to catch the attackers once they get in. (2012)

Security science, to be scientific in improving detection, must collect and utilize evidence concerning the methods and passage of attackers. Evidence is data that can be used for formal testing and comparison to assess and validate explanatory theory or hypotheses. When prediction is the goal, evidence substantiates to what degree predictions are accurate given defined conditions.

Evidence has a particular meaning in the context of predictive machine learning. Labeled data, as it is known in the machine learning context, provides evidence of a particular outcome given certain conditions, as represented in data, which can be used to train predictive models. This can take the form of an array with a flag or categorical label or may be a less structured set of records spanning across systems which trace an incident step-by-step from improper access and escalation of privilege (e.g., as recorded in an authentication log file) through to data exfiltration (e.g., as recorded in NetFlow records and proxy logs).

Similar to the fraud analytics realm, cybersecurity incidents are relatively rare in the context of the great volumes of data recording non-adversarial behaviors. There is an uncertainty principle at play, namely, that a certain proportion of incidents occur without being detected. As well, there is the challenge that the event in question may not be discretely presentable in aggregate form (i.e., as a single vector, set of variables, or integrated collection capturing a process). As has been raised previously, a major operational challenge in CSDS environments concerns data management—struggles to rationalize multiple systems, fragmented data sources, and disparate data types. This leads to challenges in representing security events as aggregated datasets capturing complex processes that occur stepwise over time.

Even when incident evidence is available, CSDS practitioners seeking to extrapolate detection models must address the resulting imbalanced nature of the evidence. Evidence frequency is typically quite minimal compared to the exponentially larger dataset of "normal noise." Statistical mechanisms must be utilized to accommodate this problem of imbalanced evidence, for instance, bolstering representation, reducing the larger sample, or applying specialized treatments during machine learning training.

Traditional approaches to explanation and prediction in the security domain often exhibit less rigor than would be suggested by formal scientific approaches. Typically, criteria for adopting a security rule or signature rests on clinical observations, expert judgment, or routine observational verification. For instance, if a malware infection generates a particular set of network behavioral measures (e.g., communicating with DGA-associated URLs), these measures may be framed by a practitioner for adoption as a new rule in a tracking and alerting system such as a security information and event management (SIEM) platform.

In security practice, it is common to derive rules-based methods to identify incidents by simple transitive reasoning, for example, because incident A involved X, Y, and Z factors and observed case B also involved X, Y, and Z, case B must also be an incident. It is less typical for formal statistical diagnostic tests, probabilistic analysis, variable significance diagnostics, feature engineering, or causal assessments to be undertaken with reference to broader "normal" events (as baselines). The result is that the de facto approach to deriving rules in security practice is largely inductive—based on direct observations.

This inductive focus, when taken to the extreme, is the source of a good deal of operational pain in cybersecurity security operations centers (SOCs). In the case above, for instance, there may be many instances of X, Y, and Z factors which are not incidents, such that when a rules-based alert is implemented, it gives rise to an overwhelming set of false positives. Failing to substantiate, individuate, and refine indicators can lead variously to model overgeneralization or over-specification.

While there is no doubt that efforts are made to test security rules and signatures in operational settings and to refine these rules, it is less frequently the case that a refined rule is formally established as a proposed hypothesis to be tested. For instance, it is useful to determine that the observation of parameters associated with rule X accurately indicates an attack of type Y in more than 80% of the investigated cases. A formal effort to track and assess the results of follow-on investigations from alerts supports an experimental process and allows for statistical and probabilistic determinations.

The circumstance of alert overload in many SOCs is, more or less, directly the result of not implementing rigorous experimental falsification efforts in operational settings. However, this is not to understate the operational challenges involved. Many incidents are the result of staged, multiphased processes. The indicators of an incursion or an attack, as recorded in data, thus may be spread across a range of repositories and systems. Many rule and signature-based systems look for focused indicators, as opposed to framing and testing complex arrays of indicators resulting from combined measures and sources.

The fallback is that, lacking evidence, many CSDS practitioners are induced to free-associate inductive and deductive modalities to support root cause analysis or to rely primarily on deductive reasoning (i.e., stock propositions based on instincts grounded in past clinical experience). Deductive reasoning initially rests upon theoretical assertions based on understanding or experience and then attempts to test and validate assertions via observations or measurements. Testing may involve direct experimentation or the application of inferential statistics to establish validity. However, to the degree that data is unavailable or otherwise unaggregated for such testing, operational judgment risks proceeding from strength-to-strength drawing upon clinical instincts alone. Research substantiates the poverty of clinical versus actuarial (statistical) reasoning (Dawes et al. 1989).

In the cybersecurity realm, deductive reasoning is often the domain of cybersecurity professionals with a great deal of operational experience applying clinical judgment to assess whether a particular set of circumstances appears suspicious or not. However, in the majority of cyber SOC settings, a cyber-analyst is accessing visualizations and distributed measures and alerts, but typically does not have the tools, time, means, or knowledge to conduct comprehensive hypothetical-deductive statistical tests.

This is not to discount the approach to rule and signature building entirely, however. It is to assert that such approaches are quasi-hypothetical and would otherwise benefit from follow-on experimental testing and validation efforts. As CSDS develops as a field, there is the risk of a backlash against both rules-based inductive and expert-based deductive approaches. This is particularly the case if there is a biased (and erroneous) standpoint that machine learning algorithms are capable of fully automating human judgment and that such methods are somehow qualitatively "better." Quite the reverse, rules and expert judgment are quite valuable, and efforts should be made to frame and experimentally assess the hypothetical propositions that result. It is mainly in remaining quasi-hypothetical and not being formally validated that rules, signatures, and expert judgment fail the test of rigor. Again, the general lack of recorded incidents to undertake validation remains a shortcoming, pending focused improvements in efforts to capture and store labeled data.

Turning to broader outlooks on adopting scientific rigor in security operations, in *Research Methods for Cyber Security*, Edgar and Manz enumerate a range of methods to undertaking primary research in the cybersecurity domain (2017):

- *Observational*, i.e., exploratory, descriptive, machine learning-based
- *Mathematical*, i.e., theoretical and simulation-based
- *Experimental*, i.e., hypothetical-deductive and quasi-experimental
- *Applied*, i.e., applied experiments and observational studies

Providing examples of security-relevant approaches to undertaking such research efforts:

- *Observational*: assessing multivariate probabilistic behavior through aggregate and summary measures (i.e., aggregate measures combining byte traffic, connections, ports, scanning behavior, etc.); deriving probability distributions and estimates for common user and device entity behavioral measures as a baseline for framing anomalous behavior (e.g., number of systems connected to, bytes sent on average, number and types of ports commonly connected to); and utilizing unsupervised machine learning to derive statistical segments as propositions for significant categories to refine analysis (e.g., cluster analysis, self-organizing maps).
- *Mathematical*: application of simulation methods—Monte Carlo, systems dynamics, discrete event, multi-agent (e.g., utilizing entity probability measures to calibrate simulation-based analysis of expected multivariate distributions for entities and categorical peer groups).
- *Experimental:* testing and staged incidents in controlled virtualized environments; utilization of Bayesian methods to guide and analyze experimental approaches.
- *Applied*: vulnerability and penetration testing (i.e., PEN testing, red teaming) to identify pathways susceptible to threat vectors (e.g., exposed ports, unpatched systems, etc.).

A related shortcoming is that many organizations lack the practical wherewithal, in terms of staff, tooling, and/or time, to stage such research efforts comprehensively. A central challenge is a lack of basic data and data collection capabilities. In interviews, 56% ($n = 28/50$) of CSDS practitioners reported a "lack of labeled incidents to focus detection" as a key challenge (CH4). Addressing this challenge is nontrivial, as a prerequisite to data collection is the design and implementation of data collection protocols and mechanisms. Building data collection protocols and mechanisms, however, requires some notion of the data required, including its experimental efficacy. There is a practical chicken-and-egg principle at work: to collect efficacious security data, one needs foreknowledge of the data required in order to design and refine data collection protocols and mechanisms.

Many organizations lack the resources and time required to stage experimental data collection exercises in order to build the requisite understanding to standardize a data collection program. A notable exception is that many organizations do stage red teaming and/or penetration testing to assess vulnerabilities and defenses. However, the approach in these exercises typically is focused on simply cataloging a list of weaknesses in order to improve defenses. However, an ancillary goal could be an effort to monitor and improve security data collection protocols and mechanisms. Forward-thinking organizations should consider the very valuable evidentiary data which issues from such staged exercises and endeavor to collect the results more comprehensively.

By staging an attack on a network and/or set of devices, evidence of the attack is manifested in data artifacts—for instance, in authentication, lookup, proxy, network flow, network device, systems access, data access, and endpoint behavior logs and records. By collecting the composite data surrounding an attack, an organization builds evidentiary details characterizing the attack that can be used subsequently to conduct statistical tests, to test hypotheses concerning the notable indicators of attack, and to train and validate predictive models. A staged attack can be considered a type of applied field experiment aimed at determining how particular attacks are evidenced in data.

To support and orient the consideration of structured efforts to operationalize evidentiary security data gathering, it is helpful to differentiate a range of evidence collection techniques:

1. Field evidence (e.g., observed incidents)
2. Data derived from field testing (e.g., local experiments)
3. Honeypots ("fake" systems staged to attract attackers and record their behavior)
4. IDSs (intrusion detection systems)
5. Simulation findings
6. Laboratory testing (e.g., examining malware behavior in a staged environment)
7. A-B testing (examining the focused effects of intervention versus nonintervention)
8. Stepwise discovery (iterative interventions to examine vulnerabilities)
9. Pen testing (attempts to penetrate the network)
10. Red teaming (staged attacks to achieve particular goals)
11. Incidents (records associated with confirmed incidents)
12. Research examples (shared datasets recording attacks from research efforts)
13. Reinforcement learning experiment (goal-oriented, self-improving machine learning)
14. Expert review (opinion, examination, determination, and guidance from experts)
15. Intelligence feed (aggregated indications from a third-party service)
16. Staged thought experiments (e.g., boundary conditions, counterfactuals)

As a focused case, concerning evidence type number 12, research examples, a recent development has been the increasing availability of datasets recording attacks or incidents made available to practitioners and researchers (Kent 2016; Mongeau 2019b). While these datasets were created within, and are thus partially unique to, specific environments, and not all relevant details are shared, such "real" datasets expose phenomena and dynamics suitable for hands-on testing and exploration.

For organizations desiring evidence specific to their own environment, evidence types 1–11 offer approaches to evoking "own" research datasets for treatment. In either case, evidentiary data obtained externally or invoked internally, there is great value in the base exercise of investigating and processing the data prior to analytical treatments and modeling. The structured process of investigating evidentiary datasets is itself insightful in terms of training, building contextual understanding, developing methodological procedures, and testing and refining operational approaches (e.g., scoping and designing ETL procedures for instantiation as data pipelines).

Efforts invested in exploring and structuring evidentiary datasets can be a helpful proxy for designing and refining operational data management protocols. Working through the iterative process of cleaning and preparing sample data helps to refine approaches to designing operational collection mechanisms, data staging, exploring contextual dynamics, identifying entities, normalizing, segmenting, labeling, time stamping, extracting features, etc. As well, and perhaps most importantly, the presence of labeled attack evidence is centrally useful in subsequently building, testing, validating, and refining analytical treatments and models.

Whereas data and context surrounding attacks and incidents are not easily available in many operational settings, practitioners can fruitfully apply evidentiary datasets to develop structured processes that lead to refined operational data collection practices. This essentially provides a bootstrapping approach to overcoming the chicken-and-egg quandary surrounding developing operational data collection protocols and mechanisms. As many production environments may not be conducive to operational data exploration and testing, practitioners can use research and/or derived datasets to develop and refine data structuring approaches prior to engineering operational ETL and data processing pipelines.

As an example, Fig. 4.26 offers a step-by-step data extraction and refinement process developed by this researcher utilizing the Los Alamos National Laboratory (LANL) research dataset as a test bed (Kent 2016). The example demonstrates that the process of refining cybersecurity data is nontrivial and worthy of isolated consideration well in advance of approaching operational environments and systems.

Each progressive step represented in Fig. 4.26 involves a procedure or set of procedures that serve to refine and structure data in preparation for analytical treatments. Decisions made during each step embed assumptions and judgments concerning the external "realities" the dataset is representing. Exploration during this process leads to iterative understanding and refinements to the judgments made. The prepared dataset thus iteratively embeds semantic assumptions concerning security phenomena and dynamics, both of a general nature (quasi-theoretical assertions) and as specific to the environment and phenomenon under scrutiny (inherent idiosyncrasies of the host environment).

As a central concluding remark, given the breadth of options for collecting evidentiary security data, it is useful to consider how evidence collection relates to two key modalities of scientific inquiry: *inductive* (primary observational) and *deductive* (based on expertise and rational extrapolation). Evidentiary data can either be

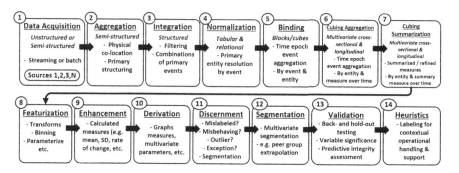

Fig. 4.26 Example of a progressive security data structuring protocol

collected directly from primary sources or *synthesized/induced* according to theoretical premises. The heuristic in Fig. 4.27 orients efforts towards gathering, framing, and testing security evidence

Connecting to previous design guidance on operationalizing discovery, once evidence is gathered, it can be fed into self-improving operational processes. Staged processes can be applied which combine methods for evidentiary discovery, for instance, unsupervised machine learning and inferential statistics (observational-inductive), with iterative approaches to detection, for instance, bolstering model efficacy in stages by applying naïve, semi-supervised, and supervised machine learning amplified by human-in-the-loop mechanisms (progressively deductive). Again, this raises several opportunities for semi-automation via staged algorithmic treatments such as online machine learning, but it is asserted that human experts maintain a dual role in reviewing cases and in periodically examining and maintaining model fundamentals, per Fig. 4.28.

4.4.4 Integrating Prediction and Explanation

In "To Explain or to Predict?," Shmueli differentiates between modeling efforts aimed at explanation versus prediction. Commenting that there is a methodological schism resident between, on the one hand, statisticians and researchers seeking to explain phenomena causally, and, on the other, practitioners seeking pragmatic predictive power, Shmueli comments: "measurable data are not accurate representations of their underlying constructs. The operationalization of theories and constructs into statistical models and measurable data creates a disparity between the ability to

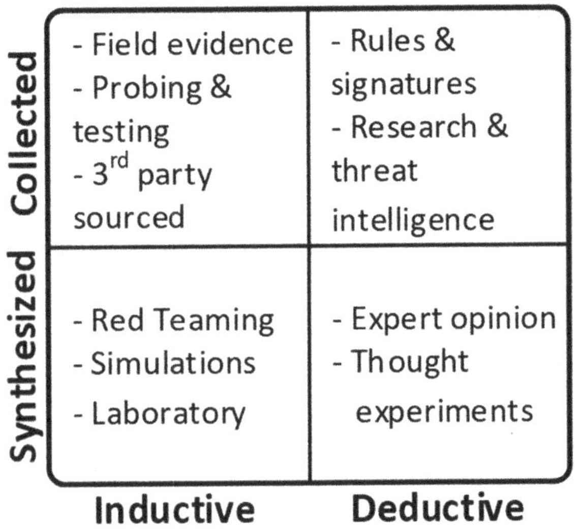

Fig. 4.27 CSDS model to orient modalities and data

	Inductive	Deductive
Collected	- Field evidence - Probing & testing - 3rd party sourced	- Rules & signatures - Research & threat intelligence
Synthesized	- Red Teaming - Simulations - Laboratory	- Expert opinion - Thought experiments

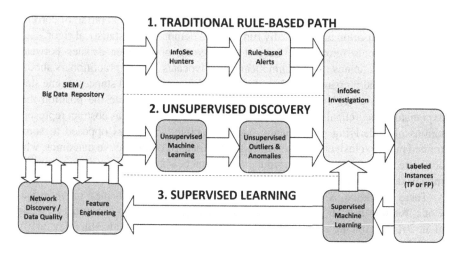

Fig. 4.28 Cyclical human-in-the-loop iterative detection process

explain phenomenon at the conceptual level and the ability to generate predictions at the measurable level" (Shmueli 2010).

The disparity between methods to explain phenomena and mechanisms for prediction raises a quandary for CSDS. The function of continuous security monitoring essentially depends upon both approaches. Focused operational identification of incidents ideally seeks raw predictive accuracy and reliability. Meanwhile, the continually shifting nature of threats and adversaries requires causal-explanatory understandings of evolving phenomena.

Models trained to predict based upon retrospective examples will likely be wrong-footed by emerging or zero-day exploits. As noted cybersecurity data scientist Joshua Neil notes in a workshop on security graph analytics: "Even if we did have labeled data, I would postulate that we don't have representative labeled data of the number of different things that the attacker could do because of complexity and combinatorics… for those reasons, (attackers) don't tell us what they do, and we don't have good labeled data and the data of today is not necessarily representative of the attack behaviors tomorrow, we typically apply unsupervised methods" (Neil 2019). Emerging threats thus require a descriptive understanding in order to extrapolate effective detection mechanisms. Based upon demonstrated understanding and efficacy, there is a stronger foundation to extrapolate predictive mechanisms.

To the degree prediction and explanation are distinct modalities yet must interoperate, designs for bridging the two approaches should be of central interest to CSDS practitioners. An exacerbating aspect of the explain-predict schism is that some practitioners start with and rely upon predictive models alone, abjuring explanatory efforts. Some abandon explanatory efforts simply due to a lack of time in demanding operational settings. Some advocates have controversially argued the obsolescence of "traditional" statistical methods and the "death of scientific theory" (Anderson 2008; Hey 2010). These advocates suggest that strong indications of correlation are enough. However, such claims invoke the classical pitfall of correlation versus causation and generally run afoul of demonstrable statistical rigor.

Disagreements remain the subject of ongoing practitioner debates between polarized statisticians and machine learning advocates. CSDS practitioners should be aware of and prepared to redress holdouts taking adversarial stances. Some statisticians, focused exclusively on explanatory models, advocate the position that explanatory theoretical and parametric understandings of data, as abstract representations of underlying phenomenon, are irreducible. This is as opposed to some trained more exclusively in machine learning, focused on predictive outcomes, who advocate that substantiating causal factors is unnecessary when complex correlative models demonstrate sufficient predictive integrity.

There are domains where a correlation-dominant approach may prove efficacious, for instance, diagnosing mechanical failure in complex machinery (Wong et al. 2014) or tracing complex hidden patterns related to public health (Hey 2010). In such cases, the phenomena are typically associated with complex systems in which root cause analysis is onerous, either because underlying phenomena are inherently complex or because direct causal elements are difficult to observe, requiring analysis through latent variables.

In such cases it is worthy to consider whether correlated variables are in fact proxy assertions of latent variables at work in complex systems. This can be problematic when spurious latent variables are invoked in the model but are not assessed for theoretical integrity. For instance, in the case of "The Parable of Google Flu," some searches which contributed to predicting flu outbreaks were, under closer examination, spurious correlations, a phantom artifact of utilizing big datasets (Lazer et al. 2014).

The appeal to complexity does not imply that causal elements do not exist but rather that the underlying phenomenon may be outside the realm of direct observation given current data gathering, analytical, and/or computational capabilities. In a general sense, social behavioral phenomena broadly may fall into this category, whereby it is difficult to uniquely identify single causes, but latent indicators can be agglomerated through sets of correlating variables. The methodological challenge arrives when practitioners propose to establish an operational predictive model based on a set of unexamined latent correlated variables that have given some indication of predictive accuracy. Lacking clear explanatory validation, there is always a risk of overfitting derived models to the training dataset, embedding spurious correlating factors (Lazer et al. 2014; Shmueli 2010).

For instance, should geolocation be correlated with security threat likelihood, it is unlikely that a region or country is the underlying cause as opposed to a set of socioeconomic, legal, and political factors that incubate hostile actors. To assume one as a proxy for the other risks either overgeneralizing (e.g., disenfranchising legitimate users in a designated high-risk region by shutting off blanket access) or, at the other extreme, overfitting (e.g., being wrong-footed when new conditions lead to the emergence of a new "threat zone"). This is the realm of theory, as opposed to routinized correlative efficacy, which, as stated, risks overfitting to local temporal conditions. The field of econometrics is otherwise well aware of such classical problems and has tests and procedures for detecting and dealing with indications of, for instance, multicollinearity, autocorrelation, and homoscedasticity.

Cybersecurity faces a special challenge concerning disassociating correlation from causation in the respect that a great deal of noise and complexity subsumes efforts to extract clear explanatory signals. Rapidly evolving events are often geographically, technically, and temporally distributed, further complicating efforts to gain clarity. Individual events are often the result of many distributed events occurring at great speeds in complex, dispersed infrastructure. This leads to large but sparse datasets. Methodologically, this raises the specter of actions occurring at a distance, events which may not always have a clear attribution or cause as recorded in the data.

One of the fundaments of the scientific method during the enlightenment was to abjure appeals to "magical" or spontaneous causes, known as "appeals to action at a distance." This was understandably considered a slippery slope to superstition and magical thinking. As an example, the "weapon salve" was a discredited medieval medical treatment whereby the weapon that caused a wound was treated, as opposed to the wound itself, under the belief that the weapon that caused the injury maintained an invisible connection to the wound.

While we might shake our heads in disbelief, the virtual world presents many temptations to attribute causes to "action at a distance." The remote, electronic, distributed, asynchronous, and often difficult-to-attribute nature of cyber events complicates efforts at direct causal attribution through empirical sensory validation. This is particularly the case when issues of doubt are raised in security environments, for instance, when an adversary is intentionally altering systems of record to erase traces of their passage.

The prospect of fundamental doubt in records and logs poses a challenge of veracity. When events may not have a trusted record, there may be a need to operate on correlative or disconnected events to frame suppositions. However, while this challenges efforts to establish direct causation through deterministic means, other methods may prove suitable substitutes. Inferential statistics, inured to working with limited, poor quality, and incomplete datasets, offers a range of statistical diagnostic approaches that are useful to substantiating rigorous conclusions even when the evidence, as recorded in data, may be spotty, interrupted, noisy, or in doubt.

In operational settings, predictive models built upon observations of complex correlation are often a fallback when explanation appears out of reach or onerous to substantiate. This is the case when many indicators and alerts are aggregated together without an attempt to sort out component contributions and interactions in a composite model for alerting. While demonstrable practical prediction of incidents is otherwise a "holy grail" for CSDS practice, pure predictive approaches risk the classical pitfall of operationalizing observations of correlation, not causation (and thus overfitting).

Highlighting the distinction, in practice there are cases where predictive models may be improved by tuning models such that they violate or ignore principles associated with parsimony in explanatory models (i.e., adding complexity or combining multiple models into an ensemble). Shmueli comments that "predictive modeling seeks to minimize the combination of bias and estimation variance, occasionally sacrificing theoretical accuracy for improved empirical precision... the 'wrong' model can sometimes predict better than the correct one" (2010). Whereas a predictive model may perform well in back-testing and generate strong efficacy metrics (i.e., ROC and AUC performance), this is no guarantee of iterative efficacy given that the primary mechanism may be local overfitting.

Especially given the nature of evolving threats and changing technical conditions (e.g., new users, devices, vulnerabilities constantly being added in expanding digital environments), overfitting may lead to missing new and emerging threats. Likewise, overgeneralization leads directly to overwhelming false alerts. It is thus advocated that predictive methods be matched with and accompanied by efforts at explanation in order to substantiate and validate production models.

The medical domain, for instance, in the areas of epidemiology and clinical studies, is well aware of the pitfalls resident in leaping to a snap diagnosis. Whereas a health epidemic such as heart disease or diabetes can often be predicted as the outcome of a host of diverse lifestyle choices and factors, efforts must be made to carefully examine the statistical aspects of proposed predictive models. It would be facile, for instance, to utilize income and zip code as primary health predictors, even as these variables may demonstrate predictive integrity in correlative models. Again, econometric model refinement techniques are well aware of such pitfalls and seek to guide the researcher towards an efficacious model maneuvering between overspecification and overgeneralization. In the predictive context, this is akin to identifying an appropriate bias-variance trade-off that holds over time.

Epidemiology seeks root cause analysis in order to frame interventions and advocacy to address improvements to public health. Thus, it may be the case that lack of exercise, poor diet, and higher patterns of smoking and alcohol consumption underly a particular public health phenomenon. It is these underlying predictors the researcher should attempt to triangulate and isolate through statistical causal analysis. In the security domain, the equivalent is to assert that a job role and education level are predictive of susceptibility to phishing attacks, when in fact there may be an underlying construct involving latent variables such as "technical sophistication" or "security awareness" that are the root, and thus treatable, factors in an explanatory model.

Several well-regarded statisticians and econometricians with firm groundings in computational methods puncture the staged debate between explanatory and algorithmic correlative methods. These advocates aver that combined approaches are essential, namely, that algorithmic methods are enriched by and made more rigorous and effective through the application of statistical techniques to ground, validate, and explain data and models (Cleveland 2001; Donoho 2018; Hastie 2016; Hastie et al. 2009; Shmueli 2010; Varian 2014).

This cooperative stance implies that data is inherently theory-laden and that algorithmic approaches have not demonstrated sufficient autonomous capabilities to frame, interpret, test, or validate general hypothetical-theoretical propositions to achieve reliable generalizability. This is not advocacy to throw the algorithmic baby out with the bathwater but to assert that efforts to explain and validate models should be invoked to substantiate rigor and assure generalizable efficacy.

In the view of those who have adopted conciliatory views and promote approaches towards methodological collaboration, to Balkanize machine learning from statistics risks establishing a highly dubious dichotomy. Statistical methods underpin and drive machine learning algorithms. Machine learning does have a claim to differentiation as methods combine complex sequential mathematical and algorithmic treatments with computational approaches (e.g., kernel density, feature hashing, random and directed sampling). However, predictive algorithmic methods should not be adopted absent efforts to consider the origins, nature, and implications of the underlying data.

Within academic literature, this viewpoint was first marked by John Tukey's seminal article "The Future of Data Analysis" (1962). Here the traditional role of the scientific researcher in framing theory and judging results based on expertise broadens to allow for guidance issuing from computationally sophisticated methods. The view asserts that algorithms can play a role in facilitating expert judgment, particularly when the data is too voluminous and/or complex to be susceptible to manual human transformation. As Tukey frames it, there are situations where "the computer makes feasible what would have been wholly unfeasible" (1962). While maintaining a central role for statistical methods, Tukey sees computation as being a facilitating tool and the researcher as guiding the process of inquiry.

The notion that an algorithm or set of algorithms can play a role in facilitating and guiding deductive theory extrapolation was and still is controversial. However, Tukey's insights have practically come to pass to a great degree. While hypothetically all data science methods could be conducted with pen and paper, the scale of problems now being framed outstrip human capabilities by vast orders of magnitude. Further, the increasingly popularity of machine learning has led to situations where machines in some cases semi- or wholly automate complex decision-making, deep learning in particular.

In "Data Science: an Action Plan for Expanding the Technical Areas of the Field of Statistics," W. Cleveland proposes and advocates the term *data science* to embody an interdisciplinary field extrapolated from statistics and focusing on data analyst practitioners (2001). He frames a curriculum for data science that balances a mix of practitioner, methodological, technical, theoretical, and pedagogical topics.

In "Statistical Modeling: The Two Cultures," L. Breiman frames two bifurcating statistical modeling cultures: statistical methods, which attempt to characterize data-generating processes, versus algorithmic modeling, which focuses on prediction and treats data generation as an unknown (Breiman 2001). Breiman advocates that statisticians increasingly make inroads to embrace the later approach and methods.

Taking inspiration from Tukey, in "50 Years of Data Science," D. Donoho further profiles the two cultures disciplinary divide by raising six functional themes associated with data science:

1. Data exploration and preparation
2. Data representation and transformation
3. Computing with data
4. Data modeling
5. Data visualization and presentation
6. Science about data science

Donoho connects the first three themes to traditions in the computer science field, thus with algorithmic methods. The fourth, *data modeling*, he characterizes as the focal point of the computational and statistical divide, with computer scientists and statisticians gleaning to their respective paradigms. The sixth theme, *science about data science*, he marks as an opportunity for an emergent discipline, mirroring Cleveland's advocacy for a new field. Donoho frames this field as being focused on evidence-based empirical research of the phenomenon of data and the many comparative methods for analyzing data (Strawn 2016).

The proliferation of networked computers and data processing have made data omnipresence itself a phenomenon. There are thus grounds, per Donoho, for viewing computationally intensive data analysis, combined with prolific data availability itself, as a discipline allied to but distinct from the traditional field of statistics (2018). Donoho offers a formal definition for such a field in a lecture entitled "Data Science: The End of Theory?": "data science concerns the recognition, formalization and exploitation of data phenomenology emerging from digital transformation of business, society, and science itself" (Donoho 2018). This view of data science is markedly more empirical and meta-referential than the computer science view, which is focused on methods and mechanisms for data analyst practitioners.

In composite, emerging views from the field of statistics advocate both distinguishing and bridging classical statistical techniques with algorithmic techniques and methods such as machine learning. Data science is positioned as a set of new statistical methods, complex algorithms included, made possible by computational power and speed combined with the ability to access, store, and process large sets of data. Statistical validity is held to be central, as is the importance of scientific standards—empirical validation and reproducibility, in particular. Significantly, allowance is made for utilizing and studying theory-independent algorithms, methods that treat data-generating processes as unknowns. Technology and tools are framed as central facilitators that make computationally sophisticated methods increasingly feasible and available.

Hastie, Shmueli, and Varian have similarly advocated that the *two culture* paradigms, although distinct, are not mutually exclusive and should be viewed as potentially complementary and additive, rather than adversarial (Hastie et al. 2009; Shmueli 2010; Varian 2014). From this view, debates concerning disciplinary exclusivity potentially reflect the difficulty of individuals to maintain deep expertise simultaneously across statistics/econometrics and the fast-expanding field of machine learning. If data science has claims to independent disciplinary status, it must be said to have interdisciplinary roots and as such to derive from and collaborate between the fields of computer science and statistics.

The principle of combining both explanatory and predictive approaches was earlier raised in the context of data management. It was advocated to apply both a version of Tukey's EDA process, aimed at explanation and aligned with the field of statistics, and some approach towards feature engineering, as aimed at prediction and aligned with machine learning. Both involve approaches to examining, preparing, and validating datasets, albeit aimed at distinct outcomes. From a process management perspective, this advocates approaches towards integrating statistical discovery, as aligned with efforts to explain, i.e., EDA, and analytics process models, as associated with machine learning model development. The following figures, Figs. 4.29 and 4.30, demonstrate two representative process-based model development approaches from the explanatory (Shmueli 2010) and algorithmic camps (SAS Institute 2018), respectively.

As more extensive representations of process-based approaches to analytics have been explicated, opportunities for integrating explanatory and predictive methods have become more clearly delineated. A key driver has been a rising consciousness concerning the need to introduce scientific rigor into unified data preparation and model development processes.

In "The Analytics Lifecycle Toolkit: A Practical Guide for an Effective Analytics Capability," Nelson explores and provides a structured guide to practitioners for operationalizing analytics (2018). In Nelson's enumeration, a number of explanatory and scientific methods are interposed. These occur in a larger context that supports the development of predictive models, albeit with rigorous efforts to substantiate, explain, and transparently frame the causal mechanisms underlying such models. Table 4.7 profiles Nelson's analytics life cycle toolkit best practices (Nelson 2018).

For the CSDS practitioner, it is essential to understand the differences between, and potential integration of, explanatory and predictive methods. Shmueli adds a third modality to the two raised, descriptive modeling, which recognizes the need for terse, pragmatic representations of phenomena. The three modalities are summarized as:

- *Explanatory modeling*: application of statistical models to data for testing causal hypotheses about theoretical constructs
- *Predictive modeling*: applying a statistical model or data mining algorithm to data for the purpose of predicting new or future observations
- *Descriptive modeling*: summarizing or representing the data structure in a compact manner (with minimal or no reliance on underlying causal theory)

Descriptive modeling may be recognizable to CSDS practitioners in the form of business intelligence efforts, for instance, a security monitoring dashboard tracking operational telemetry. This encompasses, though is not limited to, insightful

Reprinted with the permission of the Institute of Mathematical Statistics

Fig. 4.29 Explanatory-statistical modeling process

Fig. 4.30 Algorithmic—
analytics life cycle

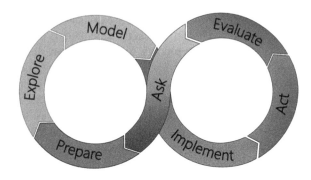

Table 4.7 Analytics life cycle toolkit best practices (Nelson)

Understanding Framing the problem	• Clarify the need • Reveal details • Generate focused hypotheses • Design a testing approach • Prioritize business cases	• Define problem • Root cause analysis • Generate hypotheses • Design question • Business case priorities
Theorization Sensemaking for data	• Specify data required for problem-solving • Extract and structure data for analysis • Explore data for understanding	• Data identification and prioritization • Data collection and preparation • Data profiling and characterization • Visual exploration
Prototyping Development of analytics model	• Testing hypotheses • Profile cause and effect • Classify analytic problem type • Uncover patterns	• Make comparisons • Measure associations • Make predictions • Detect patterns
Testing and implementation Activation of results	• Interpretation of results • Exploring alternative explanations • Communication and collaboration • Planning change activities • Evangelizing benefits	• Solution evaluation • Operationalization • Presentation and storytelling
Improvements Analytics product life cycle management	• Engagement of stakeholders • Capturing knowledge • Analyzing impact • Continuous improvements • Driving effectiveness of analytics	• Value management • Analytics life cycle execution • Quality assurance • Stakeholder engagement and feedback • Capability and talent development

descriptive visualizations, which may have interactive capabilities, allowing investigators to search for emerging and novel threats. A typical cybersecurity dashboard often aggregates a range of associated elements, for instance, combining a set of time-series views of streaming network events—byte traffic, port usage, authentication events, etc. Such descriptive dashboards are a mainstay of the typical SOC.

Having presented three distinct data analytics modalities, an idealized security monitoring use case frames a fourth. Namely, security monitoring programs ideally wish to achieve reliable, probabilistically ranked alerts accompanied by explanatory context. This frames a fourth analytical category of special interest to CSDS practitioners: detective modeling. In this proposed frame, contextual alerts combine ranked predictions with explanatory logic.

In the proposed detective mode, a prediction is made (e.g., of a security incident) and the explanatory context concerning why the alert was surfaced is also provided for the benefit of investigators. In a SOC, this might take the form of a rich alert that supplies an explanation of the chain of events that led to the surfacing of the alert as an incident of a particular type. For instance, this might combine evidence of a compromised credential, a malware infection, interactions with DGA-associated URLs, and high byte traffic to a known-bad external IP address (i.e., blacklisted or reported for abuse).

The resulting frame supports the combination of machine learning predictive techniques, descriptive background, and explanatory context in a unified view that embeds a highly contextual justification for the alert appropriate for follow-up and remediation. Adopting a notion of detective modeling, as distinct from other modalities, supports and encourages CSDS practitioners to blend explanatory and predictive modeling to drive continuous detection in context.

The heuristic in Fig. 4.31 represents how the detective modality can be distinguished from description, prediction, and explanation—namely, by noting that inductive algorithmic methods (associated with correlation) are cross-applied with deductive (theory-grounded) approaches to explain and contextualize guidance to human operators.

4.4.5 Model Context and Validation

Having framed CSDS design guidance concerning scientific discovery, evidence, and integrating explanation and prediction, it is appropriate to focus on models and approaches to validating models. Models, as theoretical vehicles, are central to CSDS practice and are the manifest outcome of efforts discussed previously.

Data science models are central artifacts through which focused CSDS security questions are addressed, be they associated with security monitoring, design, stewardship, or responsiveness. Properly specifying models in terms of context and ensuring rigor through validation are of central interest to CSDS practice. It is desirable to specify the context and use of models in practice and thus to avoid the tendency to evoke models simply as hazy abstractions.

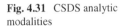

Fig. 4.31 CSDS analytic modalities

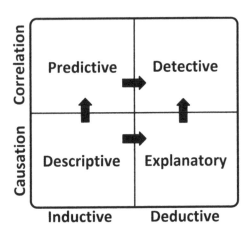

Developing operational and theoretical cybersecurity context to ground models was raised as a key challenge in CSDS practitioner interview feedback (60%, $n = 30/50$). Per the previous section, detective models were advocated as a specific vehicle for integrating security context in CSDS inquiry. If models are the vehicles for embedding context to ground CSDS insights, it is of interest to catalog the discrete *types* of models which can be specified and the methods for validating these models. This involves situating models as *theoretical vehicles* which represent and embed context to scrutinize security infrastructure, processes, or phenomena.

Invoking models as a singular species risks confusion. In practice and implementation, there are many forms and types of models. Specifying both the goals associated with modeling and the domain being modeled is central to orienting development, testing, and implementation efforts. The related specifications of models should be understood by the practitioner:

1. *Representational goal*: models are active artifacts that can be applied for a range of reasons—what is the goal of the inquiry and how does this necessitate particular methods?
2. *Domain*: what is the conceptual or systemic domain of the phenomenon being modeled—for instance, does it address technical, economic, or behavioral phenomenon or some particular conjunction thereof?

In terms of model *representational goals*, in the previous section CSDS modeling was distinguished into four types (see Fig. 4.31). To return to and summarize these modalities in terms of representational goals:

- *Descriptive*: presentation of structural dynamics and relationships
- *Predictive*: developing correlative models trained on evidence to identify new instances
- *Explanatory*: substantiating statistically validated causal theoretical constructs
- *Detective*: prediction supported and grounded by explanatory context

Concerning the model *domain*, establishing context for security environments and operations evokes both internal knowledge (e.g., local assets, infrastructure, users, vulnerabilities, etc.) and external awareness (e.g., external networks, partners, trusted external users and parties, attackers, and likely threats). Context implies that a data schema or related type of formal specification (e.g., taxonomy or ontology) exists to store, track, maintain, and update context for purposes ranging from monitoring to planning. In this sense, formal data structure specifications, be they based in entity relationships, schemas, or ontologies, are themselves models that embed theoretical suppositions regarding the presence of and interaction between key entities in a conceptual domain.

Challenges stem from conceptual confusions that develop when multiple systems or domains overlap and interact. This creates situations where paradoxical definitions arise, for instance, a "user" being an "authenticated userid" in a network context, versus being an owner and operator of several devices in a hardware context. What do we mean by user? Several equally valid answers arise, depending on the context of the domain. As a general approach, it is useful to distinguish the primary conceptual domains of technical (engineering), economic (risk), and behavioral (organizational), per Table 4.8.

Concerning the *technological* domain, the dominant engineering focus in the security field has been raised previously. While the overemphasis has been critiqued, the centrality of technical substrate is an essential foundation for modeling. To the degree economic and behavioral models are derived, they most typically directly operate off of or refer directly to structural characteristics defined by the technological domain (restrictions and constraints tied to technical substrates). For instance, cyber risk ultimately must assess vulnerabilities as tied to tangible network and device infrastructure. Frequently, this may suggest a network representation or graph model which defines the technical environment.

While behavioral dynamics open a range of complexities, in a security context, technical infrastructure imposes constraining rules, restrictions, and geographies. The technological domain in security is fundamental to modeling efforts, although practitioners should be conscious concerning monomaniacal tendencies in this regard. For instance, an overemphasis on the technological domain risks deferring to deterministic and rules-based models, whereas humans introduce behavioral uncertainty and variation which suggest probabilistic and inferential approaches.

Concerning the *economic* perspective, economic methods are a central mechanism for measuring cyber risk. CSDS as a discipline inherently opens a broad range

Table 4.8 High-level security contextual domains

Domain	Goals and methods
Technological	– Seeking patterns from inductive observations
	– Deterministic/engineering-focused representations of infrastructure
Economic	– Risk/value/efficiency analysis and optimization
	– Probabilistic and comparative understanding
Behavioral	– Substantiation of deductive behavioral propositions
	– Application of multivariate inferential statistics

of powerful approaches to measure risks. Cyber risk measurement is a prerequisite for controlling and preventing exposure, improving alerting and triage, and optimizing security detection and remediation operations. In terms of quantifying cyber risks, there are a broad range of attacks and attackers, everything from accidental incidents and company insiders, to extortion and cyber-fraud, to corporate espionage and state-sponsored cyber actors.

To understand the risks of exposure, frequency, and impact from events, it is important to develop a contextual understanding of at-risk targets and their relative susceptibilities. From here, one can match an understanding of the permissible means of attack, opportunities for staging attacks and incursions, and the corresponding motivations of prospective attackers (e.g., competitive advantage, revenge, personal gain, "just a job"). Given the behavioral aspects inherent, it is assumed such models entail inferential and probabilistic methods, as opposed to deterministic.

There are a broad range of approaches to assessing and implementing cybersecurity risk, including certifications and programmatic guidelines. The subject is voluminous, and its details are out-of-scope apart from the central notion that cybersecurity risk depends upon metrics and measurement. CSDS, at root, hosts a broad catalog of methods and processes for measuring security phenomena of all types. In this respect, CSDS is a core provider of methods that directly inform efforts to measure and monitor cyber risk.

Three works that cover the measurement of cyber risk in detail are Hubbard and Seiersen's "How to Measure Anything in Cybersecurity Risk" (2016), Freund and Jones's "Measuring and Managing Information Risk" (Freund and Jones 2015), and Wheeler's "Security Risk Management" (Wheeler 2011). All discuss and provide guidance on methods for eliciting quantitative measures relating to security across organizational, process, and technical factors. The focus is on establishing risk management programs and approaches, with the implication being that analytical methods support and amplify these efforts. Rigorous efforts to categorize, quantify, and track cyber risks have great benefits to clarifying and optimizing security operations, ranging from improved monitoring and coordination, to specifying design approaches, to informing security architectures (Muegge and Craigen 2015).

Turning to the *behavioral* domain, this identifies a frequently cited gap in approaches to security science, namely, the existing bias towards engineering and methods associated with the physical sciences, which are typically deterministic. As raised provocatively by Maxion et al., it is suggested that "better (security) models are probably the social sciences which deal with the same sorts of complex and uncontrollable environments as cybersecurity, but nonetheless have well established methodologies for performing experiments, analyzing data and reporting results so that others may build upon them" (2010). This advocacy is reinforced by Pfleeger and Caputo in terms of advocating behavioral science as a central approach to mitigating cybersecurity risk (2012).

Statistical methods common to the behavioral sciences address uncertainty by specifying probabilistic models that represent data-generating processes. Inferential statistics support diagnostics in conditions where the underlying phenomenon may not be directly measurable. This implies evoking empirical approximations of variable interactions and may infer the presence of latent variables. A key observation is that in many security environments, entities are being tracked discretely which are typically inferences to or references to an underlying phenomenon. For instance, a risk of susceptibility to phishing attempts may be discretely tracked by propensity to click on unverified links in emails. However, the more fundamental measure implies a latent user behavioral construct such as "user security sophistication" or even "security conscientiousness."

By investing efforts to identify underlying behavioral constructs, practitioners are better positioned to frame inference-based and probabilistic approaches rather than to assume deterministic dynamics. As a result, efforts to isolate and rectify underlying security shortcomings become clearer and ostensibly more efficacious. For instance, the propensity of users to click on unverified links can be treated as a mechanical policy breakdown, or it can be framed as a training theme and socialized via in-person outreach. The first, policy enforcement, positions a technical intervention, whereas the latter attempts to address a latent root phenomenon, namely, the need to socialize the importance of security awareness and risks.

The assertion made in the previous section (Sect. 4.4.4) regarding the need to combine predictive and explanatory approaches through detective analytics is relevant here. An assertion can be made that the popularity of predictive machine learning is, in effect, a symptom of attempting to extend deterministic approaches preternaturally. Given the engineering bias in the security domain, predictive machine learning offers a route to operationalize latent phenomena in correlative models while abandoning the overhead necessary to substantiate causal explanation. Correlative models have been critiqued as pure inductive approaches and thus given to missteps in navigating between overfitting and overgeneralization. Detective analytics has been advocated as a middle way for CSDS, blending predictive and explanatory approaches. To the extent that explanatory methods align closely with economic and behavioral domains via the social sciences, detective analytics is enabled as a modality.

Revising and iterating a framework proposed earlier in Fig. 2.11, a synthesis of analytic modality and model domain is posed in Fig. 4.32 to ground model development efforts. This framework is proposed to support practitioners in determining to what degree methods being applied align to the domain under scrutiny. As well it encourages explanatory insight. This framework advocates that practitioners consider to what degree economic and behavioral phenomena might underly observed technical phenomena.

As a simple example, looking at network usage dynamics on a typical corporate network, one will see a peak during local work hours and a drop late in the evening. This observes a decidedly fundamental behavioral phenomenon, namely, that most people sleep late at night. It is worthwhile to consider whether a network usage anomaly detection model should explicitly embed a notion of daily seasonality explained by this observation (i.e., the model expects that most humans sleep at night). Likewise, weekends and holidays might similarly be accommodated. Whereas a predictive-correlative model might embed such characteristics passively,

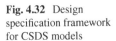

Fig. 4.32 Design specification framework for CSDS models

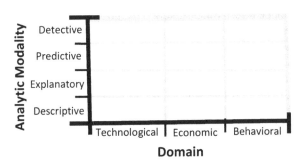

it becomes difficult to explicitly situate and resolve exceptions in gross correlative models, such as employees working in different time zones or insomniacs.

Embedding and validating behavioral explanatory understandings in models strengthens the ability to identify anomalous behaviors, namely, those behaviors which run counter to behavioral context and have no valid exceptional status. In the example above, it may be a user that is active on the network 24 h a day for 2 weeks, whereas previous usage was consistently infrequent. Depending on the particular network behaviors observed, one might hypothesize a malware infection and active exfiltration in such a circumstance.

Specifying the modality and domain of a model helps facilitate model development. As well, it provides a foundation for model validation. Clear approaches towards analytical model validation and explanation were raised in interview research as an advocated best practice (62% $n = 31/50$). This advocacy is also reflected in research (Verma et al. 2015).

Even as we recognize that models are formal, specified abstractions, validation is often elusive, especially in uncontrolled settings (i.e., outside of laboratories and controlled experiments) (Balci 1998; Pidd 2004). The objective for CSDS practitioners is thus to discretely understand where models are imperfect but useful enough in an operational setting as associated with focused goals (Mongeau 2013a). If CSDS is to be regarded as a rigorous field, it must develop a refined understanding of how CSDS models are useful abstractions, while making efforts to accommodate the resulting uncertainty.

Modeling and model validation encompasses a rich research milieu, spanning domains associated with epistemology (Morrison and Morgan 1999), economics (Morgan 2012), decision-making (Ansoff and Hayes 1973; Kaner and Karni 2004), and management (Dolk 2010). Cyber modeling in the context of risk management is its own research niche (Bevan et al. 2018; Boehm et al. 2018; Brockett et al. 2012; Hubbard and Seiersen 2016; Waldron 2019). As well, there are a broad range of frameworks and methods to support practitioners in efforts to measure and validate cyber risk from varied perspectives (Waldron 2019).

For the purpose of bolstering CSDS practitioner rigor, we focus here on providing a practical framework for validation. As an aspect of scientific rigor, Maxion frames validation as a core principle related to rigor in experimentation. He offers that experimental rigor requires:

- *Repeatability*: measurement approaches are standard and repeatable
- *Reproducibility*: experiments are capable of being repeated and provide the same results
- *Validity*: experimental design which assures internal (not subject to alternate causes) and external (generalizable) coherency, subject to tradeoffs

The implication is not that models should be treated directly as experiments but that ideally the validation of a model should assure and respect key principals of experimental rigor. A validated model should provide assurance that its measures (as inputs) are stable, regular, and understood; that the model is able to reproduce its results over time and under similar conditions (or that model degradation is understood and closely monitored); and that trade-offs between overgeneralization and over-specification are formally addressed in design.

Assuring model validity is approachable in controlled environments or under restricted circumstances. However, as practitioners attempt to represent "real-world" phenomena (ex-laboratory or uncontrolled), greater complexity and more variables are inevitably introduced. While more sophisticated computational and algorithmic methods can be applied to accommodate complexity, for instance deep learning, such approaches, as raised previously, suffer a loss to transparency (explicability). As well, it becomes more difficult to discretely identify baselines for navigating the over-specification, overgeneralization trade-off.

It is otherwise notable that models attempting to capture complex phenomena progressively deviate from the aforementioned aspects of experimental rigor: in being increasingly unable to assure the same items are being measured in the same way; in being unable to reproduce results as complex dynamics in the environment shift; and in being increasingly difficult to validate discretely (particularly as grounds for measuring over-specification and/or overgeneralization are poorly understood).

More complex models raise broader risks, including organizational factors. In considering model risk, five themes related to model complexity should be considered by CSDS practitioners:

1. *Stakeholder complexity*: complex models may cover broader domains and embed broader assertions, invoking a range of organizational stakeholders with diverse interests and goals, some of which may conflict.
2. *Probabilistic complexity*: interactions resulting from the aggregation of multiple probabilistic uncertainties lead to broader variation in projected outcomes, with some probabilities potentially overwhelming the model depending on framing.
3. *Inter-systemic complexity*: model complexity resulting from representing many interacting dynamic systems.
4. *Functional complexity*: interactions resulting from mixtures of multiple functional domains (e.g., financial, operational, technical).

5. *Technical complexity*: composite systems and technologies introducing multi-layered complexity.
Mongeau (2013b)

The rise of more sophisticated analytical methods and larger datasets increases model complexity. Rising model complexity complicates validation efforts, defined as testing which confirms that the model approximates the behavior of the system under assessment (Pidd 2004). As the systems under scrutiny broaden and expand, so does the purview of validation.

There is a growing movement advocating "explainable AI": methods for examining the actions and assumptions embedded in algorithmic processes (Samek et al. 2019). However, while the methods for validating complex models are advancing rapidly, organizational validation approaches are lagging, being typically quite rudimentary and unstructured. Business analytics model validation processes can be broadly characterized as ad hoc, consensus-based, confidence-building exercises (Sargent 2013). As such, model validation can be seen as an organizational sensemaking effort: an attempt to deem a model loosely "good enough" in its concordance with target phenomenon. Beyond this, it is generally recognized that exhaustive validation of an analytical model (ex-laboratory, uncontrolled) is not formally achievable, being at best falsifiable (Balci 1998; Pidd 2004).

There are both *practical* and *epistemological* limits to model validation rigor (Balci 1998; Kleindorfer and Ganeshan 1993; Kleindorfer et al. 1998; Morton 2001; Oreskes and Belitz 2001; Robinson 2004). Proportionally increasing overhead (time, expertise, expense, resources) offsets the proportional value gained from iterative validation rigor (assurance of concurrence, repeatability, reliability, experimental or quasi-experimental concordance). The inverse relationship between validation effort and realized value suggests a "good-enough" equilibrium point where the highest value in rigor is at the point of lowest proportional cost, as illustrated in Fig. 4.33 (Ansoff and Hayes 1973).

A model is a useful abstraction of "real phenomenon," but it is generally agreed that validity can never be fully established as a "truism," only falsified in the Popperian sense. Models are imperfect abstractions and, as such, can never be

Fig. 4.33 Model validation optimality

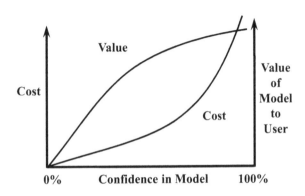

comprehensively validated against more complex reference phenomenon. As Box states: "all models are wrong; the practical question is how wrong do they have to be to not be useful" (Box 1979; Box and Draper 1987). The notion of "proof of general concordance," or even rigorous falsification, as Box alludes, is considered methodologically impossible for all but the simplest of models and even then only when pertaining to highly controlled and specific circumstances (Pidd 2004).

For business practitioners, however, on a pure overhead basis, ideal rigor is abandoned long before questions of epistemological validity are encountered. For practitioners, the establishment of rough stakeholder confidence is often cited as the ideal endpoint to achieve in the validation effort. This is understandable. Organizations face a range of pressures to rapidly deploy new and improved algorithmic decision models in order to address increasing complexity. Testing and validating complex models involves investments of time and effort. The sheer complexity of advanced algorithmic models often precludes rigorous validation efforts (Armstrong et al. 2009). Commercial organizations are pressured to take shortcuts, for instance, by proceeding based on rudimentary measures of efficacy such as ROC in predictive machine learning.

As related previously, there is a growing specific critique concerning overemphasis on the development and deployment of predictive machine learning models. Being focused on prediction, most commonly through correlation and not explanation, operationalizing predictive models without a corresponding effort to conduct diagnostics risks a range of problems, from unreliable results to an improper specification. This was raised frequently in interview research as a critique of, and discomfort with, the perceived overapplication of predictive machine learning in operational settings. Several practitioners noted organizational pressure from management to deploy predictive models based on a popular, though largely unexamined, perception that machine learning was, by nature, generally efficacious and led directly to operational efficiencies.

Tied to this was a skepticism and frustration on the part of practitioners that not only was their "hand was beings forced" by non-data scientist stakeholders to deploy predictive techniques but that there was a lack of understanding concerning what constitutes proper validation of the predictive models, calling into question the results that were then produced. A clear requirement that results is for organizations to define and apply stricter processes to not only validate but to explain the behavior and assumptions resident in predictive models. However, it also raises a more fundamental challenge concerning socializing the limits of models and of model validation with nonexpert stakeholders in organizations.

Improving organizational approaches to model validation necessitates a broader approach encompassing the full model life cycle. Organizations should clearly define standard approaches for model development, selection, evaluation, and testing. This recognizes that a model is more than an operationalized decision-making mechanism. A good model is inherently hampered by bad data, so integrity of the source data is essential. Biases in the model must be overcome, for instance, through techniques such as cross-validation. Overfitting must be addressed, the potential that the model performs well only on the dataset used to train and validate the model. This can be addressed by setting aside holdout data and testing varying subsets of the dataset and/or variant timeframes. Socializing and standardizing such approaches are crucial to bolstering rigor in CSDS modeling efforts.

Per previous discussions on data management, analytics as a process, and detective models, it has been consistently advocated that model design and validation should span a broader purview than simply instantiating an algorithm and conducting testing specific to the algorithm. Interview feedback raised this as a frequent and seductive misstep observed in CSDS practice - a tendency to fetishize algorithmic instantiations and sidestep more basic epistemological foundations.

Results from validation testing often suggest returning to more fundamental aspects of the analytics process. If a model is overspecified, there may be opportunities to aggregate measures or to seek more fundamental latent measures. If a model gives evidence of overgeneralization, the practitioner may seek new measures or transformations which present nuances regarding the phenomenon under investigation. This could include, as suggested previously, extrapolating additional evidence or attempting to identify statistically meaningful categories as novel designations. In either case, torturing existing data on-hand through the model should not be the default redoubt. Both researchers and practitioners should carefully heed subtle signals in model design and validation that the data presented to the model may be improper, misaligned, insufficient, or in need of transformation. Efforts invested in this regard are typically much more efficacious in improving performance and in improving rigor than tweaking model parameters.

At the most basic level, the results of model validation may suggest reframing the fundamental research question. Practitioners should not be seduced by algorithmic models and should be open to "killing their darlings," namely, the implied theoretical suppositions embedded across the dataset and the model as a composite. Ideally, researchers and practitioners alike undertake a theoretical introspection concerning the underlying assumptions embedded in data (as quasi-models) and in models themselves. There should be a willingness to challenge and reframe these assumptions, to frame new, alternate, or counter-hypotheses. As advocated by Banerjee and Duflo, "the only recourse we have against bad ideas is to resist the seduction of the 'obvious', be skeptical of promised miracles, question the evidence, be patient with complexity and honest about what we know and what we can know" (2019).

4.4.6   Towards Cybersecurity Theory

An overarching implication raised in the literature review was that for CSDS to advance as a profession, body of theory must evolve. *Technê*, craft, must develop a foundation of *epistêmê*, knowledge, through the vehicle of *logos*, structured reason. It has been asserted that systematically aligning data science methods with diagnosed security gaps provides an approach to iterating CSDS body of theory. A related challenge concerns the lack of an underlying theoretical tradition in the security domain. As the security domain is engineering-oriented (*technê*-focused), a strong theoretical basis on which to build CSDS body of theory (*epistêmê*-focused) is often lacking.

Design prescriptions concerning data management and scientific processes have provided a set of treatments to support theory building. However, a keystone remains, namely, the security domain itself lacks strong theoretical underpinnings (Craigen 2014; Longstaff 2012; Maxion et al. 2010; Meushaw and Landwehr 2012; Pavlovic 2012; Tardiff et al. 2016). This impedes CSDS development. However, CSDS, as a set of methods supporting both structured measurement and diagnostics, is itself a promising vehicle to rectify the shortcoming in that it is a route to substantiating theoretical propositions.

A finding in interview research was that CSDS practitioners identified an engineering and technology bias inherent in cybersecurity professional practice as a key impediment. This theme overlapped with the critique of a general overreliance on qualitative versus quantitative decision-making, a deference to experience over statistical rigor. As well, it explains the frequent invocation of methods from adjacent domains where native security domain methods are lacking. Practitioners cited eight external fields as having a strong influence on the methods they apply in CSDS work, per Sect. 3.2.3.4.

As the security domain is frequently framed in engineering terms, gaps are typically addressed as engineering requirements. This tends to diminish efforts to frame security problems in behavioral contexts (Pfleeger and Caputo 2012). For instance, despite the fact that social exploits, phishing in particular, are by far the main vector for gaining unauthorized access to systems, training users to modify their behaviors is less often invoked as a prescription than engineering treatments (i.e., blocking systems, monitoring technologies, advanced detection mechanisms, etc.).

From a security practitioner perspective, the deference to engineering treatments surfaces as a common confusion between two main forms of professional judgment, as framed by Dawes et al.: clinical, expertise aggregated from ad hoc field experience, and actuarial, empirical knowledge validated through statistical observations and testing (1989). It is the latter which builds theoretical foundations. Statistically validating hypotheses supports building a complex of connected assertions. Over time, a body of substantiated connected assertions provides a foundation for a body of theory.

The assertion that an overreliance on engineering and practitioner know-how self-limits professional development may not be immediately apparent. Technical progress provides direct physical evidence of advancement—witness the digital society. What is less apparent is when technical progress becomes self-limiting. Poorly implemented practical advancements may lead to sub-optimal outcomes. The cybersphere as a whole can be considered as a manifest lesson in this regard. Even the originator of the World Wide Web protocols, Tim Berners Lee, has repeatedly warned that web technologies have promulgated far in advance of what was intended and that sound controls and indeed more robust approaches have been pushed aside in the rush to build higher towers on shaky foundations (Hern 2019). As Barabási (2014) reflects on the growth of the World Wide Web:

> It is an excellent example of a "success disaster," the design of a new function that escapes into the real world and multiplies at an unseen rate before the design is fully in place. Today the Internet is used almost exclusively for accessing the World Wide Web and e-mail. Had its original creators foreseen this, they would have designed a very different infrastructure, resulting in a much smoother experience. Instead we find ourselves locked into a technology that adapts only with great difficulty to the booming diversity and demand imposed by the increasingly creative use of the Internet.

The implication is that pursuing *technê* without reference to *epistêmê* may lead to self-limiting, sub-optimal, or even destructive outcomes. Structured reason and general knowledge developed through quantitative means ideally accompany experiential craft. Actuarial (statistical) judgment has been demonstrated to outperform clinical (field experience-based) appeals to knowledge in professional settings (Dawes et al. 1989). Whereas professional experience can be a foundation for structured statistical analysis and evidence-based approaches, it is substandard as the sole redoubt.

Statistical judgment requires concerted efforts to gather evidence, experiment, test, and statistically substantiate assertions. Commenting on the poverty of scientific theory in the security domain, Tardiff et al. comment:

> A knowledge gap exists with respect to developing the science of cybersecurity. In other technical fields, a scientific method is typically developed and applied to address these needs. The benefit of such a science is the development and testing of theories that lead to understanding the broad sweep of cyber threats and the ability to assess trade-offs in sustaining network missions while mitigating attacks. Applying the scientific method also leads to findings and conclusions that are repeatable, and verifiable, thus providing a dependable knowledgebase to the broader cybersecurity community. (2016)

This brings the CSDS domain to the core of the challenge to individuate and professionalize: clarifying a credible path for extrapolating and validating a theoretical CSDS body of knowledge. It is asserted that CSDS can best help itself by providing the methods and means for iterating body of theory foundations in the parental security domain.

While data science, itself an emerging domain, brings a rich methodological toolbox to the table, it is not in of itself the solution to the problem. Data science, as a practitioner movement, has also suffered critiques for being lopsidedly engineering-focused (Ekbia et al. 2015; Press 2013b; Rose 2016). Developing theory is a process that can be facilitated by computational and methodological means. However, it takes a community of researchers to frame research questions, apply judgment to the suitability of experiments, and assess the theoretical implications of experimental findings (Craigen 2014; Tardiff et al. 2016). The collaborative process for sharing data, comparing models, and retesting and unifying results completes the social process of scientific research (Pavlovic 2012).

A simplification of the scientific approach can be framed as the combination of:

- *Experimental inquiry:* a process of constantly questioning assumptions underpinning understanding; proposing weaknesses and failures of current understanding; seeking alternate explanations, exceptions, and inconsistencies; and testing the limits of understanding in novel ways
- *Theoretical knowledge:* a standing *body of knowledge* that encompasses the aggregated understandings from observing, measuring, and experimenting

Theory, in this context, admits to a broad range of experimental approaches. Experimental methods may be *controlled*, comparing tightly constrained treated instances to a holdout sample and focusing on key variables to substantiate efficacy; *natural* (quasi-experiments), observing focused effects across a range of variables; and *field*, focusing on observed variables in a natural setting outside a controlled environment. Further, *observational studies*, which are not regarded as rigorous experiments, nonetheless provide insights when it is impractical to strictly control parameters. Observational studies can be instructive in framing hypotheses for more rigorous experiments and may suggest approaches to narrowing and refining data for assessments.

Generally, in an experimental approach, there is a hypothetical assertion that a dependent variable Y responds in conditions where some combination of independent factors $X_{1,2,3,N}$ are involved. The assertion is typically formalized as an explanatory construct that specifies the interactions of related variables, which may directly affect, mediate, moderate, or correlate with the target factor and/or one another. Such an assertion, focused on causal explanation, need not lead to strong predictive integrity, although predictive power may be a byproduct.

Given the complexity of operational security environments and the demands put upon practitioners, it is understandable that pursuing and attaining experimental rigor may be considered impractical. As has been covered in detail, the challenge of gathering and aggregating evidentiary data in dynamic environments poses its own challenge. However, beyond strictly controlled experiments, there are a range of approaches feasible, from observational studies to pseudo-experiments. The assertion is that rigor can be triangulated gradually and stepwise. When controlled experimentation is infeasible, looser approaches can still provide iterative value. Key is the discipline to build and validate evidence-based statistical understandings of security dynamics (actuarial judgment, in the invocation of Dawes et al.). Building an interconnected framework of statistically validated insights provides a foundation for theoretical assertions.

Given the complexity and range of unknowns in typical security environments, it is often beneficial to undertake systematic discovery and description prior to launching experimental or pseudo-experimental efforts. Explanatory-descriptive approaches are a promising approach to context-building. As context, in the form of statistically valid categorical designations, is often missing (e.g., statistically meaningful types of users and devices), primary exploration and description provide a route to sensemaking in complex environments. A hypothesis may be associated with a descriptive assertion, for instance, the assertion that an observed statistical segment of a population is a coherent, explainable, and *durable* category—and thus suitable for tracking and anomaly detection purposes. Statistical substantiation of segments as categories, in the form of demonstrated explanatory and predictive integrity (invoking detective analytics), provides great value in operational settings.

Observations producing new variables in the form of valid categories are a central approach to building structured understandings in otherwise noise-filled, complex environments with many unknowns. The process of substantiating an operational category from validated statistical segments observes a loose process of hypothesis framing and validation. For instance, tracking user patterns associated with the aggregated use of applications, devices, and networks provides a basis for segmenting self-similar groups of users.

Under scrutiny, statistically segmented groups of users can be labeled with an explanatory designation—for instance, technically sophisticated, highly active users versus unsophisticated, occasional users. The two may amount to different species in terms of security relevance, and tracking separately may expose differences meaningful in a security context—for instance, statistically unique risks, exposures, ranges of expected behavior, propensity to be involved in incidents, etc.

Beyond an observance of unique statistical patterns, a substantiated categorical designation provides a basis for forming follow-on theoretical assertions. Namely, observed statistical patterns provide a foundation for segmenting users. Descriptive analysis and operational substantiation lead to explanatory categorization, e.g., statistically categorized user roles with meaningful explanatory labels. For example, a descriptive pattern extrapolation approach such as cluster analysis (a type of unsupervised machine learning) may result in the assertion of statistically self-similar multivariate user groupings. When examined and tied to an explanatory model, for instance, through a loose theoretical construct, a hypothesis is framed concerning expected user behavior based upon key distinct measures. Statistically substantiated categories make possible a range of new discoveries and treatments.

The assertion of meaningful descriptive categories with operational import then becomes a basis for making a range of hypothetical assertions, which, if substantiated, begin to approach a theoretical observation. An example may be the observation that less active users generate smaller amounts of network traffic and generally have a smaller variability in network behavioral measures (e.g., byte traffic, port usage, multi-device authentications, network connections, scanning events). It might be asserted and verified that these users constitute the majority of users on an organizational network. It could also be asserted that, in contradistinction, a minority of very active users tend to generate the bulk of network behaviors and have a correspondingly higher variation in their range of activities.

It could be hypothesized that there is a quantifiable Pareto principle observable over time, whereby ~20% of the users consistently generate ~80% of network traffic. Further, this could be posited as a manifestation of a power law in scale-free networks. It might also be hypothesized that when network connections per user are aggregated and represented as a graph, highly active users are frequently associated with highly connected hubs in the graph (as a representation of network usage). This then observes and explains the Pareto principle observation in terms of graph theory in nonrandom scale-free networks. Namely, highly connected hubs can be expected to appear in ordered scale-free networks and that hub-propensity cleaves roughly to the Pareto principle. The theoretical assertion is that typical organizational computer networks evidence common scaling characteristics of general networks and network phenomena as substantiated in graph theory fundamentals (Barabási 2014).

Starting with a modest hypothesis regarding differences in network usage, it becomes possible to build a complex of testable hypotheses that begin to make general theoretical assertions and observations. In this case, an observation of usage patterns leads to a theoretical proposal that network usage patterns segment users according to a classical Pareto rule and that highly active users, although the minority (±20%), account for the great bulk of traffic and activities on the typical network (±80%). Theory then begins to gain substance when this is asserted as a common principle in organizational computer networks that can be substantiated statistically through observational studies. More formally socializing and empirically testing such theoretical assertions offers value to a broader community of security researchers and practitioners as body of theory—*epistêmê*.

Such chains of assertions can be examined and tested over time to substantiate the underlying hypotheses to strengthen theoretical understanding. While an observational analysis, and not a rigorous experiment, the identification of and assertion of categorical segments have inherent value as it supports refined anomaly detection. It is otherwise easier to spot deviations within a reliably segmented statistical group, the group being defined uniquely by constrained multivariate parametric behaviors. From there, greater potential value arises in proposing that more fundamental phenomenon underly the observations.

By asserting and substantiating small observations progressively, it also becomes possible to frame and experimentally test discrete interventions, for instance, through randomized controlled trials (RCT) (Banerjee and Duflo 2019). As an example, combinations of technical treatments and user training to reduce security incidents can be tested on a focused group, while a control group remains untreated. Outcomes statistically substantiate or refute treatment efficacy.

The process of segment exploration leading to categorical explanation can be more generally represented as a process of scientific discovery which involves:

- *Discovery:* a process of statistically substantiating novel categorical, taxonomic, or ontological designations. For example, observing patterns over time and applying diagnostic statistics to validate segments; instantiating proposed segments as a probabilistic benchmark to predict future behaviors; substantiating and refining key features with demonstrated integrity over time; proposing threshold boundaries for identifying anomalous, ex-segment behavior; and investigating anomalies to substantiate the predictive integrity of surfaced anomalies (e.g., as hypotheses of security threat associated behavior)
- *Theoretical*: proposing that segmented statistical patterns encapsulate and represent explanatory categories in general ways. For example, categorizing statistically distinct groups of entities; determining key features; applying explanatory interpretation (labeling users or devices into proposed explanatory categories);

and applying knowledge extrapolated from discoveries to test new operational monitoring powers and efficiencies

Resident in these approaches are two core aspects: an experimental process involving hypothesis framing and testing (involving the assertion of a meaningful categorical designation) and theoretical knowledge gained from experimental results and deductive insights (with a stipulation concerning how theoretical propositions can be further tested and falsified over time).

Combined, these principles provide a basis for demarcation, disassociating science from pseudoscience, namely, the ability to frame scientific questions as hypotheses susceptible to empirical testing that clearly establish grounds for propositional falsification. A central staple of scientific demarcation concerns testable hypothesis, the ability to test and disprove assertions (Shostack 2012). Karl Popper framed falsifiability, as opposed to verificationism, as the central criterion for demarcation from pseudoscience: "statements or systems of statements, in order to be ranked as scientific, must be capable of conflicting with possible, or conceivable observations" (Popper 1963).

Hypotheses that have been validated empirically are grounds for theoretical propositions, such as "all swans are white" (based on the observation of only white swans). This allows, in turn, for falsification: a validated hypothesis underpinning a theoretical proposition can be discredited through a subsequent empirical observation, such as identifying black swans in Australia. Theory thereafter must either be modified, shifted, or retired, for instance, proposing that "all swans are either white or black or grue" or that "swans may be many colors depending on their genetic predisposition."

It is useful to further distinguish theory in terms of its epistemological status beyond simple empirical tests. In particular, it is important to understand divisions between practice and theory and to demonstrate the value of the latter. The medical domain is a gravid example of the pragmatic and mutually reinforcing relationships between theory and practice.

The medical field has been used as an exemplar for CSDS multiple times throughout this inquiry, both as a representation of a matured profession and as a field that addresses remarkable complexity by mediating between theory and practice. Medicine embeds the integration of practice-oriented, clinical approaches with scientific, statistically, and experimentally supported findings that frame theoretical underpinnings. Whereas primitive medicine emerged as pure practice, there often were unvalidated treatments based on superstition or hunches that were not efficacious, or worse, resulted in iatrogenesis—harm caused by the healer (e.g., bloodletting) (Taleb 2014).

The rigorous application of the scientific method, particularly evidence-based methods to assess patient outcomes, resulted in improvements to and the evolution of medical treatments. Primary investigation, observation, and experimentation led to the development of general theory, for instance, proposals on the nature, action, and interaction of the various biotic systems—circulatory, respiratory, digestive, nervous, etc. The staged validation of theory led to improvements in patient outcomes and created opportunities for alternate or entirely new treatments, be they surgical, pharmaceutical, or mechanical.

To illustrate a second example of the distinction and benefit of theoretical knowledge, for thousands of years, humans fermented alcohol without a knowledge of yeast and microbiology. Engineering progress led to innovations and scale, including process industrialization through breweries and distilleries. However, it was the revelation of the role and action of yeast, supported by theory regarding fermentation, which allowed for substantial improvements to fermentation and brewing operations, even to the extent of selecting and genetically manipulating strains of yeast to optimize characteristics such as yield, efficiency, and taste. Again, in this example an experimental validation of theory led to new approaches and improvements to a previously less well-understood process. Whereas pure craft—*technê*—led to operational developments, knowledge development, *epistêmê*, made possible entirely new, highly efficacious advances.

In the case of security phenomena, it is useful to view CSDS as an attempt to model cyber phenomena for the purpose of improving understandings of fundamental phenomena. Theory in this sense involves propositions concerning the nature of interactions between and associations amongst, humans, technical devices, and infrastructure in complex networked digital environments. The potential benefits of security theory are not hard to spot. For instance, it is not difficult to identify iatrogenic outcomes in security treatments, for instance, enforcing strong random passwords which leads users to write the unintuitive string of alphanumeric characters on a Post-it note attached to their terminal. Building behavioral security theory through social science research methods offers a range of potentially beneficial insights to avoid sub-optimal outcomes and to optimize holistic security (Pfleeger and Caputo 2012).

It is advocated that security professionals consider the benefits of CSDS methods to facilitate security theory development across a range of approaches. The range of approaches ideally transcends pure controlled experiments. Controlled experiments often necessitate a set of primary exploratory findings and a priori propositions from which testable hypotheses can be derived. Environments with many unknowns require primary efforts associated with exploration and discovery. Craigen, in "Assessing Scientific Contributions: A Proposed Framework and Its Application to Cybersecurity" (2014), proposes a set of approaches to primary security theory building:

1. Defining constructs
2. Defining propositions
3. Providing explanations to justify
4. Determining scope
5. Testing through empirical research

In this perspective, pre-experimental explanatory and hypothesis-building elements are advocated. A construct constitutes a structural model of the interaction amongst key variables associated with a phenomenon.

For instance, proposing that the difficulty of memorizing and managing passwords leads to the unintended exposure of passwords might propose that "a difficult to memorize password" (as an empirically measurable variable) has a central causal connection to "exposing passwords" (e.g., measured by known incidents of incidents of exposure, such as writing the password down and leaving it in an unsecured location).

In Craigen's framework, empirical testing is positioned after primary efforts to structure concepts, propositions, and explanations. The importance of investing in up-front efforts to establish scope prior to pursuing empirical testing is otherwise a key assertion. Given the complexity of modern security environments, there is often a deep need for primary sensemaking in advance of experimental efforts. This may, as previously asserted, involve extrapolating and substantiating statistical categories and designations of distinct forms of behavior. Key is that structured efforts to understand and describe phenomena often underly and precede formal experimental efforts (*logos* facilitates *epistêmê*).

The implication is that the poverty of theory in the security field necessitates stronger and more coordinated efforts at primary sensemaking. Socializing the value of sharing general observations as a prelude to formal theory construction incentivizes security professionals to begin with humble and simple observational theory. This avoids the risk of abandoning scientific rigor with the reasoning that controlled experiments are too costly and time-consuming. Modest foundations are needed to establish and build basic understandings.

Constructs, propositions, explanations, and scope-setting are benefitted by building and socializing conceptual propositions in the form of models. Tardiff et al. (2016) raise the importance of framing conceptual models to develop and test theoretical propositions in the security domain:

> Very often, emerging topics are poorly understood... initial conceptual models support the development of research questions that can be formalized as falsifiable hypotheses and investigated with experiments (deductive reasoning—testing a general statement or theory using specific observations). The goal for applied research is to eventually develop a sufficient understanding such that mature models can be validated for operational use.

This view advocates the approach of starting with basic sensemaking propositions to build up a platform for testable assertions. The implication is that systematic exploration and discovery efforts lead to progressive knowledge generation in the form of linked theoretical assertions. Linked assertions form a theoretical foundation upon which experimentation can progress.

A focused security domain example of the need for primary sensemaking concerns efforts to clearly define the nature of key digital entities. Simple entities such as devices and users are increasingly difficult to discretize in modern security environments. The discrete nature of physical devices is increasingly degraded in that many devices host virtual, containerized, remote, or microservice-composite applications. Physicality is no longer a discrete embodiment of a digital device; securing a device physically is insufficient to assuring security when its operation inherently depends upon distributed services and systems. A device, including the common modern smartphone, is less a discrete physical hardware object and is increasingly an agglomeration of distributed networked processes and agents interacting with complex external ecosystems.

Even a discrete definition of a user in a security context can quickly become theoretically vexing. The explosion of autonomous and semiautonomous sensors, smart devices, and industrial devices are rapidly outnumbering human agents (IDC 2014; Lohr 2011; Morgan 2017a). We are increasingly faced with automated and human-machine complexes which muddy the waters concerning discrete human agency.

Indeed, the discrete "human-ness" of human agents in a digital context is called into doubt with the advent of environments in which semiautonomous digital agents act on behalf of, interact with, or act around, between, or underneath human agents. Such a circumstance evokes an age of *cyborg agents*. Cyborg agents increasingly operate alongside and with (or in tension with) human agents on networks, whether humans united with semiautonomous apps (i.e., bots) or wholly autonomous automated agents (i.e., virtual robots).

Virtual robot or bot agents and machine-driven processes are active, and increasingly so, in autonomous ways. Whereas an autonomous agent may be initially impelled by human agents, they increasingly are capable thereafter of sophisticated decision-making behaviors. For instance, virtual agents may be impelled to pursue and achieve quite sophisticated goals when given permission and motivation by human actors. Think in this case of a travel agent bot, perhaps initiated by a mobile app, which, once impelled by a human, performs quite sophisticated searches to optimize and summarize routes and prices on behalf of the human requester. Queries and calls may be made to a broad range of systems and services, routes outlined, and prices compared, leading to a set of summarized offers to the end user. Such dynamic virtual agents operate as semiautonomous expert decisioning systems. At the bleeding edge, sophisticated human-like behaviors can be simulated, such as evidenced by chatbots, industrial robots, and autonomous vehicles.

A human typically impels and motivates an agent, but agents thereafter pursue an embedded complex logic leading to sequences of behaviors that may be difficult to predict. Bot agents generate behaviors relevant in a security monitoring context, for instance, conducting dynamic web searches, aggregating data, consuming memory and processing resources, connecting across multiple systems, communicating across networks, coordinating with or negotiating with other agents and services, framing and fulfilling financial transactions, etc.

Significantly, the growing complexity of the digital ecosystem in which virtual agents interact may result in nondeterministic reactive behaviors, leading to unexpected outcomes. This phenomenon is observable in the financial industry as related to automated high-frequency trading. Sophisticated autonomous agents may produce unintended nonlinear effects in markets, such as the 2010 flash crash, which was tied to automated trading programs (Crowe 2015).

Similar phenomenon, whereby aggregations of unintended behaviors result from complexes of autonomous interacting agents, is a worrying and largely unexamined risk emerging in broader realms. Unexpected group behaviors emerging from masses of interacting autonomous agents are a potential risk to the stability of a range of physical and virtual domains, for instance, automated factories, municipal infrastructure, or military robots and drones acting in congress (Schneier 2018).

It is becoming increasingly less clear where human and semiautonomous agent behaviors can be discretely disassociated in monitoring: they both evidence variability and seemingly complex unpredictable behaviors such that one cannot easily determine where human and bot behaviors diverge. As an example, given the well-documented action of advertising bots (Pein 2017), an employee simply leaving a webpage open to a mass media news website overnight spawns and generates a stream of complex, continuous network connections and behaviors. Autonomous browser-enabled ad bots continually make calls between the browser, the underlying system, and a complex ecosystem of external online advertising intermediaries seeking to scrape user information, push dynamic content to the browser, and correspond with a complex network of third-party intermediaries.

Even hardware component and device vendors are increasingly active in seeking to scrape and monitor user behavior, pushing signals of indeterminant meaning to a complex of often unknown and unaccountable remote third parties. While not necessarily constituting a security threat, the intrusive and persistent action of autonomous agents operating via web browsers and embedded devices increasingly creates an inscrutable and complex set of network behaviors that subsume and flummox attempts to isolate noise from signal in organizational network monitoring.

The growing challenge concerning how to accommodate the presence of semiautonomous agents in security monitoring offers a demonstration of the benefits of theoretical framing over pure engineering-based approaches. In particular, it demonstrates how CSDS insinuates itself usefully during the process of theory development, proposing methods to address gaps in knowledge. Whereas a stock engineering solution might seek to, for instance, tag and register bots, it is not clear this would be feasible or efficacious in improving monitoring. Rather, first developing a conceptual understanding of the nature and action of bots offers opportunities to propose novel approaches to classify and gain knowledge about the behavior of bots through discovery and experimentation.

A theoretical approach, per Craigen's advocacy, focuses on building conceptual knowledge through observations, propositions, and explanations. To demonstrate the process, one might begin with a propositional definition for a cyborg agent as a goal-seeking collaboration between human and semiautonomous agents. We might then examine how the proposed definition connects to related theoretical research traditions.

For instance, the concept of artificial agents as proxies and partners to (and occasionally as adversaries against) "natural" human intermediaries connects to design science through Herbert Simon and his invocation of the artificial and the notion of artifacts. Simon proposes the focused study of artifacts in terms of their goal-oriented position as an interface between an inner and outer environment. He identifies artifacts as being goal-oriented and that "fulfillment of purpose or adaptation to a goal involves a relation among three terms: the purpose or goal, the character of the artifact, and the environment in which the artifact performs" (Simon 1996).

Framing cyborg agents also invokes a number of related, established theoretical traditions, including: systems theory (von Bertalanffy 1968, 1981), complexity science (Castellani and Hafferty 2010; Erdi 2008; Lewin 1993; Miller and Page 2007), computational organizational science (Carley 2011), multi-agent theory (Aart 2004; Epstein 2006; Horling and Lesser 2005), social network analysis (SNA) (Kilduff and Tsai 2003; Popov 2003; Prell 2012; Tatnall 2003; Tushman and Fombrun 1979), and actor-network theory (ANT) (Latour 2005). The latter theoretical body, Latour's ANT, in particular provides a theoretical foundation for comingling the analysis of the dynamic agency of people and systems operating in concert (Tatnall 2003). ANT would view a cyborg agent as a composite goal-seeking decision-making system. Per Walsham, "actor-network theory is concerned with investigating the social and the technical taken together" (1997).

In terms of how such phenomenon can be measured and understood through applied CSDS, we may work out a series of propositions concerning the expected action of cyborg agents, measurable behavior which can be observationally validated. For instance, asserting that a data-generating process produces measurable records of the sequence of actions between the human and autonomous agent. We can also propose that when acting autonomously, a cyborg agent may be expected to pursue a particular set of goal-seeking behaviors that can be categorized, measured, and constrained within a particular expected range. Agents may be categorized with similarly behaving agents. It thus becomes progressively permissible to track and monitor cyborg agents in terms of categorical designations tied to statistical behavioral patterns.

We might hypothesize that there are particular measures that indicate when a cyborg agent has been initiated, measures which result from subsequent actions of the bot, and measures concerning the results that are conveyed to the user. Further, we might propose that there are a set of measures that occur in a sequence which identify a cyborg agent. For instance, this might constitute measures associated with particular target and destination IP addresses, ports activated, and byte traffic generated.

We could also propose that latent cyborg agents, for instance, advertising bots embedded in browsers, activate when users visit particular well-known websites. From here, we could opt to consider whitelisting the traffic from and to the recognized sites, along with the complex of activities occurring during the session.

The thought exercise gains import when we propose malware as a particular type of cyborg agent, one impelled by external human agency for actively malicious purposes. We then would undertake a similar process of attempting to qualify and quantify the behaviors of particular types of malware, as explored comprehensively by Saxe and Sanders in "Malware Data Science" (2018). Again, CSDS acts as a powerful knowledge-building facilitator that results in insights suggesting actionable methods for prevention, detection, and remediation.

Having given an example of propositional theory building and explanation, a second example focused on building conceptual models and constructs is useful. Refocusing on human users and human agency, there is a question concerning how we frame and measure human users behaviorally through data. A definition of a human user in terms of a data-generating process is instructive. As an example of the process of forming an empirical conceptual assertion, we begin by considering a human user on a network as acting via an authenticated human userid. A userid represents a role-based mediator between a human user, devices, applications, systems, and network access. The discrete nature of a userid thus has a range of contextual connections to associated access rights, permitted actions, connected hardware, available applications, and accessible networks.

In monitoring the behavior of an authenticated userid, actions in modern digital environments typically cannot be localized to a unique device. An active userid may proxy and authenticate on several devices and may authorize applications to operate semiautonomously or fully autonomously on behalf of the user. A user may authenticate on and maintain sessions on multiple devices simultaneously, for instance, a local workstation, a local virtual host, a remote server such as a thin-client application, a remote application such as a cloud-based service, a mobile device such as a smart phone or tablet, and a smart device such as conference phone. All may have unique device indicators, IP or MAC addresses, which host activities initiated by the primary user through their authenticated userid. Where is the discrete user in such a complex? From a security monitoring perspective, it is difficult to discretely identify where the user is active geographically, functionally, and virtually over time as they transition between devices and applications.

A malware infection typically subverts authorization covertly so that it may scan, exfiltrate, and infect under the compromised volition of a titular userid, spoofing human permission and agency. As a result, the question, "what is a userid?" has many potential answers which depend upon context. An effort to conceptually clarify the multi-contextual presence of a userid is useful to explain how a userid can be active in many locales simultaneously. We might attempt to improve understanding by proposing a conceptual model for representing the complex between a human user, userid, roles, identified network devices, external systems, and network sessions. network sessions, as in Fig. 4.34.

Such a model is useful to gaining a richer understanding of the complexities and uncertainties involved in qualifying and quantifying human users on a network. Each associated entity in the chain of connections may record unique actions and behaviors. In aggregate, exposure of human agency can become quite complex as there are several mediated relationships.

However, at a certain point, we begin to realize that a purely mechanical, engineering-based description of the relational situation of a human user does not itself constitute theory. There is no assertion concerning the expected behaviors of, nor dynamic dependencies associated with, the human. We simply understand discrete uncertainty better through having positioned the human as a combination of many distributed interacting entities and events.

Iterating a conceptual model that describes the expected activities of a user, via a role-based userid, begins to approach a set of theoretical assertions. Describing the discrete behaviors of key entities along with their interactions provides a basis for working out theoretical propositions as a prelude to testing and experimentation. A

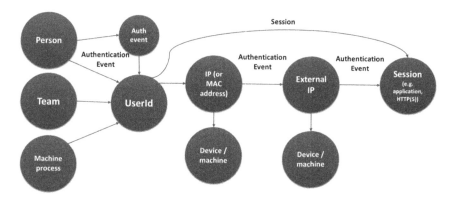

Fig. 4.34 Conceptual model of a human user interacting on a network

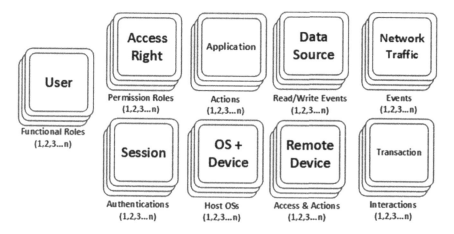

Fig. 4.35 Conceptual model of network entity actions and measures

preliminary conceptual proposition might attempt to classify the discrete elements associated with the behaviors of a user in a digital environment, per Fig. 4.35.

However, while improving an understanding of key actions associated with entities and of measurable quantities, the composite relationship between entities is missing. As an iteration of the conceptual model which begins to suggest a foundation for generating a range of testable hypotheses, it is useful to situate the embedded relational expectations amongst the key elements defined. A conceptual model, which begins to approach a pseudo-construct as a prelude to experimental propositions, would propose structural and relational connections propose structural and relational connections between entities, as per Fig. 4.36.

In this iteration, we assert a proposition concerning the nested role of a human in interacting through chains of dependencies. The proposal asserts dependencies that

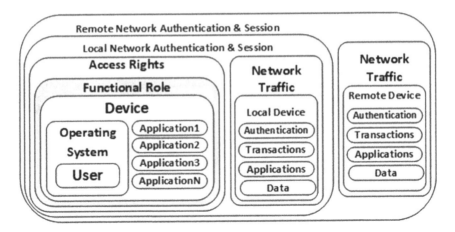

Fig. 4.36 Theoretical integration and framing

hypothesize expected processes, for instance, that a sequential set of authentication events traces the incarnation of human agency through linked network connections and devices.

Based on such proposed dependent, sequential relationships, tests and experiments can be staged either to validate and iterate the representation or to refute and reframe. Findings assist in establishing refined understandings which then can be utilized to improve monitoring, derive expected baseline behaviors, and/or frame anomaly detection methods.

By combining functional and relational context in a conceptual proposition, understanding progresses from architecture-focused engineering to knowledge of systems dynamics. The progressive conceptual working out of relational systems dynamics recalls medical diagnostics. Mechanical models are anatomical, being physical, deterministic, and discrete. Systems dynamics models are metabolic, organic, complex, interconnected, and probabilistic. This creates novel diagnostic and monitoring opportunities for characterizing and understanding complex dynamic systems behaviors.

When the discrete nature of entities, their relationships amongst one another, and the various behaviors and actions expected are outlined, grounds for encoding theoretical knowledge emerge. Namely, validated theory can be captured and operationalized via formal semantic frameworks. Proposing and experimentally substantiating that key security entities have universal aspects and behaviors can be encoded in, for instance, a descriptive ontology that specifies the formal relationships, connections, and behaviors between key entities. An encoded ontology then can serve as a semantic reference to ground taxonomies and schemas to guide and automate monitoring protocols, detection systems, and remediation efforts (e.g., investigative playbooks), per the representation in Fig. 4.37.

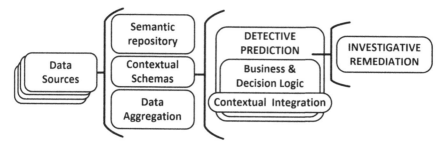

Fig. 4.37 Operationalizing theory in investigations

For instance, an anomaly and incident detection system which refers to a library of encoded semantic frameworks (e.g., encoded ontologies) has a symbolic logic for semi-automating data integration, systems interoperability, and inference-based decision-making. What results is a system that automates contextual presentation of cases to investigators for remediation.

Further, monitoring the outcomes of investigations over time provides evidence suitable for further semantic encoding, data refinement, and, ultimately, detective analytics. Operationalized semantic knowledge provides a basis for refined supervised machine learning efforts supplemented with semantic context, e.g., recommended next best actions, case context, and actionable risk scores based on experience to aid investigations. This distinguishes symbolic AI, machine learning supported by semantic context in order to engage in deductive inference and deterministic reasoning (Riley 2019a).

To conclude guidance on CSDS theory building, an iterative, integrated approach is advocated. In composite, an integrated approach spans data management, hypothesis development, and operationalization. An iterative process is assumed in which findings during one phase may suggest a return to a previous phase. Figure 4.38 summarizes a high-level perspective concerning an integrated CSDS theory building process.

4.4.7   Summary of Design Prescription II: Scientific Processes

In aggregate, the range of design requirements to inculcate scientific process rigor in CSDS frames an iterative approach. The advocated prescriptions progress from discovery and evidence extrapolation, through to integrating explanatory and predictive approaches in modeling, leading up to foundations for theory building. Scientific methods are framed as being multiple, embracing exploration and discovery leading up to experimental efforts and theory.

The unified set of processes demonstrates the flow of value-enhancing possibilities created by progressively framing and exploring theoretical propositions. Theory

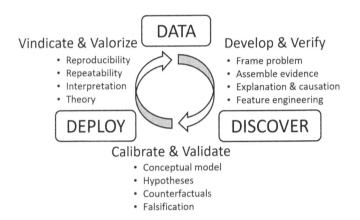

Fig. 4.38 CSDS theory building process

building offers paths to discovery and categorization, proposes experimental approaches to improve knowledge, and ultimately suggests interventions to achieve particular design goals. Theory building both frames opportunities for, and is empowered by, CSDS methods.

A central theme involves highlighting the distinction between engineering solutions and scientific advancement. The latter supports theoretical understandings of security phenomena. In the context of the iteration of CSDS as a profession, the assertion is that CSDS requires scientific methods to form, test, and falsify general theories as mediated by structured exploration, conceptual models, and explanatory constructs.

The core assertion is that CSDS enables knowledge creation by prescribing data science methodological treatments to address identified security gaps. This frames the distinction between *technê* and *logos* extrapolated *epistêmê* in applied technology. Whereas *technê*—skill, art, or craft—focuses on pragmatic results, *logos*, the expression of rational, structured thought, supports the development of *epistêmê*, understanding, knowledge, and scientific theory. It is asserted that the iteration of the latter is required for the evolution of the CSDS profession vis-à-vis maturation of domain body of theory. Scientific process design prescriptions raised in this section are summarized in Table 4.9.

4.5 Design Prescription III: Cross-Domain Collaboration

4.5.1 Problem-Solving Requirements

The final gap surfaced through triangulated literature and interview research concerns organizational aspects of CSDS management. The identified gap advocates that organizational efforts address team cross-coordination and collaboration to span the broad range of expertise covered by the CSDS domain. Key best practices

Table 4.9 Summary of scientific process design prescriptions

Framed design prescriptions: Scientific processes

General scope for inculcating scientific processes in CSDS
- CSDS is framed as the alignment of prescribed data science treatments to diagnosed cybersecurity gaps
- Scientific processes are framed as encompassing a range of treatments spanning conceptual scoping, exploration, discovery, observations, constructs, testing, experimentation, and theory extrapolation
- The security field is regarded as lacking a breadth of theoretical foundations, being largely engineering-driven
- To iterate as a profession, CSDS must develop body of theory supported by theoretical foundations accompanied by mechanisms for generating and substantiating new theory

Anomaly detection, exploration, discovery, and context-building
- Exploration and detection are two distinct processes which ideally interact and mutually reinforce cyclically
- Anomaly detection is a central approach in security environments of great uncertainty
- Anomaly detection can be approached as an iterative process of exploration and discovery leading to context-building and categorization
- Substantiating segments as durable operational categories provides a foundation for contextual sensemaking in complex environments
- Operationalizing anomaly reviews leads to establishing improved contextual understandings and evidence which can be used to drive predictive methods
- There are a range of types of evidence and methods for extrapolating evidence
- Detective analytics is advocated as a modality of special interest to security as it integrates predictive and explanatory approaches to support investigative and remediation processes.
- Predictive methods alone risk invoking correlation not causation and model overfitting

Model context and validation
- Explicitly specifying the context of a model in terms of goal/modality (descriptive, explanatory, predictive, detective) and domain (technological, economic, behavioral) supports both design and validation efforts
- Rigorous model validation is often unobtainable in complex uncontrolled environments and thus should be considered as a sensemaking activity aimed at assuring practical rigor
- Model complexity is a focused risk in the CSDS realm, especially given the growing breadth and complexity of datasets and models
- Models should be considered as a composite theoretical proposition, spanning both data and an algorithmic instantiation
- A variety of models should be compared like-to-like via champion-challenger testing

Foundations for rigor and theory
- Prior to formal experimental efforts, theory development is supported by primary efforts at discovery, scoping, conceptual modeling, and explanation
- Experimental rigor invokes reproducibility, repeatability, and validity
- A platform for experimental efforts can be approached stepwise by building primary foundations through exploration and discovery
- Theory ultimately must satisfy criteria for demarcation by framing experimentally falsifiable hypotheses
- Theory building ideally incorporates and integrates constructs, propositions, explanations, scoping, and empirical testing
- Progressive validation of contextual dynamics leads to theoretical knowledge which can be operationalized through semantic frameworks—encoded, structured models encapsulating understandings of general processes and systems dynamics

advocated by a majority of CSDS interviewees included "building process focused cross-functional teams" (76%, $n = 38/50$) and "cross-training teams in data science, cybersecurity, and data engineering" (74%, $n = 37/50$). This implies that organizations make a concerted effort to both integrate and cross-train data scientists and cybersecurity professionals.

A related theme identified in interview research concerns the challenge of the expanding breadth and complexity of the cyber domain (32%, $n = 16/50$). Literature analysis reinforced that even seasoned professionals can only hope to gain expertise in specific areas of CSDS combined with a general understanding of others. It is impractical to expect that isolated practitioners can master and perform all aspects and functions embraced by the aggregated CSDS disciplinary umbrella.

Within the data science and cybersecurity CSDS parent domains, it is already the case that most practitioners specialize in a subset of functions or methods while having a general grasp of others. The resulting challenge necessitates that institutions and organizations develop approaches to ensure cross-domain collaboration and process-driven teamwork across hybridized teams of professionals. The goal is to promote interdisciplinary understanding, a common professional language, and socialization of processes to unite the two areas of expertise.

To the degree that organizations cross-train professionals, build interdisciplinary teams composed of experts across cybersecurity and data science functions, and promote interaction through process-oriented approaches, fruitful collaboration and results follow. The facilitating mechanisms are centrally tied to organizational management, leadership, and incentives. Table 4.10 summarizes the range of themes aligned with the cross-domain collaboration gap surfaced in CSDS interview research.

4.5.2   Collaborative Science in the Enterprise

There are notable challenges to assuring scientific rigor in complex organizational settings, particularly when associated with commercially driven enterprise. A lack of focused experience supporting primary research programs can markedly hinder efforts to promote scientific collaboration. A related CSDS challenge concerns assuring rigor in analytics model validation processes. Whereas validation imposes overhead, rigor may be abandoned when judged impractical. Organizational processes safeguarding research rigor must not only assure methodological integrity but must address risks posed by diverse, potentially misaligned organizational stakeholders.

Previous discussions concerning introducing scientific processes in CSDS practice advocated a range of approaches for inculcating rigor. Model validation was raised as a central gating mechanism. It was asserted that CSDS model validation is more extensive than focused model deployment and testing. Model validation was framed more broadly to encompass both preliminary data engineering and subsequent operational implementation, as represented in Fig. 4.39.

Table 4.10 Problem-solving requirements: cross-domain collaboration

Key gap	Category	Practitioner themes
BP F2: Cross-domain collaboration	Systematic evidence	**BP16**: Participate in data sharing consortiums **CH4**: Lack of labeled incidents to focus detection
	Data structures and preparation	**BP7**: Data normalization, frameworks and ontologies **CH1**: Data preparation (access, integration, etc.)
	Uncertainty	**CH10**: Expanding breadth and complexity of domain **CH6**: Uncertainty leads to reactive stance **CH5**: Own infrastructure, shadow IT, exposure
	Management commitment	**BP18**: Upper management buy-in and support **CH8**: Ownership, decision-making and processes **CH9**: Resourcing, developing, hosting in house
	Resource coordination	**BP2**: Building process focused cross-functional team **BP3**: Cross-training team in DS, cyber, engineering

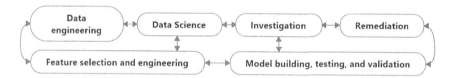

Fig. 4.39 Breadth of CSDS model validation scope

Model validation in this wider purview extends beyond assessing localized analytical model performance metrics, such as ROC and AUC (in the case of supervised machine learning). It is advocated that substantiation of problem framing, review of data suitability, and assessment of operational results be encompassed in broader validation efforts.

Model validation is defined as the "substantiation that a model within its domain of applicability possesses a satisfactory range of accuracy consistent with the intended application of the model" (Sargent 2000). The challenge for practitioners is that they are rarely afforded the opportunity to achieve rigorous validation certitude in demanding operational settings. Analytics model validation, from a practical organizational perspective, often amounts to a process of substantiating consensus-based credibility (Anderson et al. 2009; Hillier and Lieberman 2005; Law 2007).

Charged with model assurance stewardship, practitioners are often pressed to take practical shortcuts. There is a tangible operational cost to pursuing validation rigor in terms of time, resources, and opportunities foregone. Often "good enough" is a loose benchmark for model validation in commercial settings. However, this can lead to a slippery slope: validation rigor can fray and degrade as pressures to achieve economy and velocity outstrip the demands of robustness.

Practitioners, charged with stewardship of the validation process, must satisfice two key goals which may not always be aligned:

(a) to produce a model that represents true system behavior closely enough for the model to be used as a substitute for the actual system for the purpose of experimenting with the system, analyzing system behavior, and predicting system performance; and

(b) to increase the credibility of the model to an acceptable level, so that the model will be used by managers and other decision makers.

Banks et al. (2010)

With low-level validation often infeasible, particularly when associated with complex and sophisticated analytical models, a practical model validation process is tantamount to an organizational sensemaking activity. Concerning the epistemological limits of validation, per Pidd, "management scientists have a responsibility to aim at some form of validation but should begin by recognizing that this may be limited. Among many scientists it is recognized that theories and knowledge can never be comprehensively demonstrated to be true, and yet these same theories turn out to be very useful" (2004).

Substantiating "useful abstraction" and rigor need not be incompatible provided proper assurance processes are in place. Organizational model validation is not a purely mechanical process, but inherently depends upon social processes related to collaborative review, explanation, iteration, implementation, and staged adoption.

It is helpful to distinguish the collaborative aspect of assuring model rigor beyond mechanical use. While base validation aims to assure a model methodologically, there is a follow-on effort required to socially operationalize and adopt the model. This follow-on effort encapsulates the principle of *valorization*, the process of increasing the value of knowledge-based artifacts through operational productionalization. Valorization presumes a social process whereby the value of the artifact is promoted, collectively assessed, and integrated via organizational processes and mechanisms.

The necessity of valorization as an accompaniment to validation recognizes that a rigorous model is practically proven when it is adopted socially by an organization. Pavlovic (2012) comments regarding the centrality of social aspects to assuring scientific rigor in security modeling:

> The germs of a scientific approach to security, with data gathering, statistical analyses, and experimental validation, are already present in many intrusion detection and antivirus systems, as well as in spam filters and some firewalls. Such systems use measurable inputs and

have quantifiable performance and model accuracy and thus conform to the basic requirements of the scientific method. The collaborative processes for sharing data, comparing models, and retesting and unifying results complete the social process of scientific research.

It is useful for CSDS practitioners to have a clear understanding of the organizational distinction between technical validation and valorization of a model. The former assumes an engineering context, whereas the latter is aimed at assuring adoption, collaboration, and continuing knowledge exchange. Given caveats concerning the limitations of model validation, organizational processes ideally are introduced to assure model robustness and adoption through valorization.

Implementation of an analytics model typically is represented as a two-dimensional, linear process. Figure 4.40 poses a high-level, idealized representation of the application of CSDS analytics in the enterprise as a linear, end-to-end functional process. This representation recognizes that CSDS initiatives often seek to improve a focused measure or process, such as reducing false alerts, optimizing investigations, or facilitating the remediation of incidents.

However, there is a danger in over-emphasizing technology-driven, linear process designs. Namely, such approaches potentially ignore the importance of collaborative aspects of knowledge generation and valorization. Whereas data and information can be facilitated via *technê*-driven mechanisms, knowledge generation, *logos*-mediated *epistêmê,* presupposes collaborative reinforcement and iterative social exchanges.

Per earlier framing, CSDS knowledge-promoting processes are iterative and multidisciplinary. Specifically, it is assumed that continual discovery involves an intermediate collaboration between security experts and data scientists prior to operational investigation—a discovery phase. Figure 4.41 offers a reframed representation of an idealized CSDS process wherein knowledge and expertise are exchanged in a staged, preliminary discovery process between security analysts and data scientists.

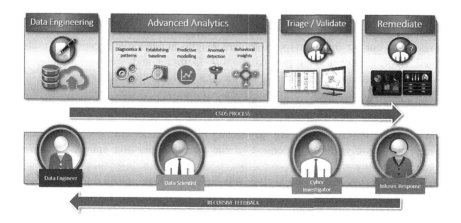

Fig. 4.40 Idealized CSDS functional-technical process

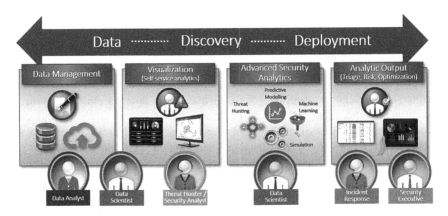

Fig. 4.41 Pragmatic CSDS functional-knowledge process

In this representation, evolving discovery-based insights are exposed to security experts prior to the invocation of alert generation and investigation (the provision of detective analytic results). Discovering insights exposes emerging threats and patterns to security experts *prior* to detection-based algorithmic processing. The preliminary review phase supports model design, refinement, and maintenance. As 76% of CSDS professionals interviewed (*n* = 38/50) cited "adversaries constantly adapting to countermeasures and innovating methods" as a key trend, the ability to assess ongoing and shifting situational dynamics is suggested as a core CSDS best practice.

The importance of facilitating organizational collaboration is reflected in analytics organizational research, namely, that organizational factors dominate process and technical factors (Henke et al. 2016). Recent industry views on the operational management of AI has promoted the importance of ModelOps, a model engineering and development function, as an intermediary between DataOps (data engineering services) and DevOps (deployment services) (Sapp 2020; Wu 2020).

As a practitioner-driven discipline, data science advocates process-driven functional approaches to refine insights from data. Analytics results must ultimately be assessed by human judgment to extrapolate insights and to make decisions. To safeguard rigor, organizational processes must facilitate and valorize a range of judgment-based assessments during the broader analytics modeling process:

- *Problem formulation*: research question framing prior to data gathering and hypotheses
- *Data collection and featuring:* grounding the data-generating process and data representation within the context of the research question and phenomenon under scrutiny
- *Hypothesis generation and testing:* formalizing testable, falsifiable hypotheses, which may range from exploratory propositions of categorical designations to causal-explanatory constructs or predictive models

- *Results interpretation and communication:* substantiating and valorizing rigor— have model errors resulted in false positives or false negatives? Does the model suffer overfitting or overgeneralization? Do stakeholders understand model assumptions and results sufficiently?

4.5.3   Aligning Organizational Incentives

The ability to extrapolate patterns and derive insights from large datasets through analytics methods has demonstrated efficacy across a broad range of domains. However, the implications of analytics-based decision-making puts fresh demands upon organizational leadership: new decision processes must be integrated within organizational structures to be viable. Teams must be orchestrated, trained, and incentivized to collaborate effectively within the context of new processes.

Viewing analytics as a process focused on improving decision-making is a perspective centrally allied to the field of MIS and associated sub-domains, particularly data mining, decision support systems (DSS), expert systems, and business intelligence (BI) (Davenport and Harris 2007; Diaz 2018; Laudon and Laudon 2017; Sharda et al. 2014; Skyrius et al. 2013). Research associated with the growth of the analytics movement focused on early adopter case studies and the value-creating benefits of analytics-based decision-making (Davenport 2009; Davenport and Harris 2007; Davenport et al. 2010).

However, given the inherent challenges of implementing analytics-based decision-making, subsequent management research has focused on organizational factors related to analytics program feasibility and quality. A range of popular business research articles focused on organizational approaches to and challenges associated with adopting analytics-based decision mechanisms (Kiron and Shockley 2011; Kiron et al. 2011; LaValle et al. 2010, 2011). Research also framed the concept of organizational analytics maturity, a measure of the readiness of organizations to adopt analytics-driven decision-making (Kaner and Karni 2004; Lovallo and Sibony 2010b).

Research has also raised the danger of deifying analytics as a decision panacea. The design and implementation of analytics solutions may itself embed biases, a topic covered by O'Neil at length (O'Neil 2013, 2016). As an adjacent, related research thread, behavioral economics research, has framed a range of inbuilt cognitive biases which commonly short-circuit attempts to improve organizational and individual decision-making (Kahneman 2011; Kahneman and Klein 2009; Miller and Page 2007; Nutt 2002). This has surfaced an awareness of challenges inherent to individual and group decision-making. From the perspective of analytics, this emphasizes threats to analytics integrity when model development and validation efforts run astray of classical cognitive biases, such as groupthink, focusing, framing, or loss aversion.

Analytics practitioners must be cognizant of the degree to which cognitive biases can hijack analytics solution and model development, for instance in missteps such as:

- *Addressing "how" before "what" and "why":* there is a tendency to rush towards technical implementations and prediction before phenomenon is understood in an explanatory context—does the model meet common sense benchmarks within the target domain?
- *Failing to bridge "what" and "why":* refinement of questions concerning the nature of phenomenon should ideally inform the framing of propositions and models—is there a rough understanding of the base mechanism and logic underlying the model?
- *Failing to address "how" adequately:* analytics decision processes should manifest as knowledge artifacts that embed contextual understanding—does the model accommodate all essential data required to analyze the phenomenon under scrutiny? Is the analytical method appropriate?

Faced with increasing complexity and data overload, beset security managers and practitioners may be tempted to retreat into the comfortable, but often misleading, domain of instinct and intuition (Bonabeau 2003). Dawes et al. frame this tendency as relying on clinical (field experience-based) rather than actuarial (statistical) judgment (Dawes et al. 1989). An associated risk is that deployment of an analytics approach or model devolves into a ceremonial band-aid or crutch.

The popularity of data science and analytics potentially leads to a misunderstanding that having an analytics technical capability is itself ipso facto sufficient to improve and safeguard organizational decision-making (Heudecker and Ronthal 2018). Such a misunderstanding might be termed the *Athena fallacy*. Athena, having sprung fully formed from the head of Zeus, metaphorically represents the notion that the blanket instantiation of an analytics program alone is sufficient to instantly and pervasively overcome organizational decision breakdowns and biases.

As has been raised and discussed, rigor in analytics solution and model design should ideally be assured through organizational processes. In both the business analytics body of knowledge (IIBA 2009; Podeswa 2009) and INFORMS analytics job task analysis (JTA) (INFORMS 2013), particular attention is paid to methodology selection, modeling, model management, validation, and the life cycle of analysis, including problem framing and communicating results.

Such best practices mirror themes surfaced in CSDS practitioner interviews related to advocating process-based rigor in problem framing, discovery, feature selection, and validation. The following themes were advocated by CSDS practitioners as key processes required to assure the integrity of analytics processes:

- Disassociating normal versus abnormal phenomenon
- Mapping and modeling environmental context
- Formal data management and data structuring approaches
- Systematic evidence gathering, framing, and labeling
- Efforts at formal risk quantification
- Formal data and model validation processes
- Linking distinct exploratory and predictive approaches

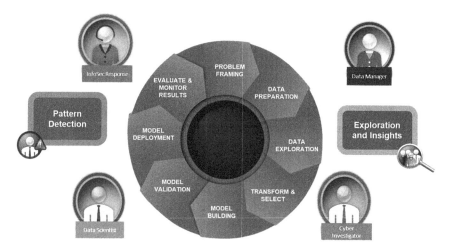

Fig. 4.42   Cyclical CSDS process

Representing these best practices within a cyclical organizational process incorporating role-based experts, Fig. 4.42 conceptualizes an ideal high-level CSDS analytics process.

In this representation, key CSDS stakeholders interact through an iterative process to drive rigor in the development, validation, and deployment of CSDS models. The scope of this process encompasses integrated problem framing, data gathering, model development, and model validation. Ideally outcomes result in cyclical improvements, staging opportunities to expand or undertake new inquiries. Efforts associated with exploration and detection are distinct yet mutually inform and reinforce one another.

Beyond process-based prescriptions, organizations must also address role-based incentives and foster an analytics culture to facilitate collaboration. Organizations are eager to understand not only how they can adopt formal processes to drive analytics but, crucially, an organizational culture that facilitates structured analytics-based decision-making. Survey research has revealed that political and cultural organizational factors are the greatest challenge to adopting advanced analytics programs (Kiron and Shockley 2011; LaValle et al. 2010, 2011). Analytics culture is cited as being a complex of fact-driven leadership, expertise (tools and skills), and processes linking analytical insight to strategic and operational decision-making (Kiron et al. 2011).

Analytics culture must overcome organizational boundaries to encourage collaborative exchange. This begins through understanding role-based conflicts that may be present. As an example, security practitioners admit that the goals of cybersecurity at times are explicitly adversarial to adjoining IT disciplines such as networking and IT infrastructure (Mastrogiacomo 2017). Security requirements are typically restrictive and constrictive by nature, having a conservative bias based in protective stewardship. Security stances may constrain strategic initiatives, for instance, seeking to veto new services or technologies due to increased threats of exposure.

Conflicts between functional mechanics may also hinder CSDS initiatives. Big data analytics for CSDS assumes the presence of large repositories of integrated security data from multiple sources. Data availability is often a challenge, and availability is not a given based on complicating technical, logistical, and/or political factors. Limited access to security-relevant data is a perennial CSDS practitioner complaint and challenge.

The data required to monitor and assess security itself must be secured due to its inherent sensitivity. Data security is a yoke on the notion that free and prolific data analytics can be conducted everywhere and at any time (Mastrogiacomo 2017). A range of agency factors arise which may disincentivize data sharing amongst distributed organizational stakeholders, ranging from fears of exposing flaws and vulnerabilities to operational concerns surrounding supplying big and fast data. For instance, network performance is impacted by streaming network flow data to CSDS practitioners for assessment. Management and network engineers may veto big data security initiatives due to the imposition of degraded network performance on the enterprise.

Political factors blocking data access can be intraorganizational, such as disputes concerning the ownership and secrecy of data, or external, such as those evidenced in strict new regulatory frameworks. Regulatory requirements may create a barrier between information stewards and cybersecurity investigators. With the growing centrality of networked computer and communication technology to facilitating labor in the enterprise, security professionals benefit from broader data beyond sources associated with systems access and usage of network infrastructure.

Having access to broader details concerning users functional job role, presence at work (e.g., work and vacation schedules, physical access through badges), and interactions in role-based professional networks (i.e., collaborative teams, functional collaborators) is potentially of great benefit to improved understandings of security behavior. Being able to track a user's work patterns throughout a typical day establishes a baseline for normal behavior, making the challenge of detecting deviations much easier. However, worker pattern profiling and tracking risks running afoul of various strictures: compliance (i.e., PID – personally identifiable information), political (i.e., European worker councils), legal (i.e., worker rights), and/or regulatory (i.e., GDPR).

In the context of the organization as a *firm* that is subject to internal and external competitive forces, implementing new analytics processes requires addressing incentives. There is a requirement to align organizational incentives to assure the cooperative operation of processes incorporating diverse functions and roles. In a CSDS context, it is useful to understand potential agency conflicts inherent to distinct role-based motivations across and between key actors:

- *Security experts:* enforce security standards to assure continuity and reduce exposure risks
- *Data scientists:* analyze diverse sources of data to extrapolate value-enhancing insights
- *Management:* optimize resource utilization and promote process efficiencies
- *Data engineers:* implementation of technical approaches to store, manage, and process data

It is not difficult to envision how these key CSDS stakeholders may develop conflicts in pursuing their respective role-based motivations. A concerted effort to explicitly facilitate collaboration through architecting incentives must anticipate and understand the potential presence of role-based conflicts. Figure 4.43 offers a unified representation of the points of collaboration and risks of disjunction amongst key CSDS roles.

Key CSDS stakeholders should be considered in terms of how they should optimally collaborate with adjacent roles to achieve common outcomes. This has a promotional aspect related to information and resource sharing associated with overcoming the fear of losing control or authority within the firm.

Incentives also frame preventative aspects in terms of avoiding adverse outcomes that may emerge as the result of perverse incentives. In the enumeration above, security experts may have a default motivation to continue "business as usual," safeguarding existing rule and signature-based monitoring systems and blocking efforts to adopt analytics programs. Collaborating with data scientists may be perceived as a loss of authority and control.

Management, being motivated to optimize human and technical resource investments, may have a bias to view analytics as an approach to automate and thus reduce labor overhead. Management may be credulous to the appeals of technology vendors who promise automation through "magic boxes." Data scientists, motivated to appease management and to promote the value of data science, may be pressured to rush predictive models into production without properly validating models. Rigor

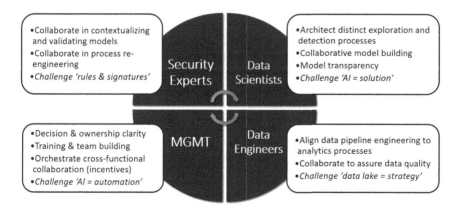

Fig. 4.43 CSDS role-based collaboration

may be abandoned to the degree data scientists are captured by management agency interests. Data engineers may have motivations to expand the authority and resourcing of their own function and thus to promote and encourage efforts to invest in facilitating technologies and to build and expand big data repositories such as data lakes.

A robust approach to planning a CSDS programmatic implementations ideally begins not with technical and architectural planning but by formally considering how to incentivize complex role-based interactions between diverse stakeholders who may have implicit or explicit adversarial dispositions to collaboration and exchange. Framing incentives such as bonuses and job tasks that explicitly frame outcomes requiring collaboration are thus advocated in the design of CSDS processes. Ideally CSDS program management frames shared metrics or KPIs which foster shared beneficial outcomes such as reducing false positives in security alerts, speeding up and improving the effectiveness of investigations, and/or improving the reported job satisfaction of security professionals.

4.5.4   CSDS Curricula for Cross-Training

The ability to cross-train professionals in CSDS fundamentals should be considered a central aspect of launching a CSDS program. A majority, 74% ($n = 37/50$), of CSDS interviewees advocated "cross-training teams in data science, cybersecurity, and data engineering" as a best practice. Given the emerging status of the domain, this is likely an aspirational objective. As CSDS body of theory is still developing, a canonical training curriculum has yet to be agreed to and broadly adopted. However, specific to particular organizations, it would be expected that local operational focus would help to refine objectives for discrete training offerings.

The larger importance of and context for CSDS training curricula ties to the aspiration of developing CSDS more broadly as a profession. Literature analysis of the emerging CSDS domain framed professionalization as a process associated with staged levels of maturity. Key thresholds for professional maturity included:

- Specialized professional training and certification programs
- Dedicated professional societies with gating criteria for membership
- Independent standards bodies
- A dedicated set of research literature and outlets
- Higher education degree offerings
- A coherent body of theory encapsulating individuated practitioner domain knowledge

As revealed in literature analysis, CSDS currently lacks these explicit, dedicated facilities outside occasional deference from within the cybersecurity parent profession. Lack of a codified body of theory has been identified as an underlying factor forestalling attainment of these thresholds. Professional knowledge individuation through a codified body of theory is a core underlying factor across these benchmarks. Each of the thresholds encapsulates, to some degree or another, vehicles for building, safeguarding, and/or conveying specialized professional knowledge.

Amongst the thresholds framed, specialized training is practically the closest at hand and achievable for CSDS. A canonical CSDS curriculum would frame and encapsulate a practical representation of CSDS body of theory. The parent cybersecurity domain has a very active and robust set of training offerings as a guide and foundation. Cybersecurity curriculum development has been the focused topic of several research contributions, providing additional guidance (Bicak et al. 2015; Conklin et al. 2014; Kim et al. 2005; Razvan et al. 2017; van den Berg et al. 2014). Curriculum design guidance from the cybersecurity field generally espouses building an awareness of key objectives, framing emerging challenges, and building awareness of associated risks, goals, roles, specialties, and methods.

CSDS training offerings ideally replicate cybersecurity curriculum development guidance with specific attention to CSDS-unique perspectives. Special attention is required to convey interdisciplinary perspectives regarding CSDS's status as a hybridized field. Verma et al. provide a focused perspective in "Security Analytics: Essential Data Analytics Knowledge for Cybersecurity Professionals and Students" (2015), referencing CSDS relevant aspects of the NIST National Initiative for Cybersecurity Education (NICE) Framework (NIST 2017). As was cited in literature analysis, CSDS training offerings are presently emerging, with O'Reilly (Givre 2017) and Udemy (Tsukerman 2019a) having hosted classes developed by CSDS practitioners.

It is asserted that CSDS training, as a focused integration of its parent domains, ideally prepares practitioners in three general respects:

- Establishing context concerning cybersecurity fundamentals
- Establishing context concerning data science fundamentals
- Understanding unique methodological and process aspects of CSDS as a hybridization

In terms of framing a draft CSDS curriculum proposal, this research effort as a whole has surfaced key themes and topics which would ideally be addressed. The literature review and derived best practices framed in *Phase I* interview research highlights central topics worthy of incorporation. This phase, *Phase III*, on design prescriptions, frames particular pedagogical points which ideally would be emphasized and reinforced.

More specifically, this researcher has designed a one-day class on cybersecurity data science and has delivered the class three times between 2018 and 2019 (Mongeau 2019a). While a full-case study or action research implementation is beyond the scope of the current inquiry, the topical curriculum designed was centrally informed by this research effort. In terms of offering design advocacy for a

CSDS curriculum, the high-level topics covered in the training offering are shared here for the benefit of interested practitioners:

1. *Introduction to the CSDS field*

    1.1. Cybersecurity basics and challenges
    1.2. Data science basics and challenges
    1.3. CSDS as a focused hybrid domain
    1.4. Differentiating analytics goals and methods
    1.5. Framing the cybersecurity analytics life cycle
    1.6. Introducing cybersecurity analytics maturity

2. *Cybersecurity data: challenges, sources, features, methods*

    2.1. Sources of cybersecurity data, research datasets, types of evidence
    2.2. Examples: log files and network traffic
    2.3. Data preparation, quality, and processing
    2.4. Statistical exploration and analysis (EDA)
    2.5. Feature engineering and selection
    2.6. Feature extraction and advanced methods
    2.7. Positioning and handling real-time and streaming data

3. *Exploration and discovery: pattern extraction, segmentation, baselining, and anomalies*

    3.1. Building contextual knowledge
    3.2. Segmentation and categorization
    3.3. Multivariate analysis
    3.4. Parameterization and probability
    3.5. Outliers and differentiating normal from abnormal
    3.6. Anomaly types, anomaly gain, and detection
    3.7. Unsupervised machine learning
    3.8. Establishing a foundation for prediction

4. *Prediction and detection: models, incidents, and validation*

    4.1. Distinguishing explanation versus prediction
    4.2. Framing detective analytics: combining explanation and prediction
    4.3. Econometric approaches
    4.4. Predictive machine learning (supervised machine learning)
    4.5. Deep learning
    4.6. Reinforcement learning
    4.7. Model diagnostics and management
    4.8. Bootstrapping detection: semi-supervised machine learning

5. *Operationalization: CSDS as a process*

    5.1. Analytics process management: integrating discovery and detection
    5.2. Human-in-the-loop: integrating investigations and investigative feedback
    5.3. Robo-automation, online machine learning, and self-improving processes
    5.4. Technical and functional architectures

5.5. Systems integration and orchestration
5.6. Cybersecurity analytics maturity recap
5.7. Cybersecurity risk and optimization
5.8. Guidance on implementing CSDS programs

The range of themes surfaced and advocated throughout this research effort are embodied in this curriculum. CSDS is framed as the application of data science to cybersecurity challenges. An integrated process-driven approach to CSDS is espoused. An effort is made to clearly distinguish explanatory from predictive approaches. Statistics-based versus machine learning methods are differentiated. A core focus on data, data management, and data preparation is resident. An effort is made to frame exploration and detection as distinct but mutually reinforcing processes. Model development and validation are raised in the broader context of data management and the operationalization of models.

Conclusions espouse approaches to bootstrapping primary context-building and discovery into refined detection. Efforts are made to explain and situate investigative processes and technical architecture with reference to the methodological approaches advocated. As a whole, the aggregated approaches frame a concept of cybersecurity analytics maturity. Advocating a process-based approach, it is asserted organizations can build CSDS maturity through the iterative stages of:

1. Discovery leading to context-building and refined anomaly generation
2. Statistical segmentation leading to security-relevant categorization
3. Prediction leading to alert refinement and investigative efficiencies
4. Cyclical model improvement through operationalized human-in-the-loop investigations
5. Leveraging results for improving risk management and optimizing operational efficiency

Further developing this curriculum and training effort through case study, design implementation, or action research initiatives would be a potential follow-on to this effort. Formalizing training through curriculum development promotes codification of CSDS body of theory. Ideally certification programs would follow, evolving CSDS professionally through the sanction of institutional and community authority.

Per research on professionalization in literature analysis, institutional support is considered necessary to foster development and maturation of professional fields. Next steps for the professionalization of the CSDS domain suggests institutional adoption, starting with sponsorship and hosting of focused training offerings. CSDS currently lacks established institutional sponsorship to adopt, ensconce, sanction, and socialize iterative body of theory development. Institutional adoption would provide a foundation for facilitating formal training, certification, organizational membership, and knowledge development through a specialized community of practice.

4.5.5   Summary of Design Prescription III: Cross-Domain Collaboration

Promoting cross-domain collaboration was the final high-level gap surfaced in CSDS interview research. Research substantiates that organizational factors are central to addressing identified requirements. Per the summary in Table 4.11, design prescriptions focused on facilitating scientific rigor through collaboration, aligning role-based incentives, and designing curricula for cross-domain training.

Table 4.11 Summary of cross-domain collaboration design prescriptions

Framed design prescriptions: Cross-domain collaboration
Facilitating collaborative scientific rigor
• Model validation ideally extends from data framing and preparation through to assessing the operational results of deployed models
• Validation rigor involves organizational sensemaking and consensus building
• Valorization encompasses the social process of organizational adoption of models
• Orchestrating collaboration transcends mechanical approaches as knowledge facilitation implies staged social exchange amongst role-based experts
Aligning incentives
• Implementing analytics-based decision-making processes pose a range of organizational challenges and considerations
• Aligning organizational culture and decision mechanisms are required to facilitate the implementation of analytics programs
• Cognitive biases, deference to intuition, agency interests, and incentive conflicts challenge efforts to assure analytics model and process rigor
• Key CSDS roles include data scientist, security expert, data engineer, and management
• Natural organizational incentive conflicts can arise when organizational goals are misaligned
• Positive opportunities and negative tendencies associated with the interaction of key CSDS roles should be formally addressed through incentive-based control systems
• Incentivizing achievement of joint metrics focused on shared outcomes is advocated
Cross-training and curriculum development
• CSDS lacks a canonical body of theory to ground curriculum design
• Curriculum development guidance is resident in research within the security domain
• Based on themes aggregated in this research, an outline for a CSDS class is offered
• Institutional sponsorship and facilitation would be a pragmatic next step to iterate CSDS professional domain development

4.6   Summary of Design Prescriptions

4.6.1   Overview of Prescribed Designs

Literature analysis revealed that immature body of theory remains an inherent challenge to CSDS professionalization. Interview research identified key gaps facing the CSDS domain. A series of design prescription gaps were surfaced through triangulated literature analysis and interview research.

Phase III has asserted that CSDS body of theory can be iterated by applying design science approaches. The three key CSDS gaps addressed were data management, scientific processes, and cross-domain collaboration. A central finding was that the nature of these three gaps is complex and overlapping, requiring systematic, multi-contextual process designs to address.

The design prescriptions advocated address key questions associated with the three gaps:

1. *Data management*: given that data is an abstract representation of underlying phenomenon, which methods should be applied to substantiate the relevance of data selected and utilized in CSDS analytical treatments?
2. *Scientific processes*: recognizing that scientific methods within CSDS practice are multiple, which methods are most appropriate during distinct phases of CSDS inquiry?
3. *Cross-domain collaboration*: given that CSDS practice entails social and organizational processes aimed at knowledge facilitation, how can organizations best foster such collaborative exchange?

A central finding was that data preparation is crucial, but time-consuming—an exacting and iterative process. This finding is at odds with the commercial excitement surrounding data science as a fast-moving and exciting discipline that focuses on algorithmic implementations. Per results from interview research, data preparation was the most frequently cited theme in interviews with CSDS practitioners, 84% ($n = 42/50$) of respondents having framed this topic as both a challenge and best practice.

There is a need indicated to set expectations with business stakeholders that a substantial up-front investment in data exploration, extrapolation, and quality is required to achieve useful CSDS results. An accompanying assertion is that time invested in data preparation, once committed, has outsized upsides in terms of inherently boosting the quality of analytics results.

Another topic addressed concerned the challenge of introducing scientific rigor into CSDS practice. The security domain is engineering-focused and generally lacks a strong theoretical, scientifically rigorous foundation. This frames the distinction between *technê-* and *logos*-mediated *epistêmê*. Whereas *technê*, skill, art, or craft, focuses on pragmatic results, *logos*, the expression of rational, structured thought, supports the development of *epistêmê*, understanding, knowledge, and scientific theory.

Process-based approaches that frame a range of scientific methods and treatments should be formally staged to address various phases of CSDS inquiry. Ideally CSDS scientific processes span from exploration and discovery through to prediction and detection, leading up to knowledge generation through general theory formulation and substantiation.

An effort needs to be made to frame data science programmatically to support CSDS efforts. The challenge is as much organizational as it is methodological and technical. As much of the collected research on data analytics organizational challenges has shown, organizational factors are the most intractable and important problems to counter in the implementation of analytics processes and programs (Kiron et al. 2014; Kiron and Shockley 2011; LaValle et al. 2010; Roy and Seitz 2018). CSDS programs ideally facilitate and incentivize social processes associated with knowledge exchange and model valorization.

4.6.2   Research Relevance

Ideally CSDS efforts position general theory before the application of data science methods. However, the security field lacks strong theoretical foundations. The de facto approach in the security profession is to lead and conclude with engineering-based practice and solutions. This results in a persistent gap in CSDS body of theory, forestalling the process of professionalization. Proposing a theory-driven approach in security practice amounts to an inversion of the status quo, per the representation in Fig. 4.44.

In order to address this shortcoming, this phase provided a capstone to previous diagnostic research by advocating a series of design prescriptions to address gaps in CSDS practice associated with building theory and rigor. This phase undertook a partial design science research cycle (Hevner et al. 2004), encompassing the development/building of theories and artifacts. Design artifacts were extrapolated from combined knowledge with reference to structured requirements.

Prescriptive problem-solving designs were extrapolated and advocated. In a practitioner research context, this addressed a partial intervention cycle, covering the phases of analysis (problem identification), diagnosis (gaps and possible treatments), and design (requirements and prescribed solutions). Application through intervention and evaluation are proposed for future action research efforts.

In the context of design science, referring to Gregor and Hevner in "Positioning and Presenting Design Science Research for Maximum Impact," design science

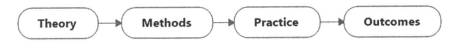

Fig. 4.44 CSDS flow of theory to outcomes

Table 4.12 Design science research contribution types

	Contribution types	Example artifacts
More abstract, complete, and mature knowledge	*Level 3.* Well-developed design theory about embedded phenomena	Design theories (midrange and grand theories)
⬍ ⬆ ⬆ ⬆	*Level 2.* Nascent design theory—knowledge as operational principles/architecture	Constructs, methods, models, design principles, technological rules
More specific, limited, and less mature knowledge	*Level 1.* Situated implementation of artifact	Instantiations (software products or implemented processes)

contribution types are differentiated in terms of their intended contributions, per Table 4.12 (2013).

The contributions in this phase focus on *Level 2* design science contributions, nascent design theory, offering a series of prescribed methods and models to address surfaced gaps in the CSDS domain. This constitutes a proposed improvement to nascent CSDS body of theory. The contributions suggest a foundation for extending to both *Level 1 - Implementations,* and *Level 3 - Well-developed design theory*.

Concerning iterating towards *Level 3*, midrange and grand theories, structured approaches supporting CSDS theory development are framed. A principle espoused here is that the cybersphere represents a new "natural" phenomenon and that its engineered nature potentially distracts from its fast-evolving organic complexity. The position that the cybersphere encompasses a new natural domain suitable for scientific interrogation opens creative opportunities for introducing rigor in inquiry and deriving general theory. Methods associated with social science research are particularly espoused to support such efforts.

Regarding iterating towards *Level 1* design contributions, instantiations, Schneider espouses principles for designing and deploying secure systems based on theoretical foundations in "Blueprint for a Science of Cybersecurity":

> The obvious alternative is to build systems whose security follows from first principles. Unfortunately, we know little about those principles. We need a science of cybersecurity... that puts the construction of secure systems onto a firm foundation by giving developers a body of laws for predicting the consequences of design and implementation choices. The laws should
>
> - transcend specific technologies and attacks, yet still be applicable in real settings,
> - introduce new models and abstractions, thereby bringing pedagogical value besides predictive power, and
> - facilitate discovery of new defenses as well as describe non-obvious connections between attacks, defenses, and policies, thus providing a better understanding of the landscape.
>
> (2012)

Implicit in Schneider's advocacy is the notion that *Level 2* design contributions, methodological rigor, per Gregor and Hevner, centrally facilitates the extrapolation of general laws (*Level 3*) required to ground improved implementations (*Level 1*).

The larger assertion concerns CSDS as a provider of structured methods and processes to promote theory development in the security domain. Promoting theoretical foundations in the security domain cyclically provides a foundation for facilitating development of the CSDS profession. In a design science research context, CSDS can thus be framed as both a methodological facilitator for iterating security domain theory (*Level 3*) as well as a meta-methodological domain in of itself which must refine its own unique theoretical foundations and value propositions (*Level 2*) to individuate as a distinct professional field.

To pursue professional individuation, CSDS, as a subdiscipline of cybersecurity, must position itself in juxtaposition to the cybersphere much as medicine addresses biological organisms. Beyond a metaphorical assertion, the methods and structures that have evolved to provide rigor in the medical profession also suggest a path and goals for development of the CSDS professional field. Namely, more systematic and rigorous approaches to integrating theoretical (research-based) and practitioner (practice-based) knowledge are required for the CSDS profession to mature.

4.6.3   Managerial Relevance

The three CSDS gaps addressed, data management, scientific processes, and cross-domain collaboration, roughly align with the MIS focus areas of technology, process, and organization. A key observation raised in framing requirements concerned the degree to which these gaps overlap. Design prescriptions framed data management and scientific processes as mutually reinforcing. The organizational context for scientific rigor was also highlighted in framing validation processes as organizational sensemaking exercises.

The design prescriptions advocated in this phase should be of central interest to managers and practitioners planning to implement CSDS programs. A selection of key management objectives were addressed as design prescriptions herein:

- Drive better security decisions
- Provide context for security events
- Extrapolate analytics-driven insights
- Improve event detection accuracy
- Lower false positive rates
- Improve staff utilization and efficiency
- Simplify security infrastructure
- Refine focus on emerging risks
- Control security risk
- Increase value of security investments

CSDS outlines a range of methods to measure phenomena, make better decisions, identify threats, control risk, and optimize processes. As a whole, the range of design prescriptions outlined by CSDS frames a basis for a cybersecurity analytics maturity model. Namely, as advocated CSDS design prescriptions are iterative and

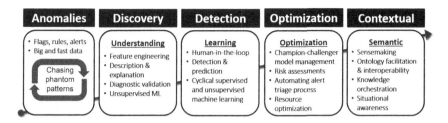

Fig. 4.45 CSDS maturity model

self-reinforcing, it is asserted that their implementation can lead to progressive improvements in operational settings. As represented in the framework posed in Fig. 4.45.

Socializing an iterative approach to CSDS effectiveness should be helpful to managers seeking to communicate a CSDS strategy and to gain buy-in from distributed stakeholders. A key aspect of building a CSDS competence involves up-front investments, especially as related to data management, model design and validation, and cross-training. The background provided here frames the focused overhead involved in building a CSDS competency. The intention is to supply managers and planners with practical background details to explain why "Rome cannot be built in a day" and to advocate that initial investments and efforts in rigor will have outsized future benefits.

4.6.4   Research Questions Addressed

The final set of research objectives were addressed in Phase III, per the summary in Table 4.13.

Table 4.13 Research questions addressed—gap-prescription designs

Management problem 3	**What guidance can be offered to orchestrate the implementation of CSDS programs?** • Address the key CSDS gaps of data management, scientific processes, and cross-domain collaboration • Address gaps through design prescriptions which accommodate MIS technology, process, and organization perspectives • Focus on the implementation of integrated processes • Address organizational process collaboration and incentives • Understand the types of up-front investments required to enable data management, rigor in model development, and cross-training
Research objective 3	*Prescribe* **design treatments based on gap analysis and extrapolation from literature** • Apply design science methods to specify data science prescriptions to treat security gaps • A range of designs, grounded by literature-derived concepts, are advocated to address the CSDS gap areas of data management, scientific processes, and cross-domain coordination

(continued)

Table 4.13 (continued)

Research question 7	**How should CSDS methodological challenges be addressed?** • CSDS experiences tension between shifting security requirements and a lack of clarity across a confusing array of data science methods • A bottom-up gap diagnostic approach addressed by design science is advocated, whereby focused security challenges are addressed by specific data science prescriptions
Research question 8	**Which data science methodological treatments are prescribed for the CSDS domain?** • It is advised to combine explanatory and predictive approaches to undertake detective analytics • Feature engineering, segmentation-to-categorization, evidence extrapolation, and self-improving detection model development methods are advocated
Research question 9	**What processes are advocated to implement prescriptions?** • Processes are advocated to facilitate improved rigor in data management, scientific processes, and cross-domain collaboration • Design science-based approaches to specify CSDS requirements and artifacts are advocated • Processes for addressing the three key gaps overlap and self-reinforce as scientific processes require both high-quality data and organizational orchestration • CSDS designs imply orchestrating collaboration between and across human and machine-facilitated processes
Research question 10	**How can CSDS prescriptions be implemented as organizational processes?** • It is recommended to adopt an analytics life cycle framework to guide the development of CSDS solution designs • Both data and scientific processes require organizational collaboration and facilitation • Careful incentive engineering is required to overcome the potential for goal-conflicts to arise across multidisciplinary teams • Scientific theory formation, as a process of knowledge generation, requires social facilitation and exchange

Chapter 5
Research Conclusions and Discussion

5.1 Summary of Research Motivations

Modern enterprise is increasingly facilitated by a pervasive global network of hyperconnected digital devices and distributed services. For many organizations, choosing not to connect to and participate in this digital cybersphere is a self-limiting option. Assuring the continuity and integrity of digital infrastructure is a growing challenge for many organizations.

Along with participation, interconnectivity exposes participants to actors seeking to damage, exploit, or otherwise misuse digital infrastructures. The exponential growth and development of networks, data volumes, and digital devices have expanded risks and vulnerabilities. Professionals charged with assuring cybersecurity are increasingly challenged by a complex of factors including expanding vulnerabilities, evolving threats, overworked cybersecurity staff, false alert overload, challenges orchestrating remediation, and difficulties integrating security systems and data.

Even as risks multiply, many organizations pursue ambitious strategies for digital advancement. The sheer pace of digital expansion frequently bypasses rigorous security oversight. Security professionals face pressures to mediate between growing demands for connectivity and increasing exposure to a range of opportunistic threats.

A central challenge in defending networked infrastructure is the growing imbalance between the capabilities of defenders and attackers. While threats and vulnerabilities expand, there are a growing lack of skilled practitioners to accommodate and monitor the risks manifested. Cybersecurity professionals increasingly struggle to keep pace with the security impacts of continual growth and technical innovation.

In aggregate, these factors have led to an interest in applying analytical, algorithmic, and machine-driven methods to semi-automate aspects of the growing security protection and assurance gap. Increasing threats, staff shortages, and an environment of growing complexity impel the need for more sophisticated, data-driven

© Springer Nature Switzerland AG 2021   317
S. Mongeau, A. Hajdasinski, *Cybersecurity Data Science*,
https://doi.org/10.1007/978-3-030-74896-8_5

decision approaches. Growing cybersecurity pressures thus organically frame requirements for cybersecurity data science (CSDS) practice.

However, as an emerging profession, CSDS lacks characteristics of mature professions. A lack of standards and codified best practices, body of theory, limits the ability to train staff, manage programmatic efforts, ensure rigor, and track effectiveness. As CSDS efforts expand, body of theory gaps will become self-limiting as practitioners increasingly encounter conflicting understandings concerning proper practice.

By improving an understanding of current challenges and best practices in the CSDS field, a foundation for advancement of the domain can be framed. It is beneficial for researchers and practitioners alike to compare where the CSDS discipline is currently (*as-is*), where it is evolving in terms of dominant trends and idealized practice (*to-be*), and where organizations need to foster development to address shortcomings (*gap-diagnosis*). A systematic comparison establishes a set of requirements and guidance for iterating CSDS professional practice.

This set of objectives motivated the practitioner problem-solving research design and diagnostic mixed methods applied. A more typical hypothetical-deductive research approach did not address the complexity and novelty of CSDS as an emerging phenomenon. Triangulated mixed methods, combining both inductive and deductive approaches, supported diagnostic gap analysis.

The often unruly reality of CSDS as a domain, emerging from dynamic practice, benefited from a mix of both top-down and bottom-up approaches to support abductive, diagnostic problem-solving. This reflects an approach more similar to medical diagnostics and reflects upon the complexity and novelty of the underlying phenomenon.

To ground and motivate the presentation of research conclusions in this chapter, it is useful to briefly convey a narrative understanding of the depth of the practice-theory gap in the cybersecurity field for the benefit of researchers. Presenting a realistic scenario: were an academic organizational researcher to visit any medium-to-large size corporate security operations center (SOC), they would be met, tensely, by a set of very busy, potentially overwhelmed, security professionals. Prior to this, securing an official invitation likely took several weeks, if not months, as the nature of the work being conducted is highly sensitive, meaning that IT, legal, compliance, and/or risk stewards had to agree to and approve the visit.

In the short interview that follows with a harried security professional, that researcher might attempt to gather background concerning the empirical basis for the work underway, the dominant work being to monitor complex digital infrastructure in order to surface potential threats and incidents amongst an ocean of noise. It will become quickly apparent that many hundreds of alerts are being tracked at any one time, and that new ones are arriving every few minutes, far many more than the professionals at work can humanly accommodate. As a result, many otherwise seemingly worrying alerts are necessarily ignored.

The researcher, seeking an operational theoretical perspective, might begin to inquire concerning the nature of several fundamental entities being monitored, for instance, asking simple questions such as "what is a user?", "what are the different types of users?", or "what types of devices are you tracking?". The security professional will, more than likely, then begin to explain that these entities are largely defined by the organization itself. That is, users are often identified as employees by HR and/or customers by sales and marketing, for instance. Further, the devices are often designated by facilities or asset management departments, working with IT operations. As a result, the definition and categorization of users and devices is not typically grounded by security relevance.

This leads the security professional to demure and qualify: the lists of users and devices the SOC receives are not necessarily accurate at any given time, being sometimes weeks or months out of date. Further, there are often stark errors, as well as misleading misunderstandings. For instance, somewhere between 1% and 3% of the registered users may actually be machines with user accounts, or in some cases may be a user account shared across a functional team. As a result, it seems that some of the users on the network are active 24 h a day, 7 days a week, which betrays the human notion of sleep.

As well, some devices may be grouped in insufficiently specific categories, for instance, simply as a "server," whereas there would be a great benefit to disassociating print, file, web, network, and application servers. These errors and misunderstandings, in aggregate (as explained with some frustration), end up causing a great deal of extra work for the SOC. Many alerts generated are the result of mis-categorizations and misunderstandings of what is being tracked and why.

What the researcher begins quickly to realize, as result of this short interview, is that there is a critical lack of fundamental theory undergirding typical enterprise security efforts. Basic questions such as "what is a user?" result immediately in an evasive answer of "it depends…." This is not to say security professionals are culpable: many, if not most, are painfully aware of this problem. However, due to the constant functional demands of their role, it is extremely difficult to "fix the engine of the automobile as it barrels down the freeway," to borrow a popular simile used in the IT professions.

An essential challenge of, and hopefully contribution of, this research is to dig deeply onto this problem: the dramatic paucity of empirically grounded, rigorous, validated theory to guide and improve security operations. As this work points out, there are more than 20,000 research articles easily at hand on the application of data science to cybersecurity (Google Scholar 2019a). However, the great bulk of this research is oriented exclusively to engineering implementations and propositions of new methods. There are very few articles or works asking fundamental theoretical questions with a grounding in organizational research concerning the scientific and empirical underpinnings of enterprise security efforts.

That CSDS demonstrably is a new field in de facto practice, there arises a natural question: how can CSDS professionals conduct their work lacking clear foundations for empirically grounded, rigorous scientific inquiry? The short answer is, apparently, "very well," in at least being able to jump *practically* into the deafening volume of data streaming forth and putting in place algorithmic and analytical treatments to surface focused security alerts.

Regarding the effectiveness of these treatments, much less is known. It is quite uncommon, for instance, for CSDS professionals to have the wherewithal (often of no fault of their own) to rigorously empirically test and track their own models in terms of the operational results, to close the loop empirically. Simply put, enterprise security, from an organizational perspective, is a complex "cyborg" combining technology, processes, and people. Reengineering such cyborg entities is a complex, staged, nearly surgical undertaking requiring integrated technical, methodological, and organizational intercessions.

This narrative digression into the practical realities of operational enterprise security frames the core research challenges this work has attempted to address. Namely, how can a discipline emerging from a dearth of theoretical foundations otherwise surface, frame, and pursue improved practices? This is, in essence, a bootstrapping problem: one that demands a comprehensive, comparative gap-based accounting of where things are (*as-is*) and where they need to be (*to-be*).

Gap analysis identifies structural gaps as requirements, and these requirements frame potential treatments, methods, and solutions. Diagnostic methods (not dissimilar from clinical medical practice in combining deductive and inductive approaches) provide a basis for this gap analysis. Solutions and treatments can then be abducted from formal requirements, supported by deductive concepts derived from literature and practice. With this motivating background established, a more formal accounting of the research objectives, methods, results, contributions, limitations, and implications follows.

5.2   Summary of Research Objectives and Questions Addressed

The central research objective of this effort is to analyze challenges in the emerging cybersecurity data science (CSDS) profession, diagnose central gaps, and prescribe design treatments to facilitate advancement. The accompanying management research objective is to strengthen CSDS body of theory in order to iterate professional maturity. The central research question addressed herein:

RQ0: What treatment designs are prescribed from diagnostic research to address gaps impeding the development of cybersecurity data science (CSDS) professional practice?

The central research question informs theoretical (TQ), empirical (EQ), and analytical (AQ) questions (Verschuren and Doorewaard 2010):

TQ0: *Phase I*—Extrapolating from diagnostic *background research* of literature, what challenges do practitioners face in the CSDS domain?

EQ0: *Phase II*—Extrapolating from interview-based diagnostic *opinion research*, what categorical challenges and best practices do CSDS practitioners perceive?

AQ0: *Phase III (triangulating Phases I and II)*—Extrapolating from mixed method diagnostic *gap analysis research*, what categorical treatment design prescriptions are recommended to address identified CSDS practice gaps?

The following three associated research objectives (ROs) are addressed sequentially across a series of ten associated research questions (RQs):

RO1: *Phase I—Analyze* the as-is state of the CSDS field based on comparative literature

RQ1: What is the basis for asserting CSDS as a nascent professional domain?
RQ2: What are the disciplinary boundaries of CSDS as a hybrid professional domain?
RQ3: Where is CSDS in a process of professionalization?
RQ4: What challenges does the CSDS domain face on the path to professionalization?

RO2: *Phase II—Diagnose* gaps impeding CSDS professionalization based on qualitative research

RQ5: What gaps can be diagnosed in the emerging CSDS field from practitioner perspectives?
RQ6: What treatments are prescribed to address gaps based on practitioner input?

RO3: *Phase III—Prescribe* design treatments based on gap analysis and extrapolation from literature

RQ7: How should CSDS methodological challenges be addressed?
RQ8: Which data science methodological treatments are prescribed for the CSDS domain?
RQ9: What processes are advocated to implement prescriptions?
RQ10: How can CSDS prescriptions be implemented as organizational processes?

RQ0-derived research objectives (**RO1–3**) and research questions (**RQ1–10**) are addressed sequentially in this research effort through three phases reflecting the key theoretical (**TQ0**), empirical (**EQ0**), and analytical (**AQ0**) questions. A summary of high-level findings associated with the key research objectives and questions follows.

5.2.1   Phase 1 (Section 1)

The first phase applied literature analysis to support diagnostic background (problem) analysis. CSDS, being an emergent hybrid professional domain, embodies new phenomenon which lacks focused coverage in literature. The integrated literature analysis synthesized sources and concepts from several domains to deliver a structured account of CDSD as an emerging field.

Literature analysis derived a model for professional maturation and assessed CSDS according to the criteria framed. A central recommendation was that institutional facilitation via professional training providers and certification authorities is a key next step to achieving *occupational closure*, whereby the CSDS specialty can assert professional specialty status.

A key outcome of literature analysis was a set of sensitizing concepts framed to serve as a foundation for subsequent opinion and gap analysis research in *Phase II*. Central sensitizing concepts extrapolated included the following: a context for assessing CSDS professional maturity, key CSDS challenge and opportunity themes, CSDS literature gaps, and a conceptual model framing the relationship between cybersecurity challenges and data science prescriptions. As well, key CSDS methodological concepts were surfaced which serve to inform design problem-solving in *Phase III*. Literature analysis thus serves multiple purposes in the scope of the larger inquiry. A summary recap of findings in the first phase encompassing *diagnostic background analysis* (**TQ0** addressing **RO1**) is offered Table 5.1.

Table 5.1 Research questions recap—Phase I literature analysis

Research objective 1	*Analyze* **the as-is state of the CSDS field based on comparative literature**
	• The CSDS field is in the process of body of theory individuation, a prerequisite for professional emergence
	• Gaps in CSDS literature include risk quantification, data management, scientific methods, and organizational management
	• There is a need for body of theory codification through institutionally sponsored training and certification
Research question 1	**What is the basis for asserting CSDS as a nascent professional domain?**
	• Active named and specialized practitioners are employed globally
	• CSDS derives sanction through its parent domains and demonstrates efficacy in focused use cases
	• The field represents a paradigm shift from traditional approaches to security assurance
	• CSDS is presently demand-driven as security professionals face growing challenges which are addressed by data science methods
Research question 2	**What are the disciplinary boundaries of CSDS as a hybrid professional domain?**
	• The locus of the CSDS field is most logically situated within the cybersecurity professional domain
	• Applied CSDS involves addressing cybersecurity challenges with data science prescriptions and implies a gap analysis is conducted
	• There are early indications of recognition of the CSDS domain as a focused specialty from a range of institutions (commercial, professional, governmental, academic)

(continued)

Table 5.1   (continued)

Research question 3	**Where is CSDS in a process of professionalization?** • CSDS is in the process of professional emergence through body of theory codification • There is a need for clearer guidelines concerning the application of scientific rigor in practice • Institutional facilitation is necessary to achieve professional individuation, particularly professional training and certification • Institutions are beginning to frame CSDS as a unique sub-specialty within the security profession • There is a growing body of focused CSDS literature
Research question 4	**What challenges face the CSDS domain on the path to professionalization?** • Both the security and data science domains are undergoing rapid technology-driven change • The methodological permissiveness of the data science domain challenges the ability to codify CSDS body of theory • The security domain lacks rigorous scientific theoretical foundations, focusing instead on applied engineering solutions • Data management is a key challenge requiring focused solutions • Commercial hype creates informational distortions compromising clarification of CSDS body of theory • Paradoxically, there is a need for institutional sponsorship, although uneven body of theory forestalls institutional adoption

5.2.2   Phase 2 (Section 2)

The second phase combined mixed qualitative and quantitative methods to surface and explore key themes raised in interview research. For the qualitative component, *diagnostic opinion research* was undertaken in the form of interviews with 50 CSDS practitioners. *Phase I* sensitizing concepts served to guide the coding of interview transcripts. This was followed by a quantitative component analyzing patterns resident in coded interview response data. Analysis served to clarify and frame key themes to inform *diagnostic gap analysis.*

Extending deductive findings from literature analysis, inductive interview analysis results supported a triangulated understanding of the relative importance of and relationships between key CSDS themes. Results suggested shared practitioner themes related to CSDS challenges and best practices. Interpretation and framing of patterns surfaced through quantitative analysis were supported by qualitative findings, bridging literature analysis, interview memoing, interview thematic analysis, and text analytics. Three central CSDS gap-prescription themes were framed: data management, scientific processes, and cross-domain collaboration. A summary recap of findings from the second phase encompassing *diagnostic opinion* and *gap analysis* (**EQ0** addressing **RO2**) is offered in Table 5.2.

Table 5.2 Research questions recap—Phase II opinion and gap analysis

Research objective 2	*Diagnose* **gaps impeding professionalization based on qualitative research**
	• An analysis of themes surfaced in CSDS interview results identified three key gap-prescription areas: data management, scientific processes, and cross-domain collaboration:
	– Interpretation and coding of interviews resulted in a set of challenge and best practice topics
	– Factor analysis of best practices and challenges extrapolated central themes
	– Correlation of factor themes surfaced key macro-gap themes associated with best practice and challenge prescriptions
Research question 5	**What gaps can be diagnosed in the emerging CSDS field from practitioner perspectives?**
	• Key challenge themes surfaced in factor analysis of interview feedback:
	– Expansive complexity
	– Tracking and context
	– Data management
	– Expectations versus limitations
	– Unclear ownership
	– Data policies
	• Key CSDS gap-prescription areas requiring focused treatments:
	– Data management
	– Scientific processes
	– Cross-domain collaboration
Research question 6	**What treatments are prescribed to address gaps based on practitioner input?**
	• Based on correlation of challenge and best factor themes extrapolated from interview feedback:
	– Improve rigor of scientific process
	– Improve approaches to data management
	– Improve cross-domain organizational collaboration
	• Details of treatments are suggested by the array of sub-themes aligned to the key gap-prescription areas

5.2.3   Phase 3 (Section 3)

In the third and concluding phase, focused guidance addressing the three gap analysis themes surfaced in *Phase II* were addressed. A design science problem-solving process was applied, framing requirements and suggesting prescriptions. Concepts derived in *Phase I* literature analysis were combined and extended to support the design of prescriptions. Standing challenges related to CSDS data management, scientific processes, and cross-domain collaboration were explored and addressed in order to provide practical guidance. Recommendations were positioned as nascent designs for future implementation and assessment.

The overarching MIS disciplinary focus framed the key gap areas within technological, process, and organizational contexts. A central finding was that the nature of the three gaps is complex and overlapping, requiring systematic, multi-contextual processes to address. Advocacy emphasized the implementation of organizational processes to facilitate multi-contextual integration. Results were presented as a potential foundation for subsequent design and action research efforts.

A key topic explored concerned the self-limiting nature of the existing engineering focus in the cybersecurity domain. The engineering emphasis in cybersecurity results in weak theoretical and scientific foundations for staging CSDS efforts. This gap was framed in terms of the distinction between *technê* and *logos* mediated *epistêmê*. Whereas *technê*—skill, art, or craft—focuses on pragmatic results, *logos*, the expression of rational, structured thought, supports the development of *epistêmê*, understanding, knowledge, and scientific theory. Process-based approaches to CSDS which frame scientific methods and treatments were advocated as an approach to addressing this *epistêmê* lacuna.

A range of CSDS scientific methodological approaches were framed, spanning from exploration and discovery through prediction and detection, leading up to knowledge generation and general theory formation. The conclusion advocated the need for programmatic process-focused approaches to CSDS, especially as related to orchestrating CSDS organizational collaboration and cross-process integration. A summary recap of findings from the third and concluding phase encompassing *design science problem-solving* (**AQ0** addressing **RO3**) is summarized in Table 5.3.

Table 5.3 Research questions recap—Phase III gap-prescription designs

Research objective 3	*Prescribe* **design treatments based on gap analysis and extrapolation from literature**
	• Apply design science methods to specify data science prescriptions to treat security gaps
	• A range of designs, grounded by literature-derived concepts, are advocated to address the CSDS gap areas of data management, scientific processes, and cross-domain coordination
Research question 7	**How should CSDS methodological challenges be addressed?**
	• CSDS experiences tension between shifting security requirements and a lack of clarity across a confusing array of data science methods
	• A bottom-up gap-diagnostic approach addressed by design science is advocated, whereby focused security challenges are addressed by specific data science prescriptions
Research question 8	**Which data science methodological treatments are prescribed for the CSDS domain?**
	• It is advised to combine explanatory and predictive approaches to undertake detective analytics
	• Feature engineering, segmentation-to-categorization, evidence extrapolation, and self-improving detection model development methods are advocated
Research question 9	**What processes are advocated to implement prescriptions?**
	• Processes are advocated to facilitate improved rigor in data management, scientific processes, and cross-domain collaboration
	• Design science-based approaches to specify CSDS requirements and artifacts are advocated
	• Processes for addressing the three key gaps overlap and self-reinforce as scientific processes require both high-quality data and organizational orchestration
	• CSDS designs imply orchestrating collaboration between and across human- and machine-facilitated processes

(continued)

Table 5.3 (continued)

Research question 10	How can CSDS prescriptions be implemented as organizational processes?
	• It is recommended to adopt an analytics life cycle framework to guide the development of CSDS solutions designs
	• Both data and scientific processes require organizational collaboration and facilitation
	• Careful incentive engineering is required to overcome the potential for goal-conflicts to arise across multidisciplinary teams
	• Scientific theory formation, as a process of knowledge generation, requires social facilitation and exchange

5.3 Summary of Research Process, Methods, and Outcomes

This effort encompasses practitioner-based partial problem-solving design research in the MIS domain supported by mixed method diagnostic research. Diagnostic background (literature), opinion (interview), and gap analysis (quantitative) research approaches are pursued to surface, categorize, and address perceived lacunas forestalling CSDS professionalization.

Mixed diagnostic methods are triangulated to abductively substantiate surfaced gaps, combining empirical and deductive approaches. Gaps frame design requirements, which are in turn treated through design science problem-solving. Results frame advocacy for prescriptive treatments to advance CSDS body of theory and practice, a necessary step in professionalization.

In pursuing diagnostic-based problem-solving research, the scientific focus of the effort is on rigorous pattern exploration, discovery, and extrapolation in a complex new domain, as opposed to framing and validating a hypothetical-deductive construct. Interview research, comparative literature analysis, and text analytics compose the core empirical elements.

Structured diagnostic gap analysis findings (*Phase II*) frame design requirements which are used to formulate conceptual design recommendations (*Phase III*) supported by literature-based, deductively derived concepts (*Phase I*). Scoped as a partial problem-solving design cycle, results are not pursued to implementation and assessment, being framed as recommended nascent designs for future implementation (Gregor and Hevner 2013).

The resulting manuscript embodies a comprehensive three-phase report intended to provide prescriptions to advance CSDS professional maturity. A high-level summary of key research phases, methods, and results is presented in Table 5.4.

In terms of the staging and integration of the three research phases, structured results from diagnostic literature analysis (*Phase I*) sensitize semi-grounded interview research of CSDS practitioners (*Phase II*). CSDS domain gaps surfaced in mixed method analysis (*Phase II*) are framed as requirements for design problem-solving (*Phase III*). Literature analysis (*Phase I*) also surfaces a series of key CSDS conceptual models to support design problem-solving (*Phase III*).

Table 5.4 Summary of research process and results

Chapter	Phase	Methods	Results
2. CSDS as an Emerging Profession	I. Diagnostic background analysis	Multi-faceted literature analysis	CSDS maturity gaps CSDS demand model Literature corpus Literature gaps Sensitizing concepts Analytical methods
3. CSDS Practitioner Interviews & Gaps	II. Diagnostic opinion research	Qualitative interview research	Key challenge & best practice themes
	II. Diagnostic gap analysis	Quantitative analysis of themes	Diagnosis of CSDS gap-prescriptions
4. CSDS Gap-Prescription Designs	III. Design problem-solving	Design science	Design requirements & prescriptions

Fig. 5.1 Interrelationship of key research phases

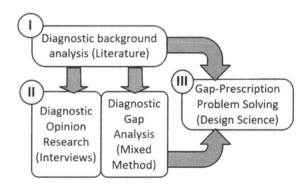

Figure 5.1 represents the interaction between the main research phases, demonstrating how *Phase I* diagnostic background (literature) analysis provides a foundation for multiple aspects of the inquiry. The literature analysis component (*Phase I*) supports both semi-grounded, inductive interview research (*Phase II*) and requirements-driven deductive design problem-solving (*Phase III*).

An overview of the three main research phases follows, summarizing main goals, methods, approaches, and results produced.

5.3.1 Phase I: Diagnostic Background (Literature) Analysis— CSDS as an Emerging Profession

The first phase applies literature analysis to support diagnostic background (problem) analysis. CSDS, being an emergent hybrid professional domain, embodies a new phenomenon which lacks focused coverage in literature. The integrated literature analysis synthesizes sources and concepts from several domains to deliver a structured account of CDSD as an emerging field. Table 5.5 provides an overview of *diagnostic background (literature) analysis* research conducted in Phase I.

Literature analysis develops an understanding of challenges facing the emerging CSDS field. CSDS professionalization is situated in the context of the cybersecurity and data science parent domains. A structured assessment of the professional maturity level of the *as-is* CSDS domain results. A central observation is that growing cybersecurity challenges organically impel the CSDS field. Growing threats, personnel shortages, and environments of increasing complexity frame a demand for data-driven approaches to security assurance.

A key corpus of 33 manuscript-length CSDS works is provided and summarized. A comparative content analysis of the corpus provides insights into the relative coverage of key themes across the domain. Combined with themes from the corpus,

Table 5.5 Overview of Phase I—Diagnostic background analysis

Phase	I. Diagnostic background analysis
Subtitle	CSDS as an Emerging Profession
Summary	*Diagnostic background analysis* of CSDS-related professional and academic literature leading to a maturity assessment of the as-is CSDS profession and sensitizing concepts
Methods	Literature analysis to profile key concepts and gaps in the CSDS domain (Hart 2000; Levy and Ellis 2006; Onwuegbuzie et al. 2012; Sammon et al. 2010); Professionalization theory supports a semi-structured maturity assessment (Greenwood 1957; Beer and Lewis 1963; Vollmer and Mills 1966; Freidson 2001; Muzio and Kirkpatrick's 2011)
Approach	Literature analysis outlines challenges surfaced by the hybridization of cybersecurity and data science within CSDS in terms of a professional maturity model. Comparative literature analysis surfaces key methods, literature gaps, and sensitizing concepts
Results	1. Frames CSDS professional maturity gaps and challenges inhibiting CSDS professionalization in a structured maturity assessment 2. Summarizes forces driving CSDS practice in a supply-demand model 3. Produces and summarizes a corpus of 33 manuscript-length works 4. Conducts a thematic gap analysis of the CSDS corpus, identifying the gaps: risk quantification, data management, scientific methods, organizational management 5. Analyzes and extrapolates sensitizing concepts to inform subsequent *Phase II* interview and gap analysis research 6. Positions key methodological concepts to support *Phase III* design prescriptions

analysis across professional and academic sources results in a set of CSDS sensitizing concepts to guide interview research and interpretation in *Phase II*. Finally, several treatments available from the field of data science are framed to support design problem-solving in *Phase III*.

5.3.2   Phase II: Diagnostic Opinion Research and Gap Analysis—Interviews and Mixed Method Analysis

The second phase combines mixed qualitative and quantitative methods to surface and explore key themes raised in interview research. For the qualitative component, *diagnostic opinion research* is undertaken in the form of interviews with 50 CSDS practitioners. This is followed by a quantitative component analyzing patterns across interview response data. Analysis serves to clarify and frame key themes to inform *diagnostic gap analysis*. Incorporating findings from literature analysis, results triangulate a structured understanding of the relative importance of and relationships between key CSDS themes.

Results suggest shared practitioner themes related to CSDS challenges and best practices. Interpretation and framing of patterns surfaced through quantitative analysis are supported by qualitative findings, bridging literature analysis, interview memoing, interview thematic analysis, and text analytics. Three central CSDS gap-prescription themes are framed: data management, scientific processes, and cross-domain collaboration. Table 5.6 offers an overview of *diagnostic opinion* and *gap analysis* research conducted in Phase II.

Phase II undertakes and analyzes over 25 h of interviews with 50 global cybersecurity data scientists. A practitioner-based understanding of perceived CSDS challenges and best practices is extrapolated, supported by sensitizing concepts

Table 5.6 Overview of Phase II—Diagnostic opinion research and gap analysis

Phase	II. Diagnostic opinion research and gap analysis
Subtitle	CSDS Practitioners
Summary	*Diagnostic opinion* and *gap analysis* surfacing and framing key CSDS gap-prescriptions
Methods	Interviews with CSDS professionals to diagnose perceived gaps impeding CSDS practice (Glaser and Strauss 1967; Kim and Mueller 1978a; Krippendorff 2019; van den Hoonaard 1996)
Approach	Themes raised in interviews with 50 CSDS practitioners are analyzed via mixed qualitative and quantitative methods to extrapolate key patterns across responses
Results	• Diagnostic opinion and gap analysis results in identification of key CSDS gaps • Key challenge and best practice themes extrapolated from interviews are framed • Key gaps impeding CSDS practice are extrapolated via quantitative thematic analysis • Three central CSDS gap-prescription themes are surfaced: data management, scientific processes, and cross-domain collaboration

derived from literature analysis. A structured analysis of interview results offers insights into current limitations of the CSDS field as compared to idealized visions of future practice. Quantitative analysis of interview results leads to the framing of three central CSDS practice gaps: data management, scientific methods, and cross-domain collaboration. A central finding relates to the need to inculcate greater scientific and methodological rigor in CSDS practice.

5.3.3   Phase III: Design Science Problem-Solving—CSDS Gap-Prescription Designs

In the third and concluding phase, the three gaps identified in *Phase II* are utilized to frame problem-solving requirements, which in turn support the development of design gap-prescriptions. Design gap-prescriptions address the gaps: data management, scientific processes, and cross-domain collaboration. Prescriptions are framed as recommendations for improving CSDS practice.

Concepts derived from literature in *Phase I* support design problem-solving. Through the design process, standing challenges in the cybersecurity profession and methodological debates in the field of data science are raised and worked through in order to provide practical guidance and insights.

As the larger research effort is constrained to partial problem-solving, design recommendations are framed at the level of *nascent designs*: constructs, methods, models, principles, and rules (Gregor and Hevner 2013). Designs are not implemented and assessed. Results are framed as advocacy for practitioners, contributions to CSDS body of theory, and foundations to guide future research efforts. Table 5.7 provides an overview of *design science problem-solving* research conducted in Phase III.

Table 5.7 Overview of Phase III—Design science problem-solving

Phase	III. Design science problem-solving
Subtitle	CSDS Gap-Prescriptions
Summary	*Design prescriptions* are framed and advocated to address surfaced CSDS gaps
Methods	MIS domain design science research (Dresch et al. 2015; Gregor and Hevner 2013; Hevner et al. 2004; Wieringa 2014)
Approach	*Design science* requirements are framed and prescriptions advocated to address gaps surfaced, combining guidance from literature and practice
Results	• Gaps surfaced in *Phase II* are refined in problem-solving requirements • Concepts raised in *Phase I* literature analysis are leveraged to inform designs • Addressing the three *Phase II* gaps, *nascent designs* are framed as unimplemented recommendations composing integrated sets of methods, models, and principles I. Guidance is provided concerning improving CSDS *data management* processes II. Methodological best practices are advocated to improve CSDS *scientific rigor* III. Organizational guidance is issued to improve CSDS *cross-domain collaboration*

Practical recommendations aim to serve managers seeking to implement CSDS programs, practitioners seeking guidance on refined practice, and researchers seeking to frame design and action research initiatives. In particular, results suggest fruitful opportunities for follow-on design, action, and/or case study research. Nascent designs framed suggest a foundation for future MIS domain design science implementations and assessments.

Design prescriptions are more broadly advocated as approaches to advance the process of CSDS professionalization. Of interest to CSDS body of theory advancement, a central meta-theoretical topic addressed concerns structured approaches to improve scientific rigor in CSDS practice.

5.4 Academic, Theoretical, and Methodological Contributions

As practitioner problem-solving research supported by diagnostic and design efforts, this work extrapolates systematic understandings of challenges and best practices in the emerging field of CSDS. A set of prescriptions are extrapolated and recommended to support advancement of the emerging field through body of theory refinement. Central mixed methods include integrated literature analysis, interview research, quantitative analysis, and design science problem-solving.

An effort to summarize the discrete academic contributions of the three-phase research effort starts with *Phase I*, literature analysis. In aggregate, *Phase I* results address a research lacuna by producing an MIS-focused organizational research perspective on the emerging CSDS profession. The literature analysis also serves an important function to ground and motivate the larger research effort.

Several unique contributions are achieved in *Phase I*, as summarized in Sects. 2.6.1–2.6.6: (1) a profile of gaps in CSDS professional maturity, (2) a model of supply-and-demand factors propelling the CSDS domain, (3) identification of a corpus of key literature in the CSDS domain, (4) a thematic gap analysis of the CSDS corpus, (5) sensitizing concepts to support interview research (conducted in *Phase II*), and (6) key data science methodological concepts to support design problem-solving (conducted in *Phase III*). The resulting six outcomes are offered as novel contributions addressing an MIS research lacuna on the topic of CSDS.

A central byproduct resulting from *Phase I* of interest to organizational scholarship concerns the framing and application of an organizationally focused professional maturity assessment model derived from Greenwood (1957). This participates in a scholarly debate concerning the nature and status of emerging postindustrial, information age job roles.

As raised in the literature analysis (Sect. 2.2.7), some modern professionalization scholars, holding a countervailing sociological view, have critiqued the more traditional organizational view of the professions. The modern information-driven workplace is rife with fast-emerging, relatively ill-defined new job roles (e.g., data scientist, social media manager, chief data officer). Many of these roles purport lofty responsibilities which, under careful scrutiny, reveal insubstantial powers in the context of organizational decision architectures (e.g., the data scientist rarely has the power to enforce standards for scientific rigor in analytical inquiry; the chief data officer rarely has the power to mandate a master data schema across the organization). There is, therefore, a cynical view that a number of the broadly aspirational IT-related job roles which have emerged in the past decade may have more to do with attracting young talent than positioning workers for success.

In this respect, the maturity analysis undertaken in *Phase I* can be considered as a potential rebuke to those asserting that organization-based assessments of the professions are passé. The exercise undertaken to assess CSDS professional maturity surfaced focused gaps which impede advancement of the field as a profession. In particular, some of these gaps are revealed as being somewhat paradoxical, for instance, the need for institutional facilitation to codify standards in the form of training and certification programs, yet being unable to resolve key body of theory to frame training and certification programs due to uneven institutional sponsorship. Next steps suggest the need for pragmatic bootstrapping approaches.

In the case of CSDS, and by comparison other fast-emerging information age roles, critical focus on the process of professionalization vis-à-vis institutional mentorship and sponsorship is otherwise sorely needed. This is a difficult but crucial problem to frame in an age dominated by free-market views and skeptical of any forces signaling potential macroprudential stewardship over the professions or, a particular bête noire, any implication of pseudo-regulatory oversight (e.g., government labor bodies, noncommercial industry organizations, or nonprofit professional associations attempting to establish ground rules for professional practice and/or repercussions for lapses in professional conduct).

Regarding the motivation for applying Greenwood's organizationally grounded view of professional maturity, the organizational view of professional maturity was lightweight, simple to ground with examples, and produced results of contemporary relevance and pragmatic interest. Further, the maturity model revealed a set of easily understandable gaps to focus subsequent CSDS gap analysis and gap-prescription efforts in the larger research effort. Of particular importance, highlighting the crucial role *body of theory* plays in the advancement of professions was central to orienting the larger research inquiry.

The results of the maturity assessment exercise reveal key gaps inhibiting advancement of both the data science and CSDS fields. In particular, the strong case made by Greenwood concerning the centrality of a coherent body of theory to the emergence of new professions is instructive. It is therefore suggested that the assessment exercise undertaken is a useful and novel contribution to the MIS domain in the respect that many other emerging information technology-related professions potentially would benefit from similar systematic maturity assessments. The working out of the local example of CSDS in this regard serves to demonstrate and highlight challenges and opportunities that face other similarly fast-emerging new information age professions.

A key outcome of the maturity assessment exercise served to raise consciousness concerning the body of theory gap in the CSDS domain. Findings have meta-theoretical implications in terms of encouraging an understanding of how MIS-associated professions are self-limiting to the degree that they focus on engineering methods and outcomes exclusively. In the case of CSDS, *Phase I* (per Sect. 2.6.2) frames the symbiotic relationship between cybersecurity challenges (demand) and data science prescriptions (supply), per Looijen and Delen's information systems management model (1992).

Specific to CSDS practice, framing a symbiotic supply-demand relationship between data science and cybersecurity situates the practical role of CSDS body of theory. In a supply-demand context, CSDS body of theory is insinuated as a super-visory agent in diagnosing cybersecurity challenges and specifying matching data science prescriptions. In a theory building context, this frames the theoretical inter-action between engineering-focused cybersecurity practice (*technê*), data science treatments-as-artifacts (*logos*), and CSDS domain body of theory (*epistêmê*).

In hypothetically conjoining cybersecurity *techne* with data science *logos*, CSDS body of theory is positioned as a mediator of integrated techno-logical (*techne-logos*) synthesis. This situates CSDS in a functional clinical diagnostic role not dissimilar from medical practice: diagnosing cybersecurity gaps and prescribing subsequent data science treatments. This focused positioning of the practice-based application of professional body of theory (*praxis*), especially as related to diagnostic practice, is submitted as a concept of potential wider interest to the development of IT professions.

Concerning the contributions of *Phase II*, empirically grounded CSDS practitioner interviews, interview findings are broadly of interest to the emerging CSDS profession. CSDS practitioner themes and gaps, as extrapolated through quantitative methods from coded interview data, are a central knowledge artifact produced.

Through quantitative analysis of coded interview results, *Phase II* surfaces three key CSDS domain gaps from the perspective of CSDS practitioners: data management, scientific processes, and cross-domain collaboration. This constitutes a novel contribution to researchers and organizational stakeholders, addressing a research lacuna by extrapolating a structured, empirically grounded profile of gaps in CSDS practice. These findings are of interest to promoting a larger understanding of gaps impeding CSDS professionalization.

The identification of sub-themes linked to the three key themes is also a unique and valuable contribution. A set of gap sub-themes (challenges and best practices attached to the three key gaps) are extrapolated from coded interview results. Results pinpoint the focused topics that underlie macro-gaps in the CSDS domain. The surfacing of key CSDS gaps with accompanying, statistically extrapolated sub-themes frames requirements that empirically ground subsequent *Phase III* design problem-solving. As gap sub-themes frame problem-solving requirements, an empirical basis is provided for the development of problem-solving prescriptions.

Interview research is less typical than survey research in MIS and organizational research. The inherent benefits of interview research were highlighted, namely, the ability to surface unexpected and hidden (latent) insights. Survey research, by nature, largely constrains results within the confines of the survey instrument, potentially disguising researcher bias in the data collection instrument itself. While interview research must address the risk of bias in researcher interpretations of interview results (i.e., the coding process), this can be addressed by properly detailing and substantiating the staged effort. A detailed description of the coding effort undertaken here is submitted in order to substantiate assertions of rigor and reproducibility (as covered in Sects. 3.2.2 and 3.2.3, along with an exposition on methodological underpinnings in Sect. 3.2.1.4).

More broadly concerning the interview effort, the novel mix of quantitative methods applied are of interest in demonstrating new approaches to analyzing interview data. The application of mixed logistic regression, correlation analysis, factor analysis, and factor-to-factor analysis to examine latent patterns in coded interview data was a unique approach.

Another contribution concerns the use of mixed methods to substantiate the results from the quantitative analysis of interview data. Mixed qualitative and quantitative findings are used to cross-validate the CSDS professionalization gaps surfaced. Quantitative results, being inductively and empirically grounded, but open to potential charges of bias (per the semi-grounded approach taken), are cross-validated through mixed deductive and inductive findings. Per the summary in Sect. 3.4.1, triangulated mixed methods applied to cross-validate quantitative results include the following: (1) literature gap analysis (Sect. 2.6.4), (2) interview memos (Sect. 3.2.2.3), (3) text analytics (Sect. 3.2.3.7), and (4) theme frequency in interviews (Sects. 3.2.3.2 and 3.2.3.3). The cross-validation of inductive and deductive findings constitutes *abductive reasoning* in the service of pragmatic satisficing.

In combination, diagnostic research undertaken across *Phases I* and *II* results in structured problem-solving requirements presented for design treatments in *Phase III*. *Phase II* results frame discrete problem-solving requirements in the form of the three CSDS practice gaps along with supporting sub-themes. Models and concepts surfaced in *Phase I* literature analysis serve as a foundation for the extrapolation of conceptual design gap-prescriptions, nascent designs, per Gregor and Hevner (2013).

The resulting three gap-prescriptions, framed as *processes* which group concepts, methods, models, and principles (nascent design artifacts), are advocated for practitioner and researcher consideration as problem-solving solutions. The design process utilizes abductive reasoning to extrapolate practical insights from mixed inductive and deductive findings surfaced in *Phases I* and *II*.

Phase III gap-prescription problem-solving invokes discrete findings enumerated previously in the research, but also utilizes general concepts surfaced in *Phase I* literature analysis. Such general concepts from literature include insights surfaced from sources such as technical research, industry reports, government studies, and case insights raised in the CSDS corpus (per Sect. 2.6.3).

Design prescriptions are advocated to address the focused gaps surfaced in *Phase II*. These are asserted as professional body of theory principles to underpin and advance the emerging field. Gap-prescription designs surfaced in *Phase III* are advocated as a set of key best practices as well as foundational meta-theory to advance the CSDS domain. Conclusions suggest fruitful opportunities for follow-on design, action, and case study research implementations.

The result is a set of advocated conceptual artifacts to guide the implementation of CSDS best practices. Guidance aims to serve managers seeking to implement CSDS programs, practitioners pursuing development and improvement, and researchers seeking to frame follow-on implementation efforts. Based upon advocated design requirements and suggestions, guidance concerning possible future research avenues is provided.

As the integrated research design is novel and complex, it is useful to briefly recap the discrete connections between the three research efforts. *Phase I* frames the basic research domain, motivates the research problem, and surfaces key concepts to support later problem-solving efforts. *Phase II* extrapolates key gaps in CSDS professionalization through quantitative analysis of interview results. Triangulated diagnostic methods substantiate and enrich an understanding of the key gaps surfaced.

Gaps identified are resolved into requirements that are addressed through design gap-prescriptions in *Phase III*. *Phase III* provides the capstone contribution in consolidating results from *Phases I* and *II*. Design problem-solving frames requirements based upon the gaps surfaced and proposes nascent design gap-prescriptions to address these gaps. Results are framed as practical processes to strengthen CSDS organizational, process, and methodological rigor. The prescriptions resulting are asserted as novel MIS research contributions of interest to advancing CSDS practice.

Having profiled the discrete research contributions in detail and having summarized the integration of research findings, it is appropriate to highlight the empirical foundations of the research effort. Concerning empirical grounding, there were several inductively focused elements of note. These include the following: (1) the *Phase I* comparative literature thematic gap analysis (a review of content across 33 manuscript-length CSDS works, per Sect. 2.6.4), (2) the substantial interview effort in *Phase II* (25+ h of interviews with 50 CSDS professionals and subsequent quantitative analysis of the coded results, per Chap. 3), and (3) thematic text analytics conducted on raw interview transcripts in *Phase II* (algorithmic extrapolation of themes supported by natural language processing and array-based key-term proximity analysis, per Sect. 3.2.3.7).

Having noted the empirical foundations, it is appropriate to substantiate claims to scientific rigor. As practitioner-focused diagnostic problem-solving research, outside the empirical foundations, the appeal to scientific rigor rests in the suitability, application, and integration of the mixed methods applied within the context of the novel research design. To substantiate this, a detailed exposition on the motivations, traditions, and application of the mixed methodologies and research design applied will be expounded upon.

In terms of research background and methodological motivations, the research effort is focused on the management of information systems (MIS) domain (Haag and Cummings 2012; Hsu 2013; Laudon and Laudon 2017; Pearlson et al. 2016; Sousa and Oz 2014). While both data science and cybersecurity individually align to the domain of MIS research, there is a lacuna regarding their conjoint interrogation which the research addresses centrally.

The MIS focus approaches CSDS from the combined perspective of organizational, process, and technological factors (ITIL 2019; Keen 1981; Leavitt 1964; Looijen and Delen 1992). The core effort encompasses a partial design science cycle within the MIS domain (Dresch et al. 2015; Gregor and Hevner 2013; Hevner et al. 2004; Wieringa 2014).

A systematic research plan was designed based upon methodological guidance derived from Verschuren and Doorewaard (2010). The research design is based on the practitioner-problem-solving tradition, with a core focus on mixed diagnostic research efforts to analyze new and complex phenomena. The effort follows a partial problem-solving cycle, encompassing analysis (problem identification), diagnosis (gaps and possible treatments), and design (requirements and prescribed solutions). Diagnostic research encompasses the sequential application of *background analysis*, *opinion research*, and *gap analysis* (Verschuren and Doorewaard 2010; Groenland 2009).

Mixed methods support gap-diagnostic analysis and include literature, interview, and quantitative methods. Literature research (*Phase I*) frames sensitizing concepts by profiling key themes and CSDS gaps. Interview research supported by quantitative analysis (*Phase II*) extrapolates and integrates a set of key themes and gaps to be addressed from the practitioner standpoint. Combined, the results establish a foundation for problem-solving design (*Phase III*). Design prescriptions (latent designs) are advocated to address the key gaps identified. The design process is supported by concepts surfaced in literature analysis (*Phase I*).

Having framed the research method and process in detail, assertions of scientific validity and rigor can be framed and substantiated. To ground an assertion of scientific validity, it is important to reinforce that a more traditional hypothetical-deductive research design was not followed. The research did not pursue development of hypothetical conceptual models, nor empirical testing of such models.

The newness and complexity of the domain advocated the application of triangulated mixed methods aimed at exploring and extrapolating hidden patterns as opposed to pursuing hypothetical-deductive approaches. There being no target variable or rigid theoretical model to test, the methodological approach taken focuses on rigor in pattern extraction and substantiation. The scientific basis rests in demonstrating rigor in the diagnostic process, particularly through mixed and multiple methods, triangulating inductive evidence and deductive concepts sourced from literature.

Concerning possible critiques of rigor from the perspective of the hypothetical-deductive tradition, there is risk of false equivalences. Critiques of not having applied or followed hypothetical-deductive methods amounts to the chicken critiquing the goose for not being chicken enough. A proper critique must assess the goose according to criteria for goose-ness, like-to-like. As such, the focus here will be to address topics of rigor, repeatability, and thoroughness in the context of "goose-like" diagnostic problem-solving research.

The scientific basis of diagnostic problem-solving research rests in demonstrating systematic rigor in the diagnostic process. This can be substantiated by a clear and well-executed research design, cross-comparing complementary mixed and multiple methods, and triangulating inductive and deductive findings. Confronting complex, novel, fast-emerging phenomena, this effort focused on problem definition and diagnosis via diagnostic background (literature), opinion (interview), and gap analysis (mixed/quantitative) research approaches.

The manuscript format supported an integrated practitioner problem-solving research approach, linking the results of diagnostic research directly to design prescriptions. It is asserted that the complexity and novelty of the domain under examination benefited directly from mixed method analysis. This approach has addressed a call from Aguinis et al. to advance approaches in organizational research through the application of innovative mixed qualitative and quantitative research methods (Aguinis et al. 2007).

A related critique that must be politely countered is the notion that because a hypothetical-deductive methodology was not applied, there is a lack of empirical grounding and/or scientific rigor (or validity). Such charges are based on a false syllogism as the hypothetical-deductive method is certainly not the only valid scientifically rigorous method (a point discussed in Sect. 4.4). As well, not applying the hypothetical-deductive method does not ipso facto mean that research lacks an empirical element, nor inherent systematic rigor.

A variant criticism from the hypothetical-deductive perspective is the absence of a focused, rigid theoretical framework to orient empirical material to advance theory. Producing and testing a master theoretical framework was not in scope for this research from the outset. As practitioner-focused diagnostic problem-solving research, mixed methods were focused on triangulating and abducting a patterned understanding of a complex new phenomenon (in the form of problem-solving requirements) in order to generate a series of practical recommendations. However, a number of theoretical vehicles sourced from literature were utilized along the research path, and several proto-theoretical concepts (nascent designs, in Gregor and Hevner's conception) emerge from the design process.

With the core effort focused on systematic diagnostic efforts, the research was aimed at extrapolating patterns and substantiating structure regarding hitherto largely unexplored phenomenon (organizational factors underpinning CSDS practice). Rather than operational proof of a hypothetical construct, scientific effort was focused on rigor in diagnostic exploration, discovery, description, substantiation, categorization, and gap analysis.

A demonstrative appeal to broader traditions of scientific diagnostic reasoning is appropriate to emphasize this position. In this appeal, we can observe the application of similar diagnostic methods in, for instance, clinical medical practice, epidemiological social health research, biological taxonomic classification, and anthropological and archeological field research. In each of these cases, mixed inductive and deductive methods drive abductive reasoning. A key focus is on inferential techniques and efforts to rigorously qualify and quantify operational categories for the purpose of rationalization and treatment. Examples include identifying and classifying a novel ailment with reference to an array of symptoms and conditions, pinpointing the mechanisms of disease transmission, classifying a new species in a taxonomic framework, identifying a particular ritual in the context of sociocultural traditions, or hypothesizing the probable use and function of an ancient artifact unearthed at an excavation site.

An earmark of rigor in diagnostic analysis is holistic, systemic thinking, a modality which admits and integrates empirical evidence and primary deductive reasoning. This notion otherwise serves to frame the holistic organizational context of research findings, findings that address CSDS practice as a complex, multilayered organizational problem. Grounded in the MIS discipline, *Phase III* results are presented as integrated technological, process, and organizational prescriptions. A central insight is that the nature of the three CSDS prescriptions surfaced are complex and multi-contextual, requiring overlapping and integrated processes to address. Concluding advocacy emphasizes the importance of holistic organizational perspectives to implement and realize such multi-contextual integration.

The final contribution, in this holistic sense, is meta-theoretical in terms of espousing body of theory for improving CSDS theory building practice. By extension, this also establishes an approach to improving empirical and theoretical foundations in the parent cybersecurity domain. A core assertion is that CSDS ideally facilitates knowledge creation by prescribing data science methodological treatments to address diagnosed security gaps. This highlights the distinction between engineering solution design and scientific diagnostic inquiry. It is asserted, through the CSDS professional maturity assessment, that more rigorous scientific diagnostic approaches are necessary to promote the maturation of CSDS body of theory in this regard.

A central conclusion is that for the CSDS profession to advance, a tighter and clearer integration of practice with theory must be pursued. This posits that CSDS, as an emerging profession, can advance through a tighter symbiosis between practice (*technê*) and logic-guided theory (*logos* driven *epistêmê*). The facilitation of this synthesis is framed in a multimodal MIS context, espousing a unified approach to technology (improved data management), process (scientific methods), and organization (cross-domain collaboration). Conclusions of the *Phase III* design gap-prescription exercise frame, motivate, and position treatments in this direction.

Concluding the discussion on contributions, it is asserted that a particular set of thresholds related to the suitability of research offerings, as adapted from Wilson (2002), have been addressed:

1. *Are problems discussed of substantial interest and would solutions to these problems materially advance knowledge of theory, methods, or applications?*

   - Literature analysis emphasized the importance of the growing problems facing cybersecurity, the promise of CSDS prescriptions in this regard, and remaining challenges resident in the as-is approach to CSDS.

- Interviews and mixed method gap analysis clarified the nature of challenges facing the CSDS domain. Triangulated findings framed a structured understanding of the as-is state of the CSDS domain, addressing a research lacuna.
- Design science requirements and advocated problem-solving approaches have advanced an in-depth, integrated understanding of ideal CSDS practice.

2. *Does the work either solve these problems or else make contributions towards a solution that improves substantially upon previous work?*

- Analysis identified data management, scientific methods, and organizational management to be relatively less represented in CSDS literature.
- No single existing work focused on CSDS has addressed these topics in a comprehensive and integrated fashion.
- It is asserted that the design prescriptions resulting from gaps identified in diagnostic analysis are substantial new contributions to understanding the emerging CSDS field.

3. *Are the methods of solution new and can the proposed solution methods be used to solve other problems of interest?*

- The application of mixed methods to diagnose a triangulated understanding of challenges in the emerging CSDS field were unique in their combination.
- In particular, an effort to combine inductive and deductive methods, both qualitative and quantitative, improved rigor in substantiating and interpreting themes surfaced.
- Given the complexity and newness of the emerging CSDS domain, the methods applied could be reproduced to improve understandings of other emerging professions.

4. *Does the exposition help to clarify understanding of this area of research or application?*

- A concerted effort has been made to satisfice interests across academic, management, practitioner, industry, and public sector audiences.
- Literature analysis (*Phase I*) results should be of interest to managers and practitioners, whereas the structured mixed method analysis of interview results (*Phase II*) should be of greater interest to management research academics and strategic planners.
- Design prescriptions (*Phase III*) are offered as contributions to all audiences as advocacy for improving CSDS practice and as foundations for potential follow-on design, action, and case research implementations and assessments.

5.5 Limitations of Research

The research purview was ambitious and natural compromises were made to balance the competing interests of scope and rigor. A core principle of this effort from the beginning, and reinforced by mentors at the host institution, was that when undertaking practice-based research on new and complex phenomenon, often a firmly grounded explanation and diagnosis of the problem is quite enough if done rigorously.

There is a potential argument that, given the resulting scope, length, and complexity of the three-phase effort, that, with modifications, the research might have remained focused on *Phases I* and *II* exclusively (spanning diagnostic background analysis, opinion, and gap analysis research) and still have evidenced acceptable contributory value.

On this point, there was a strong motivation to offer results of potential benefit to practitioners, a central audience. The capstone partial design problem-solving phase (*Phase III*) addressed this goal. Early feedback from CSDS practitioner-researchers concerning the value of *Phase III* was positive, reinforcing the value of this component.

Limitations of the research relate to complications presented by the atypical mixed method research design, the constrained amount of empirical content, and the absence of theoretical constructs. Methodological motivations, factors, and rebuttals concerning related critiques were covered in detail in the previous section. Limitations of research that occurred as a result will be further discussed here.

Having pursued practitioner problem-solving research, a hypothetical-deductive approach was not applied. As a result, an operational conceptual model was not framed or tested. Outside this, there were challenges in the effort to establish theoretical grounding and to form a foundation for theory generation. A major limitation in this regard related to the stark lack of theory and empirically grounded research across the central practice-oriented domains of interest: cybersecurity, data science, and CSDS.

As a result of this theory gap, there was a lack of theoretical models for determining which data science methods to apply in CSDS practice, as well as a lack of empirical foundations for validating (or comparing) CSDS methodological outcomes. Establishing a foundation for improving these circumstances was challenged by a larger lack of empirically grounded research in both the cybersecurity and data science domains.

While engineering-focused research is well-addressed in both cybersecurity and CSDS literature (see Sects. 2.3.3 and 2.5.3, respectively), there is a lacuna of theoretically and empirically grounded research (see Sect. 4.4). The theory-poverty of CSDS is tied to a strong engineering bias in practice, inherited both from the cybersecurity and data science parent domains.

A resulting challenge in framing the research effort was that, in colloquial terms, CSDS is, from a theoretical grounding standpoint, an abandoned child, chaotic and manic due in large part to parental negligence. The poverty of literature and tradition left few forebears to support the development of operational theoretical concepts for empirical testing. A hypothetical deductive approach would have required, to some extent, spinning raw theory from what, as the *Phase I* literature analysis demonstrates, amounts to rather insubstantial foundations.

The general lack of theory in the CSDS domain, combined with the complexity and novelty of the field as emerging phenomenon, motivated the application of diagnostic problem-solving research, rather than a hypothetical-deductive approach. As practitioner-oriented research, combined diagnostic and design problem-solving approaches matched well with the objective of developing a structured set of recommendations to advance practice.

Diagnostic research presented the need to conduct fundamental discovery related to a new, complex domain. This called for a set of mixed methods associated with exploration, pattern analysis, and pattern substantiation leading to recommendations. The resulting nonstandard research design and mixed methods applied presented challenges in substantiating rigor and scientific merit, topics broached and addressed in detail in the previous section.

The larger challenge presented by the CSDS domain concerns the reality that the domain is emerging from actual practice, not from a supervisory effort, nor from academic theory building. This fact influenced the approach to design recommendations suitable for professional practice. Rather than rigid theoretical prescriptions, recommendations advocate a series of integrated organizational processes to improve practice. This was a pragmatic accommodation. To attempt to prescribe grand alterations or dramatic new theoretical approaches would be to invoke the *Athena fallacy*: the misapprehension that a professional field can be synthetically birthed, and, as a result, emerge fully formed from the head of Zeus, so to speak. To wit, dynamic fields of practice require practice-focused prescriptions.

This principle reflects research into the emergence of professions undertaken at the outset of the inquiry, namely, that fields typically evolve stepwise from practice into professions with well-developed theoretical foundations, often mentored by institutional entities along the path. To presume the reverse, that a rigid theoretical platform could be transplanted as a composite into a new profession, is to ignore the stepwise, communal sensemaking that goes on during the evolutionary process of professional emergence. The medical field is used at several points in this inquiry as a point of comparison, a field that emerged historically from technical practice and then, stepwise, through theory building and fundamental research.

The challenge of a lack of theoretical and empirical foundations both in the cybersecurity and CSDS domains also presented opportunities. In particular, it presented a very clear gap that needed to be addressed. As raised in *Phase I* concerning requirements for iterating CSDS professionalization, development of a body of theory in the form of a set of focused theoretical and methodological practices is a necessary next step for maturation of the field. A related goal of this research thus became to frame and motivate a foundation for theory building in CSDS practice. To the degree that CSDS improves methodological rigor, cybersecurity research itself is strengthened via more rigorous data-driven methods.

However, there is some disappointment that the amount of effort that was required to isolate, explore, substantiate, and address CSDS methodological and theoretical gaps left a little remaining scope for follow-on implementation research. It is clear that much more could be done in the future to broaden empirical efforts. This was, in the final analysis, a limitation of scope. That this research provides a fruitful foundation to stage such efforts is addressed in the section following concerning future research opportunities.

The scope of this research, constrained by time and overhead, was projected as a partial practitioner problem-solving design cycle. The scope focused on defining, framing, and diagnosing a set of refined problem-solving requirements. The conclusion proposes a set of nascent design artifacts to address the focused requirements. However, the artifacts are not implemented, nor are implementations assessed.

Design recommendations were framed at the level of *nascent designs*: constructs, methods, models, principles, and rules (Gregor and Hevner 2013). It was determined that nascent designs were appropriate to and sufficient for the novel and complex phenomenon being scrutinized.

Fully developed theoretical vehicles are not specified for empirical testing. The design process benefited from ad hoc practitioner experience from the researcher. However, field experience is inadequate as a rigorous basis for theory framing, testing, and assessment. Future implementation and testing would complete a full problem-solving research cycle in the design science, action, and/or case study traditions. Results presented here are restricted to advocacy for practitioners, contributions to CSDS body of theory, and foundations to guide future research efforts.

Given the newness and the complexity of the phenomenon under examination, along with the lacuna in the research tradition, it is asserted that a partial problem-solving design cycle, supported by a mix of inductive and deductive efforts, is a worthy and sufficient contribution. Guidance is offered as a basis both for improved professional practice and for future research efforts, a topic expanded upon in the following section.

5.6   Future Research Foundations

Concerning future research opportunities, a range of promising follow-on possibilities are framed by results achieved, methods framed, and outcomes surfaced. Key research opportunities encompass four particular categories: (1) direct extensions of this effort to address limitations of scope, (2) follow-on efforts extending findings, (3) applications of the meta-theoretical principles framed, and (4) broader opportunities implied by the extended conclusions.

In the first regard, recognizing this effort as a partial design cycle, diagnostic research results from *Phases I* and *II* are framed as problem-solving requirements. Problem-solving design prescriptions extrapolated and explored in *Phase III* are framed as recommendations. However, due to scope, these recommendations were not carried through to implementation and assessment. A foundation for follow-on design, action, and case research is thus suggested. This would provide empirical feedback regarding the efficacy and nature of the gap-prescriptions framed in *Phase III*.

Completing a design science problem-solving research cycle would involve undertaking the stages of intervention and evaluation. Implementations might include the design, enablement, and assessment of new CSDS associated processes and/or solutions in organizational settings. Key areas advocated for follow-on implementation research include CSDS data management, scientific methods, and cross-domain collaboration. For example, a follow-on action implementation could be a detailed working out of CSDS data preparation and exploration best practices in an operational setting.

Concerning the second opportunity, follow-on efforts extending findings, a promising next step for the mixed qualitative-quantitative interview component covered in *Phase II* would be to leverage interview themes surfaced to design and conduct survey research. Interview findings are suitable for framing and designing a survey instrument. Survey research presents the opportunity to gather additional results at scale.

In addition to surfacing CSDS challenges and best practices in interviews, practitioner feedback on domains adjacent to CSDS and adversarial trends were also collected. While reported, results were not explored in greater depth due to limitations of scope. This suggests the opportunity for follow-on survey research, for instance, refining an understanding of practitioner perspectives on adversaries, or research studying focused adversarial methods in greater depth. The topic of machine learning-driven (and -targeting) adversarial approaches is particularly interesting.

Interview feedback concerning adjacent domains utilized by CSDS practitioners would be a fruitful area to pursue in greater depth. Practitioners provided both *adjacent methods* (i.e., network graph analytics, natural language processing, time-series analysis/process analysis, and deep learning) and *adjacent domains* (i.e., fraud analytics, epidemiology/medical diagnostics, social sciences, and risk management).

It would be of interest to refine and explore these two distinct adjacent categories in greater depth. For instance, an effort to profile *adjacent methods* surfaced in research would be of interest along with example implementations to demonstrate the approaches. Concerning *adjacent domains*, it would be useful to understand the range of methods and practices underlying the domains advocated to frame guidance on CSDS best practices. This might also suggest an effort to formally extrapolate and dissociate discrete methods from domains in order to unify a comprehensive list of distinct methods utilized in (or available to) CSDS practice.

Findings specific to the importance and centrality of organizational factors to CSDS implementations suggest the potential for follow-on organizational case study research. The visibility and centrality of organizational factors in cybersecurity is reflected in a growing call to focus research in this area (Forsythe et al. 2013). Topics of particular interest to CSDS include the themes of cross-disciplinary coordination and cross-training. As well, the finding of a pronounced gender imbalance in the nascent CSDS profession would be a worthy focused research topic, especially if arriving at a set of recommendations to rectify this disparity.

Ideally, CSDS will see the future emergence of a CSDS-specific analytics process model as domain best practices become codified. An iteration of an existing analytics life cycle model specific to CSDS, leveraging advice framed herein, would be a fruitful direction. As well, the notion of a CSDS analytics life cycle model presents opportunities for framing a CSDS organizational maturity model. An effort to frame a comprehensive CSDS maturity model would be a valuable research avenue.

A range of more focused topics are touched upon throughout this work and could similarly be expanded upon. The application of text analytics to focused CSDS text artifacts suggests that additional research in this direction could lead to the development of CSDS controlled vocabularies, taxonomies, and/or ontologies. Bechor and Jung have demonstrated promise in this line of research by deriving a CSDS topic model through text analytics performed on relevant research articles (2019).

Derivation of CSDS taxonomies and/or ontologies through text analytics would have numerous potential benefits, including support for cybersecurity policy analysis (Kolini and Janczewski 2017), vulnerability assessments (Shuai et al. 2013), threat analysis (Samtani et al. 2015), domain document classification (Bechor and Jung 2019), and guidance for curriculum development (Bicak et al. 2015; Conklin et al. 2014; van den Berg et al. 2014).

Beyond pure research opportunities, academic instructors and students should find this work a valuable reference to inspire teaching-focused discussion and inquiry into the emerging CSDS field. The MIS domain focus supports those teaching or studying cybersecurity, data science, and/or IT management. The extensive references, including a corpus of 33 key CSDS works, should provide ample background for follow-on self-study. A number of technical case studies are well-represented in the CSDS corpus (a strand not explicitly profiled here). Student-researchers may opt to leverage the methodological guidance issued herein, particularly as related to improving scientific rigor in security research, as a frame of reference for critically evaluating, or even undertaking, case study research.

Of interest to academic and commercial instructors, the conclusion to *Phase III* includes guidance on CSDS curriculum development. An excellent outcome would be to see additional development of a focused CSDS curriculum, perhaps as an action implementation for a new CSDS course offering. Codifying CSDS methods and theory in training offerings is framed in *Phase I* as a key step towards professionalization, particularly as a prelude to professional certification offerings. The next steps for the professionalization of the CSDS domain suggest more formal institutional sponsorship, starting with the framing and hosting of focused training and certificate offerings. Research, teaching, and curriculum development on this topic would have great value in this regard.

Concerning the third set of opportunities, applications of the meta-theoretical principles framed, *Phase III* advocates methods to inculcate greater rigor in CSDS research (and by extension, cybersecurity research). A working out of the functional role of CSDS through the Looijen and Delen (1992) IT supply-demand model positions CSDS as a supervisory diagnostic mechanism in prescribing data science treatments to address cybersecurity gaps. As emphasized at several points, this positions CSDS as conjoining cybersecurity *technê* with data science *logos*. CSDS body of theory, as *episteme*, is positioned as a mediator of integrated techno-logical (*technê-logos*) synthesis.

This situates CSDS in a functional-clinical role not dissimilar from medical practice: diagnosing cybersecurity gaps and prescribing subsequent data science treatments. Leveraging this proposed meta-theoretical model for CSDS practice (theory pertaining to body of theory as *praxis*), a range of design, action, and case research opportunities are suggested. Whereas technical and engineering research is well-addressed, advocacy reframes and encourages greater efforts towards establishing empirically based theoretical frameworks and inculcating scientific rigor in such research. The focused positioning of the need for more rigorous theoretical foundations (body of theory), especially as related to diagnostic practice, is submitted as a meta-theoretical concept of potential wider interest to the advancement of adjacent IT professions, a topic positioned for future inquiry.

Concerning the fourth opportunity, broader opportunities implied by the extended conclusions, these range from innovative technical research to policy-related interdisciplinary research opportunities. Concerning technically oriented advanced research opportunities, CSDS ontologies are a fast-emerging area worthy of greater attention (see Sects. 2.3.5 and 4.3.3). It would be worthy to demonstrate the focused deployment of CSDS ontologies to facilitate the management and interoperability of CSDS processes, systems, and models. Ontologies, as discussed in the literature review in *Phase I*, not only are useful in orienting organizational cyber responses but also have a growing value in facilitating knowledge-driven security operations, including systems integration and automation (Riley 2019a).

There is great promise in pursuing the design and implementation of semantic (or symbolic) machine learning solutions (those which formally situate and leverage semantic models), as proof-of-concept implementations. For instance, there is promise in refining anomaly and incident detection systems through applied computer-driven semantic reasoning. Such an approach leverages a semantic framework (i.e., ontologies supported by interpretive mechanisms) for assessing complex threats. This might entail a system that automates triage decisioning and the presentation of contextual reasoning (i.e., evidence + logic) to investigators for rapid case remediation. Such solutions demonstrate an operational approach to enable the concept of detective analytics, framed and advocated herein (Sect. 4.4.4).

Monitoring the outcomes of investigations over time provides evidence suitable for the continual improvement of semantic encoding, data refinement, and detective analytics. Operationalized semantic knowledge (i.e., encoded evidence-based reasoning logic) provides support for refined supervised machine learning efforts supplemented with explanatory context (e.g., recommended next best actions, case context, and actionable risk scores based on experience to aid investigations). Together, this frames opportunities for designing and staging cyclical human-in-the-loop, self-reinforcing detection solutions. The design, implementation, and assessment of self-reinforcing, human-in-the-loop, semantic analytics (symbolic AI) orchestrated "cybersecurity monitoring and remediation cyborg-solutions" would be a compelling topic to pursue. This includes opportunities for operationalizing reinforcement learning and human-computer gamification processes.

Another topic related to advanced technical and engineering research regards the growing specter of AI cyber warfare, including growing indications of an emerging machine learning-facilitated cybersecurity cold war. This includes the topics of the use of machine learning to semi-automate attacks and defense, attacks on machine learning systems (a healthy and growing research area), and broader aspects concerning the use of machine learning to orchestrate public disinformation campaigns (so-called fake news and related use and abuse of the Internet as a platform for manipulating public opinion and behavior). Research into related engineering trends is of strong interest, particularly as related to semantic analytics, as are the broader military, intelligence, and government policy implications and reactions. That such phenomenon is not simply technical, but emerges as techno-economic-behavioral phenomenon, frames such research as interdisciplinary.

Finally, the broader implications of this research effort suggest the possibility of more outré opportunities. Advocacy for greater theoretical, empirical, and scientific grounding in cybersecurity research opens the possibility that a range of nontechnical prescriptions to address cybersecurity challenges emerge. This implies hybrid techno-economic-behavioral research which might frame novel techno-economic, economic-behavioral, sociocultural, or sociopolitical recommendations.

A case has been made that only by strengthening CSDS as a scientific undertaking, principally in framing cybersecurity challenges as phenomenon susceptible to scientific investigation leading to theory-grounded solutions, can a strategically advantageous foothold be gained in cybersecurity practice. However, this principle may lead to demonstrable macro-optimal conclusions that challenge the very rules of the game, for instance, advocating stricter regulatory standards leading to requirements for stronger authentication to access Internet carriage.

While most every nation requires a license to operate an automobile, anonymous access to Internet carriage in many cases is unquestionably provided. As a result, trustworthiness is a recognized and growing gap in securing cyberspace (Amla et al. 2012; Lowry et al. 2018). As observed by de Bruijn and Janssen, as the cybersphere becomes progressively more central to all aspects of social, political, and economic exchange, and thus increasingly more dangerous in misuse and abuse, loose regulatory and oversight practices will increasingly be called into question (2017).

Chapter 6
Managerial Recommendations

6.1 Management Problems

The practical intention of this research is to provide insights and guidance of interest to a range of organizational stakeholders, including managers, practitioners, researchers, educators, strategists, planners, and analysts. In mapping CSDS challenges and prescribing practical treatments, this work offers guidance to those interested in advancing CSDS professional practice and general effectiveness, including stakeholders planning programmatic operational implementations.

Three practical management problems (MPs) are addressed via the three research components:

MP1: What gaps challenge the effectiveness of the CSDS field and impede professionalization?

MP2: What prescriptive treatments address categorical CSDS gaps?

MP3: What guidance can be offered to orchestrate the implementation of CSDS programs?

Summarizing the three management problems in terms of the three key phases:

1. *Phase I—Diagnostic Literature Analysis*
   MP1—What gaps challenge the effectiveness of the CSDS field and impede professionalization?

   CSDS is asserted as a hybrid domain and nascent profession. The field has emerged from the fusion of cybersecurity needs and data science methods in response to the growing economic and organizational pressures associated with the effects of rapid digital expansion and evolution. From a managerial perspective, a structured understanding of the broad dynamics shaping the emergence of CSDS provides an ability to control and foster this transformation.

   Being a practitioner-led discipline, managerial stewardship is essential to the maturation of the CSDS field. Analysis seeks to inform managerial actors concerning the landscape and issues resident in the current-versus-idealized state of

© Springer Nature Switzerland AG 2021
S. Mongeau, A. Hajdasinski, *Cybersecurity Data Science*,
https://doi.org/10.1007/978-3-030-74896-8_6

the field. Managers, as institutional mediators, are essential to iterating this evolution by facilitating the codification of a clear *body of theory*, which in practitioner terms constitutes operational *best practices* united with *theoretical context*.

Literature analysis frames the major theoretical, practical, and structural elements necessary for transitioning from CSDS *body of theory* knowledge individuation to *occupational closure* via institutional codification. Specifically, institutional facilitation via professional training providers and certification authorities is proposed as a key next step in achieving *occupational closure*, whereby the CSDS field can assert professional specialty status.

2. *Phase II—Diagnostic Opinion (Interview) and Gap Analysis (Mixed Methods)*
   MP2—What prescriptive treatments address categorical CSDS gaps?
   Guidance based upon interviews with active practitioners advocates the need to integrate organizational and process-driven treatments, especially to align human and technical resources through organizational processes. Key enabling processes framed center around data management and applied scientific inquiry. Results suggested that resource coordination, data management, and scientific methods must together conjoin to drive CSDS as a functional organizational process. This presents CSDS as necessitating multiple, overlapping processes, particularly cyclical data discovery and scientific inquiry.

3. *Phase III—Design Prescriptions.*
   MP3—What guidance can be offered to orchestrate the implementation of CSDS programs?
   The three CSDS gaps addressed, data management, scientific processes, and cross-domain collaboration, roughly align with the MIS focus areas of technology, process, and organization. A key observation raised in framing requirements concerns the degree to which these gaps overlap. Design prescriptions frame data management and scientific processes as mutually reinforcing.

The design prescriptions advocated relevant to managerial interests:

1. *Organizational*: Ensure that management is engaged and knowledgeable concerning the inherent complexities associated with CSDS. Organizations should introspect concerning how the breadth and complexity of the CSDS domain complicates and obfuscates ownership, decision-making, team collaboration, and cross-domain knowledge.
2. *Process*: A drive towards process-driven approaches to integrate data management and scientific methods is advocated. Data management and applied scientific processes are recognized as having distinct overlaps in addressing the questions of "what to track," "what data to collect," "what is normal," and "how to analyze." These are topics that require the application of concerted scientific methods to frame, extrapolate, and substantiate.
3. *Technology*: Although technology topics were raised by practitioners, it was notable that their explicit appearance was muted in surfaced themes. This does not imply technology is unimportant to CSDS best practices, but rather that prac-

titioners indicated that technology is viewed as an enabler and that organizational framing and process alignment goals take precedence.

6.2   Management Problems Addressed

This research frames and analyzes CSDS as an emerging field in the early stages of professional maturation. Being in the early stages of professionalization, there are key gaps in CSDS organizational practice that impede growth and effectiveness. As a nascent practitioner-driven discipline, many aspects of mature professions—standards, certifications, and a codified body of theory—are yet-to-emerge.

CSDS hybridizes two existing professional fields, cybersecurity and data science, each hosting focused bodies of theory and associated best practices. However, the root parent fields themselves are evolving amid rapid technological change. The recency of the emergence of the CSDS field combined with rapid change in the parent domains has resulted in a lack of consensus concerning key theoretical foundations. This research fulfills an academic and managerial need for systematic inquiry into the challenges and opportunities facing the nascent CSDS discipline.

Three associated practical management problems (**MPs**) were addressed herein:

MP1: What gaps challenge the effectiveness of the CSDS field and impede professionalization?

MP2: What prescriptive treatments address categorical CSDS gaps?

MP3: What guidance can be offered to orchestrate the implementation of CSDS programs?

Table 6.1 provides an overview and summary of findings, beginning with management problem 1 (**MP1**), diagnostic literature analysis (*Phase I*), as aligned with **TQ0** and addressing **RO1**.

CSDS is asserted as a hybrid domain and nascent profession. The field has emerged from the fusion of cybersecurity needs and data science methods in response to the growing economic and organizational pressures associated with the effects of rapid digital expansion and evolution. From a managerial perspective, a

Table 6.1 Management problem 1 (MP1) summary

Management problem 1	**What gaps challenge the effectiveness of the CSDS field and impede professionalization?**
	• Immature CSDS body of theory hinders professionalization
	• CSDS lacks clear standards for driving scientific rigor in practice
	• Uneven institutional sponsorship forestalls professionalization
	• Data science as a profession lacks clear disciplinary boundaries
	• Cybersecurity as a profession is increasingly challenged by the rapid expansion and evolution of threats
	• Cybersecurity practice is engineering-focused and lacks strong theoretical foundations

structured understanding of the broad dynamics shaping the emergence of CSDS provides an ability to control and foster the growing field.

Being a practitioner-led discipline, managerial stewardship is essential to the maturation of the CSDS field. Managers, as institutional mediators, are essential to iterating this evolution by facilitating the codification of clear *body of theory*, which in practitioner terms constitutes operational *best practices* united with *theoretical context*. This mirrors the classical distinction between *technê*, knowing how to do, and *epistêmê*, knowing what to do and why.

Although driven by a growing pressure and demand, development of CSDS as a profession is challenged by a lack of clarity concerning core body of theory. When considered as a hybrid of its parent domains, CSDS faces inherited challenges. Namely, CSDS faces a metaphoric challenge of having an overly didactic parent (cybersecurity) and an overly permissive parent (data science). Between the two, there is a risk of misguidance and delinquency. Design prescriptions framed through this effort advocate a series of structured approaches to ground development of professional practice.

The results of background analysis frame the major theoretical, practical, and structural elements necessary for transitioning from CSDS *body of theory* knowledge individuation to *occupational closure* via institutional codification. Specifically, institutional facilitation via professional training providers and certification authorities is proposed as a key next step to achieving *occupational closure*, whereby the CSDS specialty can assert professional specialty status. Phase II addressed management problem 2 (**MP2**), diagnostic opinion and gap analysis, as aligned with **EQ0** and addressing **RO2**, per summary in Table 6.2.

Practitioner interview guidance advocated the need to integrate organizational and process-driven treatments, especially to align human and technical resources through organizational processes. Key enabling processes framed centered around data management and applied scientific inquiry. Results suggested that resource coordination, data management, and scientific methods must together conjoin to drive CSDS as a functional organizational process. This presented CSDS as necessitating multiple, overlapping processes, particularly cyclical data discovery and scientific inquiry.

A summary listing of CSDS practitioner interview findings follows, framed within the three central MIS categories of organization, process, and technology:

Table 6.2 Management problem 2 (MP2) summary

Management problem 2	**Interviews and Mixed Methods Gap Analysis:**
	What prescriptive treatments address categorical CSDS gaps?
	• A range of organizational, process, and technical best practices are advocated by CSDS practitioners to address focused challenges surfaced in interview research
	• Key CSDS gap-prescription areas surfaced for treatments:
	  – Data management
	  – Scientific processes
	  – Cross-domain collaboration

1. *Organizational:* Ensure that management is engaged and knowledgeable concerning the inherent complexities associated with CSDS. Organizations should introspect concerning how the breadth and complexity of the CSDS domain complicates and obfuscates ownership, decision-making, team collaboration, and cross-domain knowledge. The following key themes were advocated by practitioners as CSDS organizational enablers:

   - Management commitment
   - Data-driven paradigm shift
   - Uncertainty reduction/risk management focus
   - Operational optimization
   - Multidisciplinary team orchestration
   - Cross-domain training

2. *Process:* A drive towards a process-driven approach to enable both data management and scientific methods is advocated. Data management and applied scientific methodological processes are recognized as having distinct overlaps in addressing the questions of "what to track," "what data to collect," "what is normal," and "how to analyze." These are topics that require concerted scientific inquiry to frame, extrapolate, and substantiate. Scientific processes were surfaced as multifarious, spanning problem framing, discovery, feature selection, categorization, validation, detection, prediction, and optimization. The following key themes were advocated as CSDS process enablers:

   - Data management and structuring
   - Mapping and modeling environmental context
   - Systematic evidence gathering, framing, and labeling
   - Disassociating normal versus abnormal phenomenon
   - Risk quantification
   - Data and model validation
   - Linking distinct exploratory and predictive methods

3. *Technology:* Although explicit technology topics were raised in interview feedback (i.e., cloud, honeypots, endpoint detection, SIEMs, data engineering, unsupervised vs. supervised ML), it was notable that their direct appearance was muted in surfaced gap-prescription themes. This does not imply technology is unimportant to CSDS best practices, but rather that practitioners seemed to indicate that technology is an enabler and that organizational framing and process alignment goals take precedence. This suggests technological solutions are viewed as the outcome of organizational direction and process goals. Given the preponderance of engineering-driven approaches in the broader security field, the muted focus on technology was both surprising and laudable.

Phase III addressed management problem 3 (**MP3**), design science gap-prescriptions, as aligned with **AQ0** and addressing **RO3**, per the summary in Table 6.3. The three CSDS gaps addressed, data management, scientific processes, and cross-domain collaboration, roughly align with the MIS focus areas of technology,

Table 6.3 Management problem 3 (MP3) summary

Management problem 3	**What guidance can be offered to orchestrate the implementation of CSDS programs?** • Address the key CSDS gaps of data management, scientific processes, and cross-domain collaboration • Address gaps through design prescriptions which accommodate MIS technology, process, and organization perspectives • Focus on the implementation of integrated processes • Address organizational process collaboration and incentives • Understand the types of up-front investments required to enable data management, rigor in model development, and cross-training

process, and organization. A key observation raised in framing requirements concerned the degree to which these gaps overlap. Design prescriptions framed data management and scientific processes as mutually reinforcing. The organizational context of scientific rigor was also highlighted in framing validation processes as organizational sensemaking exercises.

An effort needs to be made to frame data science programmatically, as an effort that is as much organizational as it is methodological and technical. As much of the collected research on data analytics organizational challenges has shown, organizational factors are the most intractable and important problems to counter in the implementation of analytics processes and programs. CSDS programs ideally facilitate and incentivize social processes associated with knowledge exchange and substantiation.

The design prescriptions advocated in *Phase III* should be of central interest to managers and practitioners planning to implement CSDS programs. A selection of key management objectives which are addressed by pursuing the design prescriptions advocated:

- Drive better security decisions
- Provide context for security events
- Extrapolate analytics-driven insights
- Improve event detection accuracy
- Lower false-positive rates
- Improve staff utilization and efficiency
- Simplify security infrastructure
- Refine focus on emerging risks
- Control security risk
- Increase the value of security investments

CSDS outlines a range of methods to measure phenomena, make better decisions, identify threats, control risk, and optimize processes. As a whole, the range of design prescriptions outlined by CSDS frames a basis for a cybersecurity analytics maturity model. Namely, as advocated CSDS design prescriptions are iterative and self-reinforcing, it is asserted that their implementation can lead to progressive improvements in operational settings.

A summary of key prescriptions surfaced of interest to managerial stakeholders:

- Investments in data quality and feature selection are central to improving the results of CSDS efforts.
- Big data by itself is not a solution or goal for CSDS analysis as the results of data analysis and engineering efforts often result in compressed and focused datasets ("smart data").
- Combine both explanatory and predictive methods: traditional statistical diagnostic techniques should accompany the application of predictive machine learning approaches (correlation is not causation).
- Continually revisit assumptions trapped in data and models, particularly in the sense that challenges encountered in models often imply the need for transformed or new data.
- Test "fancy algorithms" (e.g., deep learning) like-to-like against "boring techniques" (e.g., logistic regression)—as there is no "best" algorithm fit to address all problems, match the algorithm to the particular problem and accompanying data.
- Machine learning is applied computational statistics—involve someone with a background in statistics when pursuing machine learning solutions.
- There is no magic box—if it sounds too good to be true, it likely is.
- Results begin and end with the organization—listening, communicating, and collaborating is key, especially given the breadth and complexity of the CSDS domain.

The concluding design gap-prescription, cross-domain collaboration, advocated organizational orchestration of distributed CSDS processes. The dominance of engineering approaches in cybersecurity was framed as being self-limiting. It is asserted that a rigorous scientific approach to CSDS requires organizational collaboration to validate and valorize CSDS theory. This proposes a paradigmatic distinction between engineering- and orchestration-based approaches to CSDS. Framed through Herbert Simon's three-phase approach to management decision-making (Simon 1960), the two management paradigms are counterposed in Table 6.4.

Table 6.4 Security engineering versus CSDS management paradigm comparison

	Security engineering	Security orchestration
1. Information gathering (framing)	Architecture driven Rules-based Big data Deterministic Prediction	Sensemaking Inferential Feature engineering Pattern identification Evidence-based modeling
2. Design (solutions)	Engineering solutions Didactic Linear systems design Upgrades	Behavioral Collaborative Cyclical detective analytics Self-improving
3. Choice (decisions)	Rules and algorithms Engineering-driven technê Policy and rule focused Technology and device target	Optimizing and satisficing Logos-driven epistêmê Theory-based and theory building Human behavioral target

6.3   Summary of Stakeholders Served

The results of this research will be of interest to a range of organizational stakeholders spanning professional, commercial, public, and academic domains. It is useful to profile major groups of stakeholders in terms of their domains of interest in order to orient interested parties to areas of this work which may be particularly compelling.

A central intended audience for this work is security, data science, and CSDS professionals. These parties will be interested to understand the evolving nature of the CSDS domain as participant-practitioners. This work has outlined paths for *getting from here to there*, an overview of current challenges in CSDS practice and how these limitations can be addressed and overcome. As central participants in the evolving CSDS profession, practitioners are both suppliers and consumers. It is intended that the views espoused will be seen as a call-to-action and exhortation to advance CSDS professional rigor. Additional discussion is offered in Sect. 6.4.

Concerning stakeholders with commercial interests, this group also represents both consumer and supplier perspectives. Commercial consumers include organizational managers, planners, and strategists seeking guidance on implementing CSDS programs and solutions in the firm. The potential pitfalls covered herein will be instructive. Guidance concerning best practices exposes collective practitioner learnings, and hopefully will lead to efficiencies and insights to optimize approaches. Additional perspectives are provided in Sect. 6.5.

Commercial suppliers are a focused audience that will find the results presented of potential value. The analysis of the CSDS field and advice for improved practice offers both a focused overview of the market and perspectives on product and solutions design. Software and services firms should find solid insights concerning framing, packaging, marketing, and hosting offerings. Consulting firms should find the approaches covered in Sect. 6.4 of particular interest. There are multiple opportunities presented for framing consulting value propositions, spanning technical, methodological, and organizational frames.

There is an overlapping relationship implied between CSDS consumers, firms seeking CSDS solutions, and suppliers, either internal professionals or external service providers (i.e., software, services, or consulting firms). That is, as CSDS is a very new domain, and thus a function of new potential value to firms, the cost efficiencies of insourcing versus outsourcing various CSDS functions is not yet clear. There is an open question related to the benefits and costs of hosting an internal CSDS function versus outsourcing. As this is an important topic of interest to both organizational managers and commercial suppliers, this is addressed in greater detail in Sect. 6.6.

Lastly, but certainly not least on a scale of importance, governmental organizations and NGOs should find this work of interest in terms of understanding how CSDS will impact the cybersecurity discipline as the CSDS field develops. Military and intelligence services will be interested in understanding prospects for both the defensive and offensive applications of CSDS methods to support various campaigns and initiatives. Regulators will be interested in an early perspective on how CSDS will likely alter the cybersphere in the coming decade, requiring new rules and regulations. Focused insights concerning national and regulatory interests are offered in greater detail in Sect. 6.7.

NGOs and other institutions focused on cybersecurity will be interested in how they might participate in the evolution of the CSDS field. *Phase I* framed a central point in this regard: next steps in the formalization of the CSDS profession will require the sponsorship of industry and professional organizations. In particular, formalizing training offerings (curricula), codifying best practices (body of theory), and hosting certification programs will be essential to iterating and formalizing the CSDS profession.

Research-oriented professionals with a potential interest in this work include professional academics, industry researchers, instructors, and students. This work provides a foundational reference to the emerging CSDS field, including an overview of the state of the art, a structured evaluation of key challenges, and a rich bibliography. Students and instructors will benefit from this work as a CSDS domain reference and subject-matter primer, as it includes both fundamentals and detailed examples of applied approaches.

Professional academics, academic researchers, and postgraduate students with a specific interest in the research findings are directed to the material covered previously in Chap. 5. Of particular interest, Sect. 5.4: *Theoretical Conclusions* provides a summary of the main theoretical findings. A central intended contribution of this inquiry is to establish a foundation for follow-on research. Section 5.6 addresses opportunities in this regard: a range of indications are provided. The results presented in *Phase II* frame opportunities for follow-on survey research. The design contributions advocated in *Phase III* present a foundation for case, design, and/or action research implementations and assessments. A focused action opportunity of great potential practical importance is the furtherance of CSDS curriculum design and teaching efforts.

6.4   Guidance for CSDS and Related Practitioners

CSDS, data science, and cybersecurity practitioners are a main audience served by this work. CSDS practitioners will gain a better understanding of the current status and ideal direction of the emerging profession. Security practitioners will benefit from a focused perspective on the emergence of and critiques associated with the data-driven security paradigm. Data scientists seeking an understanding of the application of data science within the cybersecurity domain will be presented with guidance on best practices and focused examples.

An overview of the main phases as related to general practitioner interests:

- *Phase I*: introduces the CSDS domain, including a focused assessment of challenges

- *Phase II*: insights from 50 global CSDS practitioners concerning challenges and best practices
- *Phase III*: detailed guidance concerning best practices centered on three key CSDS gaps: data management, scientific methods, and cross-domain collaboration

CSDS is presented as a recent development, having emerged as a focused professional title in the last 3 years. The domain is identified as a functional fusion of the cybersecurity and data science professional fields. Participating professionals have mainly entered either from a cybersecurity or data science background and build cross-competencies in the junction between the two domains. This highlights the need for clearer, more explicit cross-disciplinary bridges.

Pressures afflicting cybersecurity professionals frame use cases for data science solutions. Data science offers a range of practical methods to address focused cybersecurity challenges. Big data analytics, data engineering, pattern analysis, anomaly detection, predictive algorithms, and prescriptive methods address aspects of noted cybersecurity gaps. Industry research has reinforced assertions of the efficacy of applied security analytics.

An overview of the CSDS profession is presented in *Phase I*, framing the field as an emergent hybridization of cybersecurity and data science. *Phases II* and *III* frame and explore CSDS best practices. In aggregate, these best practices address the question: what does a CSDS professional do? The following offers a derived high-level summary of key CSDS professional functions:

1. *Manage data*

    - Gather, aggregate, treat, analyze, and interpret large volumes of cybersecurity relevant data
    - Ensure data quality through data cleansing and transformation
    - Transform, combine, and extrapolate variables and focused datasets
    - Integrate, structure, and rationalize unstructured data in preparation for analytics
    - Perform statistical and diagnostic tests validating the importance and explanatory significance of variables (i.e., EDA)
    - Extract and select features conducive to modeling (i.e., feature engineering)
    - Validate the suitability of datasets for modeling
    - Frame and maintain focused data models for structured storage and extraction (i.e., SQL calls to databases) and/or for extracting on-demand from less-structured big data repositories (i.e., schemas for reading data from cloud repositories)
    - Frame, test, and adopt approaches to aggregate, reduce, compact, and focus datasets
    - Assist in the design, testing, and deployment of data pipelines (i.e., ETL processes)

2. *Apply data science methods*

- Marshal, explore, and describe security-relevant data
- Apply security domain knowledge to refine statistical, analytical, and algorithmic treatments
- Apply innovative methods from a range of disciplines to the goal of assuring security
- Conduct process analytics to understanding and track various adversarial approaches
- Undertake discovery-focused pattern analysis and extrapolation to develop a contextual understanding of the environment and situational awareness of vulnerabilities and threats
- Gather and validate evidence
- Develop a variety of data analysis methodologies, tools, and techniques using statistical approaches and machine learning
- Set and validate objectives related to descriptive, diagnostic, predictive, and prescriptive analytics
- Apply both deductive and inductive approaches
- Design and validate both explanatory and predictive models
- Supply cybersecurity professionals with informative and efficacious "detective analytics" results, aggregating explanatory and predictive indications
- Develop and apply CSDS-focused data analytics life cycle processes and best practices
- Evaluate and formally validate model performance
- Test, analyze, compare, and promote varying algorithms and modeling techniques (i.e., champion-challenger approaches)
- Conduct forensic analysis to extrapolate detection approaches
- Adapt predictive mechanism and leverage past incidents to support real-time detection
- Enable security operations performance monitoring and optimization efforts

3. *Organizational collaboration*

- Determine and validate CSDS organizational objectives working with a range of stakeholders
- Support and facilitate approaches to optimizing security operational efficiency
- Frame new and expand existing big data analytics tools and infrastructure
- Work cross-functionally with data management, infrastructure, and software teams to facilitate analytics model validation and deployment
- Collaborate cross-functionally to enable rapid predictive detection remediation
- Work with implementation/support teams and end-users to investigate and resolve production model and data problems
- Advocate scientific methods and rigor in the conduct of CSDS
- Facilitate the development, testing, and validation of theoretical propositions related to security phenomena

6.5   Guidance for Managers and Security Stewards

Managers and security stewards will be interested in this work for a broad range of practical reasons, including:

- Preparing a strategic set of goals related to CSDS for proposal to executive management
- Developing and delivering a CSDS training program to prepare and upskill staff
- Understanding the relevance of CSDS within the context of security technical architectures
- Simplifying security infrastructure by rationalizing and optimizing the utilization of data
- Increasing the value of existing security investments by improving decision-making
- Pursuing the design and deployment of CSDS systems and solutions
- Extrapolating analytics-driven insights to understand emerging threat trends and risks
- Utilizing CSDS to improve cybersecurity metrics and/or risk management
- Framing and facilitating better security decisions
- Establishing context for security events relevant to investigation and remediation
- Setting objectives for data-driven security in terms of "what to track," "what data to collect," "what is normal," and "how to analyze"
- Improving staff utilization and efficiency
- Improving the accuracy of security event detection mechanisms through the application of scientific principles in order to reduce false-positive alert rates
- Developing a CSDS programmatic implementation strategy
- Developing discrete goals and metrics for the performance of a CSDS program
- Undertaking organizational change management to realize a CSDS function, including cross-training, collaboration, and aligning incentives

Phase I profiles the discrete objectives that impel adopting a data-driven security paradigm. These drivers include realizing cost savings from automation, accommodating shortages in trained personnel, operating efficiently at great speed and scale, deriving value from large sets of security data, and improving operational decision-making in environments of fundamental complexity.

Phase II brings managers into contact with practicing CSDS professionals to understand both focused challenges and best practices in the domain. The results of interview research provide focused insights into where CSDS professionals experience frustrations and limitations. To address these challenges, guidance on ideal CSDS best practices are offered. These results are integrated and summarized within three key areas: data management, scientific methods, and cross-domain collaboration. The latter area profiles organizational and management factors that influence the success of CSDS efforts.

Phase III provides guidance concerning the three gap areas. This will be of central interest to managers planning to implement CSDS programs. The central importance of organizational factors to the success of CSDS programs and implementations is emphasized. Guidance highlights the importance of management support, clarity in organizational ownership, and a structured approach to cross-training and collaborative teaming. A primary goal identified concerns the importance of clarifying and strengthening organizational processes for the rigorous application of scientific methods within CSDS practice.

Socializing an iterative approach to CSDS effectiveness should be helpful to managers seeking to communicate a CSDS strategy and to gain buy-in from distributed stakeholders. A key aspect of building a CSDS competence involves up-front investments, especially as related to data management, model design and validation, and cross-training. Discussion frames the focused overhead involved in building a CSDS competency. The intention is to supply managers and planners with practical background to explain why "Rome cannot be built in a day" and to advocate that initial efforts and investments in rigor will have outsized future benefits.

The CSDS best practices profiled herein outline a range of methods to measure phenomena, make better decisions, identify threats, control risks, and optimize processes. The aggregation of best practices frames a vision concerning ideal CSDS practice. Collectively, prescriptions advocated outline a basis for a CSDS maturity model. Maturity indicators are useful to managers in benchmarking their own *as-is* capabilities. This provides a basis for gap analysis and strategic planning. An ideal vision for CSDS should be of interest to managers seeking to communicate a CSDS strategy and to gain buy-in from distributed stakeholders. Formalizing and socializing a CSDS maturity model as a benchmark establishes a series of strategic goals to achieve, manages executive expectations, establishes training and development targets, and frames metrics for performance tracking.

6.6 Guidance for Firms and Service Providers

A central concept raised and explored in this inquiry concerns a principle of reciprocity between vulnerabilities and defenses created by digital technical innovations. As each new technical innovation creates vulnerabilities and dangers, avenues for leveraging the innovation in defense are created. CSDS as an innovative new approach to security follows suit: a range of offensive and defensive treatments are framed.

The proposition that CSDS creates permeable opportunities between actors and interests on both sides of the security divide raises strategic existential questions for both firms and industry. From an economic value perspective, the notion of reciprocity frames a circular and self-reinforcing aspect to the security industry. An optimistic perspective is that this is simply a cost of doing business, that the cybersphere writ large is a powerful, albeit flawed, innovation, and that there are embedded costs associated with its use and growth.

A more pessimistic interpretation would be that there are potentially better macro-systemic approaches to delivering the key services offered while lowering the associated risks and costs. The existential question posed concerns potential sunk-cost and status-quo biases: the possibility that fundamental aspects of the current situation are negligently sub-optimal. This poses a question as to what degree "not fixing the problem" is comfortably facile, and that there are a number of vested parties benefiting from the relatively poor and degrading state of broad cybersphere macro-security.

This pessimistic interpretation frames CSDS potentially as much more than a Band-Aid in improving security investigations. Whereas CSDS can be implemented as an engineering vehicle to patch and improve the status quo, it potentially also offers approaches to examine and critique systemic issues of sub-optimality, risk, and value destruction.

CSDS demonstrably assists in improving threat detection; it also potentially offers more rigorous quantitative and scientific approaches to examine macro-systemic efficiencies. CSDS should also be considered as a mechanism that can facilitate a range of higher-value offerings, both at microeconomic and macroeconomic scales. From a microeconomic perspective, CSDS offers methods to improve security investment decision-making in the context of the firm.

CSDS also potentially offers macroeconomic perspectives that might suggest improved macro-systemic prescriptions. More rigorous theoretical foundations and scientific practices provide approaches for charting strategic guidance, for instance, to identify optimal public-private collaboration structures or to proffer perspectives on improved and expanded regulatory approaches.

As a foil, adopting an optimistic perspective, given the growing proliferation and inherent multi-contextual complexity of the cybersphere, there is a valid critique that no demonstrably optimal state of security is possible. A central economic benefit of the cybersphere, which has both technical and social aspects, concerns its fundamental openness.

As markets themselves create value, increasing value accompanies growth and proliferation. Larger and more valuable markets also inherently open opportunities for abuse, misuse, and capture. A case can be made that the cybersphere evidences greater and more serious vulnerabilities precisely in lockstep with its increasing reach and proliferation. However, here too CSDS also offers macro-prescriptions to the extent it can facilitate the identification of satisficed equilibrium points between opportunity and risk.

In either case, the assertion is that the CSDS field, as a vehicle for scientific inquiry and rational decision-making, offers approaches to improve economic and firm decision-making. In the case of improving firm decision-making, organizations seek to address growing security exposures as an optimal investment between opportunity and risks assumed.

To assess firm decision-making, it is necessary to understand the context and structure of the firm vis-à-vis economic factors associated with security decisions. The socio-structural perspective views firms as complex, dynamic information processing entities which act and react to information based on unique structural characteristics and conditions.

Embracing the notion of an organization as a structural-relational complex of interacting roles and incentives allows for practical and discrete comparisons between organizational structure and firm behavior, including performance. This viewpoint is supported by notions associated with *theories of the firm*, the *knowledge-based view* (KBV) of the firm in particular (Grant 1996, 1997; Jensen and Meckling 1992), itself a refined specification of the *resource-based view* (RBV) (Barney 1991, 1999; Conner 1991; Penrose 1959; Wernerfelt 1984; Williamson 1975).

The KBV recognizes the value-enhancing capabilities of information systems in expediting firm knowledge to make informed decisions. The KBV core concept asserts the importance of incentives, acting in concert with information and decision rights, to understanding the capabilities and behavior of organizations. The importance of decision rights, information access, and incentives in orienting organizational actors connects KBV to considerations of *organizational design* (Burton et al. 2011), *management control systems (MCS)* (Merchant and van der Stede 2003), *organizational culture* (Cameron and Quinn 2011), and *change management* (Hiatt 2006; Kotter and Cohen 2002).

In the broadest theoretical frame, the socio-structural perspective views organizations as *sensemaking bodies* (Klein et al. 2006; Klein et al. 2010; Weick 1995, 2001, 2009). Sensemaking in this context encompasses working out decisions regarding optimal investment between opportunities created and vulnerabilities accepted. If we are to assume that there is no perfectly optimal solution *satisfying* all stakeholders, *satisficing* (an optimal compromise) is necessary amongst the array of competing firm stakeholders (e.g., sales and strategy vs. security and risk functions).

In this purview, CSDS offers treatments, yet also involves foregoing opportunity costs associated with capital outlay and embedded overhead in maintaining a CSDS function. Costs raise particularly in the context of substantiating and enforcing rigor in CSDS practice. As discussed in this inquiry in *Phase III*, whereas ironclad CSDS model validation is practically infeasible, firms must decide on an optimal investment outlay regarding rigor. The process of validating CSDS treatments is, at some fundamental level, a firm sensemaking and satisficing exercise, especially when CSDS treatments are understood as organizational decision-making artifacts (Pidd 2004).

A related question arises concerning the focused economic value proposition of CSDS professionals. This does not appear to have a single answer given the range of host industries in which CSDS professionals work. For a large corporate host, the CSDS professional offers risk reduction services in terms of cyber incidents detected and prevented (hypothetical, but ideally as measured tangibly).

Implicit is that the CSDS professional is also improving efficiency over traditional manual security investigations which creates economic value through cost savings. This may take the form of facilitating and improving the efficiency of security investigators, providing indicators, tools, and orchestration to traditional investigators, or potentially directly reducing or circumventing the need for traditional investigations.

An allied question concerns cost economics in the firm as related to the CSDS role as an opportunity investment versus anticipated returns. Instructive in this regard is a consideration of Porter's five forces model (Porter 2008), which in this context poses a strategic decision to the firm concerning whether to host, outsource, offload, or ignore the risks associated with cybersecurity threats.

For a larger firm with a serious and pressing need to maintain digital security, an in-house solution may be optimally cost-effective. We see this as related both to the size of and core functions associated with host firms. For instance, large financial services and telecommunications firms centrally depend upon the integrity of their digital information and infrastructure, suggesting outsized demand for security assurance.

Concerning the future of the CSDS domain, a core question is framed concerning the costs versus benefits of hosting internal CSDS functions versus contracting a managed service or contracting the implementation of a solution, potentially involving a mix of consulting and software tool provision.

Given the substantial up-front and continuing embedded costs of staging an end-to-end programmatic CSDS function for individual firms, broad future uptake of homegrown CSDS functions in mid-to-small enterprises is unlikely. More likely, large-scale enterprises will be early adopters and innovators, possibly working in consortia and with national and NGO institutions. This leaves an opportunity for CSDS product and service providers to address gaps via a range of solution, tool, and service offerings.

This frames a growing security industry opportunity for CSDS service and solutions providers. Concerning packaged solutions, a likely trend will be subscription-based services, observing the general growing preference for subscriptions over licensed software (Thomason 2018). Cloud-based services in particular will be increasingly attractive, given the embedded cost and overhead of maintaining own infrastructure and storage. This assumes providers can maintain confidence in security and address data volumes by reducing raw data requirements to refined, focused datasets. This implies that service providers themselves stage efforts to conduct primary research on feature efficacy and to undertake rigorous model validation.

Although machine learning methods hold great promise, statistical validation will continue to be a requirement for substantiating methodological efficacy. The ability to create and stage effective remote service offerings thus implies a serious up-front capital investment in substantiating efficiency and efficacy through primary research.

In summary, while implied CSDS market opportunities are substantial, in keeping with the new and emerging nature of the CSDS domain, rigorous efficacy demands substantial up-front capital investments, particularly in research and development. Novel business models might, for instance, subsidize R&D capital outlay by allocating the assumed embedded economic value of handling data at scale, whereby providers gain pan-systemic strategic power through analytical insights gained as data hubs (i.e., as a security data ecosystem intermediaries).

6.7 Guidance for Nation-States and Regulators

The advent of the cybersphere and related innovations have been rapidly adopted globally at scale, often with limited critical security oversight. Due to many unanticipated effects from this broad adoption, the notion that the unmitigated proliferation of network-mediated computer and communication technology is inherently and unquestionably pro-social is increasingly being called into question. Whereas immense economic benefits have accreted to a select few, a range of negative externalities increasingly abound. As social and economic costs expand with the proliferation and impact of the cybersphere, increased regulatory oversight may become a necessity.

This work has framed a number of approaches whereby CSDS can improve the currently dire and degrading standards for public security and safety on the Internet. However, the efforts and overhead necessary for establishing CSDS programmatic capabilities are substantial. Economic cost factors to host robust CSDS centers of excellence in organizations may be beyond the means of all but nation-states and large transnational conglomerates and financial institutions. In terms of supporting ambitious CSDS development efforts, this raises four possibilities:

1. *Free market responses*: consortia of private interests collaborating with third-party service providers (as-a-service market) to host CSDS offerings to foster improved cybersphere control and safety.
2. *Centralized governmental responses:* cascading failures lead to a growing appreciation of the value of regulatory control from governmental agencies (some model involving taxes, subsidy, and greater regulatory oversight).
3. *Public-private collaboration*: a combination of approaches 1 and 2, assuming the mechanisms and traditions for facilitating such collaborations are in place (i.e., as present in Western Europe and relatively absent in the United States).
4. *Do nothing:* maintain the status quo and accept the outcomes, albeit with the assumption that the proliferation of digital exposure will continue to expand exponentially, risking increasingly perverse "tragedy of the commons" scenarios and outcomes.

Currently, although there are some movements across the range of these responses, the dominant response has been largely "do nothing" (or little). That a burgeoning commercial security industry benefits economically from a degenerating status quo does not bode well for a rapid reversal. However, whereas macro-systemic responses on the defensive side have been sluggish, this, unfortunately, has not been the case on the adversarial side of the equation. The market for so-called contract hacking is in rude health. This comes in part due to passive and active support from national governmental interests.

An ironic trend in state-sponsored cyberattacks are indications that former hard-line anti-capitalistic regimes give every indication of being active innovators in employing free markets to undertake arms-distance campaigns and actions. Russia and China are both believed to actively collaborate with semi-private and contractor-fronted outfits to undertake cyber mischief and to incentivize actions vis-à-vis ostensibly independent proxies.

Contracting arms-distance cyber mischief achieves a number of goals, the first being plausible deniability of attribution in the event of exposure. Second is the advantage of agility—private outfits typically being geographically, virtually, and technically more dexterous than ensconced, bureaucratized military outfits.

The combination of deniability and agility finds a third benefit, that of a wider purview concerning what constitutes mischief and flexibility concerning motivations. Namely, high-level state sponsors have focused sociopolitical and socioeconomic remits to probe and destabilize adversaries. For mercenary outfits, an implied creative license is granted, whereby economic theft and sewing economic discord are inherently profitable activities to the independent agents while also achieving broader strategic objectives associated with wrong-footing adversarial institutional targets.

A vibrant dark web ecosystem, empowered by strong encryption and the anonymity of cryptocurrency transactions, facilitates and enables such contracted pseudo-state actors. The stark reality is that there is a thriving economy of skilled mercenary actors willing to take on particular actions or to produce specialized tools under contract and anonymously with few questions asked (The Economist 2019c).

As an example, bot-net armies can be assembled by private hackers by exploiting zero-day exploits and/or unpatched systems. These massed compromised machines can then be bid-out to take on attacks both overt, such as mass DDoS attacks, and covert, such as "rentable" arms-distance, obfuscated command and control platforms for complex and subtle incursions.

Beyond semi-private agents of discord, there are also indications of increasing public-private collaborations whereby state interests may incentivize (or mandate the enlistment of) otherwise privately presenting foreign technology manufacturing and service firms. As an example, in recent news coverage, doubts have been leveled concerning the degree to which global mobile and telecom equipment providers exhibit fundamental independence from state interests.

Beyond large nation-state-associated firms, there are many hundreds of smaller global independent manufacturing and technology firms developing networked mobile devices, embedded components, and consumer products. These smaller players in the global technology supply chain are at risk of hosting or otherwise acting as carriers for intrusion gateways or infections, perhaps even inadvertently if their supply chain is compromised. That technical devices assembled for public, private, and consumer use often aggregate a complex array of third-party sourced microprocessors and embedded controllers means that there are many opportunities for inadvertently hosting covert monitoring or interference mechanisms.

As raised in interviews and several times during this inquiry, CSDS meta-challenges abound on the horizon. There are active indications that data science-associated methods are being examined and applied by state-level adversaries, opening a hall of mirrors scenario in which methods of attack encompass adversarial machine learning and attacks facilitated by machine learning mechanisms (Chebbi 2018, 2019; Chio and Freeman 2018; Forrester 2020; Gopalakrishnan

2020; Yin et al. 2019). Implicit is that effective countermeasures must encompass and adopt machine learning vectors (Johnson 2019). The future points towards staged algorithmic offensive and defensive platforms, each attempting to gain minute advantages over the other: *a CSDS cold war.*

A threat to many western democratic countries is that many lack strong regulatory and technical controls over their own national cyberinfrastructure. As opposed to countries that restrict and control the use of the Internet, this leaves countries which have embraced openness-as-a-virtue, or "religious deregulation," at risk of being wrong-footed in adversarial actions from less democratized players. From a technical-epidemiological perspective, open networks and prolific access are a hotbed of vulnerabilities for adversarial actors to explore, compromise, damage, or commandeer. The standoff in this sense is asymmetric, the west being inherently vulnerable for its sheer accessibility.

The most worrying trend goes beyond technical vulnerabilities and attacks, encompassing the vulnerability of democratic media itself. State-sponsored disinformation campaigns find accommodating and fruitful targets in online social media channels and the permeable diaspora of online independent media. Adversarial states with a deep history of strict media control and propaganda have many decades of experience concerning the subtle mechanisms of spreading disinformation and controlling public opinion through webs of indirect channels, including innuendo, rumormongering, fomenting division, incentivizing mob violence, etc.

One need not look far to see evidence of these factors appearing in current events, for instance, opportunistic disinformation campaigns related to the 2020 coronavirus crisis (Beaumont et al. 2020). The results gained for state actors in such campaigns are the attrition of sociopolitical stability and sowing general institutional discord in rival states, all at a relatively low cost.

In terms of prevention, socializing basic digital literacy is the best proven defense. This includes direct education of the public concerning digital threats and outreach to the private sector to actively promote patching and basic defenses. Beyond this, larger organizations are prompted to introspect concerning making cyber incursions and mischief time-consuming and expensive. For all the resources of state actors and their seeming willingness to sponsor semi-independent outfits, at base, undertaking an adversarial cyber campaign or action is an economic effort, and, at some point, economic fundamentals are restrictions. To the degree that defenders present difficult and time-consuming-to-circumvent countermeasures, adversaries will turn away and seek more fruitful paths and targets. That CSDS-driven adversarial mechanisms and strategies bring about automation and, thus cost savings, should be a central point of worry.

Concerning larger existential issues, the drama playing out in western democracies involves a renewed soul-searching regarding the wisdom of open, deregulated communications infrastructure, telecommunication carriage, and unfettered interactive media. As recent events have revealed, the real sources of danger may simply be laissez-faire openness itself, and efforts to draw back towards modest regulatory oversight of national telecommunications security and even online media communications may be increasingly on the table as an effective palliative.

Younger citizens in western democratic countries likely have lost appreciation of the fact that telephones were once state-owned devices lent out to private citizens under strict conditions. While telecommunication industry deregulation in the late 1970s and early 1980s led to miraculous technical innovation and the proliferation of media technologies, we are returning to a sober understanding of the true value and purpose of telecommunication regulatory oversight.

From an economic perspective, there is some case for viewing the growing centrality of the cybersphere as staging elements of a tragedy of the commons scenario. Whereas the cybersphere offers carriage for a vast array of services and efficiencies, it is also generating a growing host of negative externalities and risks. The challenge of the status quo relates to adverse and perverse outcomes resulting from broadly absent regulatory oversight.

When free markets are left to their own devices, sub-optimal outcomes are not unusual, especially when incentives are misaligned. As Herbert Simon notes concerning economic optimality: "the real economic actor is in fact a satisficer, a person who accepts 'good enough' alternatives, not because less is preferred to more but because there is no choice" (1996). Increasingly, the array of challenges presented by the largely deregulated nature of the cybersphere is resulting in fewer and less optimal choices for individuals across a range of measures.

In conclusion, pursuing more rigorous scientific theoretical foundations in the security domain may lead to surprising, even potentially disruptive, insights. This is especially the case when stronger models for satisficing broader principles of social and economic value are embraced. The status quo is increasingly leading to devolution to substandard equilibriums, lowest common denominator scenarios. The libertarian and anti-regulatory roots of the Internet are appropriate for revisiting as the scale of the exponentially expanding cybersphere increasingly presents disappointing and, at times, dystopic outcomes.

The great potential of CSDS to empower bad actors is a real concern. It is not yet clear that CSDS-driven defenses can match and effectively stalemate CSDS offensive capabilities. Beyond securing against and hunting threats, the broader value for CSDS as an emerging field may unexpectedly be to facilitate improved perspectives on and approaches to macroprudential regulatory stewardship of the cybersphere.

References

Aart CV (2004) Organizational principles for multi-agent architectures. Birkhauser, Berlin

Abbott A (1988) The system of professions: an essay on the division of expert labor. University of Chicago Press, Chicago

Acito F, Khatri V (2014) Business analytics: why now and what next? Business Horizons 57(5):565–570. https://doi.org/10.1016/j.bushor.2014.06.001

Adams N, Heard N (eds) (2014) Data analysis for network cyber-security. Imperial College Press, London

Adams N, Heard N (2016) In: Adams N, Heard N (eds) Dynamic networks and cyber-security. World Scientific Publishing Ltd., London

Aggarwal CC (2017) Outlier analysis, 2nd edn. Springer, Cham

Aggarwal CC, Sathe S (2017) Outlier ensembles: an introduction. Springer, Cham

Aggarwal CC, Wang H (2010) Managing and mining graph data. Springer, London

Aguinis H, Pierce CA, Bosco FA, Muslin IS (2007) First decade of organizational research methods: trends in design, measurement, and data-analysis topics. Organiz Res Methods 12:43. https://doi.org/10.1177/1094428108322641

Akoglu L, Tong H, Koutra D (2015) Graph based anomaly detection and description: a survey. Data Min Knowl Discov 29(3):63

Alazab M, Tang M (eds) (2019) Deep learning applications for cyber security. Springer, Cham

Albright SC, Winston W, Zappe C (2011) Data analysis and decision making, 4th edn. South-Western Cengage Learning, Mason, OH

Allaire JJ (2018) Machine learning with R and TensorFlow. In: RStudio conference 2018. YouTube: rstudio::conf

Allemang D, Hendler J (2011) Semantic web for the working ontologist, 2nd edn. Elsevier, Waltham

Alsmadi IM, Karabatis G, AlEroud A (eds) (2017) Information fusion for cyber-security analytics. Springer, Cham

Alter SL (1980) Decision support systems: current practice and continuing challenge. Addison-Wesley, Reading, MA

Amla N, Atluri V, Epstein J, Greenspan S, Muhlberger P, Piotrowski VP et al (2012) Toward a secure and trustworthy cyberspace. The Next Wave 19(4):5. Retrieved from https://www.nsa.gov/Portals/70/documents/resources/everyone/digital-media-center/publications/the-next-wave/TNW-19-2.pdf

Anderson C (2008) The end of theory: the data deluge makes the scientific method obsolete. Wired. Retrieved from https://www.wired.com/2008/06/pb-theory/#

Anderson DR, Sweeney DJ, Williamson TA, Wisniewski M (2009) An introduction to management science: quantitative approaches to decision making. Cengage Learning, Beijing

© Springer Nature Switzerland AG 2021
S. Mongeau, A. Hajdasinski, *Cybersecurity Data Science*,
https://doi.org/10.1007/978-3-030-74896-8

Ansoff HI, Hayes RL (1973) Roles of models in corporate decision making. Paper presented at the IFORS international conference on operational research, Amsterdam, Netherlands

Antonakakis M, April T, Bailey M, Bernhard M, Bursztein E, Cochran J et al (2018) Understanding the Mirai Botnet. Paper presented at the 26th USENIX security symposium

Antoniou G, Groth P, van Harmelen F, Hoekstra R (2019) A semantic web primer, Kindle edn., 3rd edn. The MIT Press, Cambridge

Armstrong RC, Mayo JR, Siebenlist F (2009) Complexity science challenges in cybersecurity, Albuquerque, New Mexico. Retrieved from https://wiki.cac.washington.edu/download/attach-ments/7478403/Complexity+Science+Challenges+in+Cybersecurity.pdf

Ashford W (2019) Perfect storm for data science in security. Computer Weekly. Retrieved from https://www.computerweekly.com/news/252464071/Perfect-storm-for-data-science-in-security

Asparouhov T, Muthen B (2010) Simple second order chi-square correction. Retrieved from https://www.statmodel.com/download/WLSMV_new_chi21.pdf

Aucsmith D, Dixon B, Martin-Emerson R (2003) Threat personas. Microsoft. Redmond, WA

Azevedo AIRL, Santos MF (2008) KDD, SEMMA and CRISP-DM: a parallel overview. Paper presented at the IADIS European conference data mining 2008

Baarda B (2010) Research this is it! Noordhoff Uitgevers bv, Groningen

Baesens B (2014) Analytics in a big data world. Wiley, Hoboken, NJ

Baesens B, Van Vlasselaer V, Verbeke W (2015) Fraud analytics using descriptive, predictive, and social network techniques. Wiley, Hoboken, NJ

Baglin J (2014) Improving your exploratory Factor analysis for ordinal data: a demonstration using FACTOR. Pract Assess Res Eval 19(5):2

Balci O (1998) Verification, validation and testing: principles, methodology, advances, applications, and practice. In: Banks J (ed) Handbook of simulation. Wiley, New York

Banerjee A, Duflo E (2019) Good economics for hard times. PublicAffairs, New York

Banerjee A, Bandyopadhyay T, Acharya P (2013) Data analytics: hyped up aspirations or true potential? Vikalpa 38(4). https://doi.org/10.1177/0256090920130401

Banker RD, JKauffman RJ (2004) The evolution of research on information systems: a fiftieth-year survey of the literature in "management science". Manag Sci 50:17

Banks J, Carson JS, Nelson BL, Nicol DM (2010) Discrete-event system simulation. Pearson Education, Inc., Upper Saddle River, NJ

Bara G, Backfried G, Thomas-Aniola D (2019) Fake or fact? Theoretical and practical aspects of fake news. In: Bosse E, Rogova G (eds) Information quality in information fusion and decision making, Kindle edn. Springer, Cham

Barabási A-L (2014) Linked: the new science of networks. Perseus Publishing, Cambridge, MA

Barlow M (2013) The culture of big data. O'Reilly Media, Sebastopol, CA

Barney J (1991) Firm resources and sustained competitive advantage. J Manag 17(1):21

Barney J (1999) How a firm's capabilities affect boundary decisions. Sloan Manag Rev 40(3):9

Barreno M, Nelson B, Sears R, Joseph AD, Tygar JD (2006) Can machine learning be secure? Paper presented at the ACM symposium on information, computer, and communication security, Taipei, Taiwan

Barrett PT, Kline P (1981) The observation to variable ratio in factor analysis. Personal Study Group Behav 1(1):10

Barros A, Chuvakin A (2018) Preparing your security operations for orchestration and automation tools. Retrieved from https://www.gartner.com/en/documents/3860563

Bartholomew DJ, Steele F, Moustaki I, Galbraith J (2008) Analysis of multivariate social science data, 2nd edn. CRC Press, London

Bartlett R (2017) A practitioner's guide to business analytics. McGraw-Hill, London

Baskerville R, Pries-Heje J (1999) Grounded action research: a method for understanding IT in practice. Account Manag Inform Technol 9(1):23. https://doi.org/10.1016/S0959-8022(98)00017-4

Bayrak T (2015) A review of business analytics: a business enabler or another passing fad. Proc Soc Behav Sci 195:230–239

Beaumont P, Borger J, Boffey D (2020) Malicious forces creating 'Perfect Storm' of coronavirus disinformation. The Guardian

Beaupérin T (2019) Cyber security is an enterprise risk, FERMA tells the European Commission

Bechor T, Jung B (2019) Current state and modeling of research topics in cybersecurity and data science. Syst Cybernet Inform 17(1):27

Beer JT, Lewis WD (1963) Aspects of the professionalization of science. The MIT Press 92(4):20

Bejtlich R (2013) The practice of network security monitoring. No Starch Press, San Francisco, CA

Benbasat I, Zmud RW (2003) The identity crisis within the IS discipline: defining and communicating the discipline's core properties. MIS Q 27:2

Berman DS, Buczak AL, Chavis JS, Corbett CL (2019) A survey of deep learning methods for cyber security. Information 10(4). https://doi.org/10.3390/info10040122

Bernard HR, Ryan GW (2010) Analyzing qualitative data: systemic approaches. Sage, Thousand Oaks, CA

Bevan O, Boehm J, Manocaran M, Riemenschnitter R (2018) Cybersecurity and the risk function. Retrieved from https://www.mckinsey.com/business-functions/risk/our-insights/cybersecurity-and-the-risk-function

Bhattacharyya DK, Kalita JK (2013) Network anomaly detection: a machine learning perspective, Kindle edn. CRC Press, Boca Raton, FL

Bhattacharyya DK, Kalita JK (2014) Network anomaly detection: a machine learning perspective, Kindle edn. CRC Press, Boca Raton, FL

Bicak A, Liu M, Murphy D (2015) Cybersecurity curriculum development: introducing specialties in a graduate program. Inform Syst Educ J 13(3):12

Bishop M (2003) What is computer Security? IEEE Security & Privacy Magazine 3:67–69

Blenko MW, Mankins MC, Rogers P (2010) The decision-driven organization. Harv Bus Rev

Bloor R (2013) A data science rant

Bodeau DJ, Graubart R (2011) Cyber resiliency engineering framework (MTR 110237). Retrieved from https://www.mitre.org/sites/default/files/pdf/11_4436.pdf

Boehm J, Merrath P, Poppensieker T, Riemenschnitter R, Stähle T (2018) Cyber risk measurement and the holistic cybersecurity approach. Retrieved from https://www.mckinsey.com/business-functions/risk/our-insights/cyber-risk-measurement-and-the-holistic-cybersecurity-approach

Bollobás B (2001) Random graphs, 2nd edn. Cambridge University Press, Cambridge

Bolón-Canedo V, Sánchez-Maroño N, Alonso-Betanzos A (2015) Feature selection for high-dimensional data. Springer, Cham

Bonabeau E (2003) Don't trust your gut. Harv Bus Rev

Booz Allen Hamilton (2019) 2019 Cyber threat outlook. Retrieved from https://www.boozallen.com/content/dam/boozallen_site/sig/pdf/publications/booz-allen-2019-cyber-threat-outlook.pdf

Box GEP (1979) Robustness in the strategy of scientific model building. Paper presented at the robustness in statistics

Box GEP, Draper NR (1987) Empirical model-building and response surfaces. Wiley, Minnesota

Boyd D, Crawford K (2014) Critical questions for big data. Inform Commun Soc 15(5):17

Breiman L (2001) Statistical modeling: the two cultures. Stat Sci 16(3):199–231

Brockett PL, Golden LL, Wolman W (2012) Enterprise cyber risk management. In: Emblemsvåg J (ed) Risk management for the future—theory and cases. IntechOpen, London

Brundage M, Avin S, Clark J, Toner H, Eckersley P, Garfinkel B et al (2018) The malicious use of artificial intelligence: forecasting, prevention, and mitigation. Future of Humanity Institute, London. arXiv: 1802.07228

Bryman A, Burgess RG (1994) Developments in qualitative data analysis: an introduction. In: Bryman A, Burgess RG (eds) Analyzing qualitative data. Routledge, London

Brynjolfsson E, McAfee A (2012) Big data's management revolution. Harv Bus Rev. Retrieved from http://blogs.hbr.org/cs/2012/09/big_datas_management_revolutio.html

Buczak AL, Guven E (2016) A survey of data mining and machine learning methods for cyber security intrusion detection. IEEE Commun Surv Tutorials 18(2):23

Bureau of Labor Statistics (2018) 2018 Standard occupational classification system. Retrieved from https://www.bls.gov/soc/2018/major_groups.htm

Burnard P (1991) A method of analyzing interview transcripts in qualitative research. Nurse Educ Today 11:5

Burr M, Endicott B (2020). Coronavirus will permanently change how we work. The Wall Street Journal. Retrieved from https://www.wsj.com/articles/coronavirus-will-permanently-change-how-we-work-11584380290

Burton RM, Obel B, DeSanctis G (2011) Organizational design: a step-by-step approach, 2nd edn. Cambridge University Press, Cambridge

Caltagirone S, Pendergast A, Betz C (2013) The diamond model of intrusion analysis (ADA586960). Retrieved from https://apps.dtic.mil/dtic/tr/fulltext/u2/a586960.pdf

Cameron KS, Quinn RE (2011) Diagnosing and changing organizational culture: based on the competing values framework, 3rd edn. Jossey-Bass, San Francisco, CA

Cannon SF (1978) Science in culture: the early Victorian period of medicine. Dawson and Science History Publications, London

Carley KM (ed) (2011) Computational and mathematical organization theory, vol 17. Springer, Heidelberg

Carnegie Mellon University & John Hopkins University (2019) Cybersecurity maturity model (CMMC). Retrieved from https://www.acq.osd.mil/cmmc/draft.html

Carrascosa IP, Kalutarage HK, Huang Y (eds) (2017) Data analytics and decision support for cybersecurity: trends, methodologies and applications. Springer, Cham

Castellani B, Hafferty F (2010) Sociology and complexity science: a new field of inquiry. Springer, Heidelberg

Chakraborty G, Pagolu M, Garla S (2013) Text mining and analysis: practical methods, examples, and case studies using SAS. SAS Press, Cary, NC

Chalvatis D (2017) The conundrum of big data—why organizations still struggle with it. Information Management. Retrieved from https://www.information-management.com/opinion/the-conundrum-of-big-data-why-organizations-still-struggle-with-it

Chandler N, Hostmann B, Rayner N, Herschel G (2011) Gartner's business analytics framework (G00219139)

Chapman P, Clinton J, Kerber R, Khabaza T, Reinartz T, Shearer C, Wirth R (2000) CRISP-DM 1.0: step-by-step data mining guide

Charmaz K (2014) In: Seaman J (ed) Constructing grounded theory. Sage, London

Chaturvedi A (2019) Securing microservice architectures. Retrieved from https://www.ibm.com/downloads/cas/JY6LNAWX

Chebbi C (2018) Mastering machine learning for penetration testing. Packt, Birmingham

Chebbi C (2019) How to attack artificial intelligence systems. Retrieved from https://www.peer-lyst.com/posts/how-to-attack-artificial-intelligence-systems-chiheb-chebbi

Chen WWS (ed) (2005) Statistical methods in computer security. Marcel Dekker, New York

Chen CLP, Zhang C-Y (2014) Data-intensive applications, challenges, techniques and technologies: a survey on big data. Inform Sci 275:314–347

Chen H, Chiang RHL, Storey VC (2012) Business intelligence and analytics: from big data to big Imact. MIS Q 36(4):23

Cheng Y, Deng J, Li J, DeLoach SA, Singhal A, Ou X (2014) Metrics of security. In: Kott A, Wang C, Erbacher RF (eds) Cyber defense and situational awareness, vol 62. Springer, Cham, p 32

Chio C, Freeman D (2018) Machine learning & security: protecting systems with data and algorithms. O'Reilly, Sebastopol, CA

Cisco Systems Inc. (2017) The Zettabyte era: trends and analysis. Retrieved from https://www.cisco.com/c/en/us/solutions/collateral/service-provider/visual-networking-index-vni/vni-hyperconnectivity-wp.html

Cisco Systems Inc. (2018) Cisco 2018 annual cybersecurity report, San Jose, CA, USA. Retrieved from https://www.cisco.com/c/dam/m/digital/elq-cmcglobal/witb/acr2018/acr2018final.pdf

Cisco Systems Inc. (2019) Cisco visual networking index: global Mobile data traffic forecast update, 2017–2022 white paper, San Jose, CA, USA. Retrieved from https://www.cisco.com/c/en/us/solutions/collateral/service-provider/visual-networking-index-vni/white-paper-c11-738429.html

Cleveland WS (2001) Data Science: an action plan for expanding the technical areas of the field of statistics. Int Stat Rev 69(1):6

Coffeen L (2009) If you can't measure it, you can't manage it. Retrieved from https://perma.cc/3VVC-SGHT

Cohen Z, Marquardt A (2020) They are trying to steal everything. US coronavirus response hit by foreign hackers. CNN

Collins M (2014) Network security through data analysis. O'Reilly, Sebastopol, CA

Conklin WA, Cline RE Jr, Roosa T (2014) Re-engineering cybersecurity education in the US: an analysis of the critical factors. Paper presented at the 47th international conference on system science, Hawaii, US

Conner KR (1991) A historical comparison of resource-based view and five schools of thought within industrial organization economics: do we have a new theory of the firm? J Manag 17(1):33

Connors ES, Endsley MR (2014) Foundation and challenges. In: Kott A, Wang C, Erbacher RF (eds) Advances in information security, vol 62. Springer, Cham

Corbin J, Strauss A (2015) Basics of qualitative research: techniques and procedures for developing grounded theory, 4th edn. Sage, London

Craigen D (2014) Assessing scientific contributions: a proposed framework and its application to cybersecurity. Technol Inform Manag Rev 4:5–13

Craigen D, Diakun-Thibault N, Purse R (2014) Defining cybersecurity. Technol Innov Manag Rev 4:13–21

Crowe P (2015) The trader blamed for the 'flash crash' tried to blow the whistle on other traders. Business Insider, 14 May 2015. Retrieved from https://www.businessinsider.com/navinder-singh-sarao-blew-the-whistle-on-other-traders-2015-5?international=true&r=US&IR=T

Cudeck R, MacCallum RC (2007) Factor analysis at 100: historical developments and future directions. Lawrence Erlbaum Associates, Mahwah, NJ

Curry S, Kirda E, Schwartz E, Stewart WH, Yoran A (2013) Big data fuels intelligence driven security. Retrieved from https://www.scribd.com/document/183089601/big-data-fuels-intelligence-driven-security-io-pdf

Cybersecurity Data Science (2019)

CyberSeek (2019) Cybersecurity supply/demand heat map. Retrieved from https://www.cyberseek.org/heatmap.html

D'Amico A, Buchanan L, Goodall J, Walczak P (2009) Mission impact of cyber events: scenarios and ontology to express the relationships between cyber assets, missions, and users. Paper presented at the 5th international conference on information warfare and Security, Dayton, OH, US.

Data Science Association (2019) Data science code of professional conduct. Retrieved from https://www.datascienceassn.org/code-of-conduct.html

Data Science for Security Professionals (2017)

DataScience.Community (2019) College & University data science degrees. Retrieved from http://datascience.community/colleges

Davenport TH (2009) Make better decisions. Harv Bus Rev

Davenport TH (2013) Analyics 3.0. Harv Bus Rev

Davenport TH (2014) Big data @ work. Harvard Business Review Press, Boston, MA

Davenport TH, Harris JG (2007) Competing on analytics: the new science of winning. Harvard Business School Press, Boston, MA

Davenport TH, Patil DJ (2012) Data scientist: the sexiest job of the 21st century. Harvard Business Review

Davenport TH, Harris JG, Morison R (2010) Analytics at work: smarter decisions, better results. Harvard Business Review Press, Boston, MA

Dawes RM, Faust D, Meehl PE (1989) Clinical versus actuarial judgement. Science 243:6

de Bruijn H, Janssen M (2017) Building cybersecurity awareness: the need for evidence-based framing strategies. Gov Inf Q 34:7

De Villiers MR (2005) Three approaches as pillars for interpretive information systems research: development research, action research and grounded theory. Paper presented at the South African Institute of Computer Scientists and Information Technologists (SAICSIT), White River, South Africa, 20–22 Sept 2005

Deerwester S, Dumais ST, Furnas GW, Landauer TK, Harshman R (1990) Indexing by latent semantic analysis. J Am Soc Inf Sci 41(6):16

Deshpande S, Kim E, Canales C, Kish D, Contu R, Wheeler JA et al (2018) Forecast: information security, worldwide, 2016–2022, 2Q18 update

Diaz A (2018) Why data culture matters. McKinsey Quart 3(3):16

Dipert RR (2013) The essential features of an ontology for cyberwarfare. In: Yannakogeorgos PA, Lowther AB (eds) Conflict and cooperation in cyberspace: the challenge to national security. CRC Press, Boca Raton, FL

Dolk D (2010) Structured modeling and model management. In: Sodhi MS, Tang CS (eds) A long view of research and practice in operations research and management, vol 148. Springer, New York, p 25

Domingos P (2012) A few useful things to know about machine learning. Commun ACM 55(10). https://doi.org/10.1145/2347736.2347755

Donoho D (2015) 50 Years of data science. Paper presented at the Tukey centennial workshop, Princeton, New Jersey, US. http://courses.csail.mit.edu/18.337/2015/docs/50YearsDataScience.pdf

Donoho D (Producer) (2018) Data science: the end of theory? What is data science? Retrieved from https://www.youtube.com/watch?v=PekBM76z2qE&feature=youtu.be

Dresch A, Lacerda DP, Antunes JAV (2015) Design science research, Kindle edn. Springer, Cham

Drinkwater D (2018) 6 Ways hackers will use machine learning to launch attacks. ComputerWorld

Dua S, Du X (2011) Data mining and machine learning in cybersecurity. CRC Press, London

Dykstra J (2016) Essential cybersecurity science. O'Reilly Media, Sebastopol, CA

Edgar TW, Manz DO (2017) Research methods for cyber security, Kindle edn. Elsevier, Cambridge, MA

Efron B, Hastie T (2016) Computer age statistical interence: algorithms, evidence and data science. Cambridge University Press, New York

Egan M (2019) What is the dark web & how to access it. Internet How-To

Ekbia H, Mattioli M, Kouper I, Arave G, Ghazinejad A, Bowman T et al (2015) Big data, bigger dilemmas: a critical review. J Assoc Inform Sci Technol 66(8):22. https://doi.org/10.1002/asi

Epstein JM (2006) Generative social Science: studies in agent-based computational modeling. Princeton University Press, Princeton

Erdi P (2008) Complexity explained. Springer, Heidelberg

Erdős PR, Rényi A (1959) On random graphs. Publ Math 6:7

Ettinger J (2019) Cyber intelligence tradecraft report: the state of cyber intelligence practices in the United States (DM19–0447)

EUROSTAT (2007) Handbook on data quality assessment methods and tools. Retrieved from https://unstats.un.org/unsd/dnss/docs-nqaf/Eurostat-HANDBOOK%20ON%20DATA%20QUALITY%20ASSESSMENT%20METHODS%20AND%20TOOLS%20%20I.pdf

Fayyad U, Stolorz P (1997) Data mining and KDD: promise and challenges. Fut Gener Comput Syst 13(2–3):16

Fayyad U, Piatetsky-Shapiro G, Smyth P (1996) From data mining to knowledge discovery in databases. AI Mag 17:3. Retrieved from https://www.aaai.org/ojs/index.php/aimagazine/article/view/1230/1131

Ferguson S (2019) A vision of the role for machines in security. Retrieved from https://www.bankinfosecurity.com/vision-role-for-machines-in-security-a-12124

Find a Masters (2019) MSc degrees in United Kingdom (cyber security). Retrieved from https://www.findamasters.com/masters-degrees/msc-degrees/united-kingdom/

Fisk M (2019) Data-driven decision making for cyber-Security. In: Heard N, Adams N, Rubin-Delanchy P, Turcotte M (eds) Security science and technology. World Scientific, London

Foroughi F, Luksch P (2018) Data science methodology for cybersecurity projects

Forrester (2020) The emergence of offensive AI: how companies are protecting themselves against malicious applications of AI. Retrieved from https://www.darktrace.com/en/resources/research-forrester-offensive-ai.pdf

Forsythe C, Silva A, Stevens-Adams S, Bradshaw J (2013) Human dimension in cyber operations research and development priorities. Paper presented at the foundations of augmented cognition: 7th international conference, Las Vegas, NV, USA. https://www.researchgate.net/publication/268512910_Human_Dimension_in_Cyber_Operations_Research_and_Development_Priorities

Fourer R (2010) Cyberinfrastructure and optimization. In: Sodhi MS, Tang CS (eds) A long view of research and practice in operations research and management science, vol 148. Springer, New York, p 297

Freidson E (2001) Professionalism: the third logic. Polity Press, Cambridge, MA

Frenkel KA (2017) Can characteristic-based AI fight malware? CIO Insight

Freund J, Jones J (2015) Measuring and managing information risk, Kindle edn. Elsevier, Butterworth-Heinemann

Gallagher R, Bloomberg (2020) Hackers 'without conscience' demand ransom from dozens of hospitals and labs working on coronavirus. Fortune

Gandomi A, Haider M (2014) Beyond the hype: big data concepts, methods, and analytics. Int J Inf Manag 35:7

Garande P (2019) The gap within the skills gap: what does cybersecurity really need. Information Management. Retrieved from https://www.information-management.com/opinion/the-gap-within-the-skills-gap-what-does-cybersecurity-really-need

Georgescu TM, Smeureanu I (2017) Using ontologies in cybersecurity field. Inform Econ 21(3):5–15. https://doi.org/10.12948/issn14531305/21.3.2017.01

Gerhardt B, Griffin K, Klemann R (2012) Unlocking value in the fragmented world of big data analytics. Retrieved from http://www.cisco.com/web/about/ac79/docs/sp/InformationInfomediaries.pdf

Gilbert EN (1959) Random graphs. Ann Math Stat 30(4):4

Giles M (2018a) AI for cybersecurity is a hot new thing—and a dangerous gamble. MIT Technol Rev

Giles M (2018b) Cybersecurity's insidious new threat: workforce stress. MIT Technol Rev

Givre C (2017) Data science for security professionals. O'Reilly, Live Online Training

Glaser B (1978) Theoretical sensitivity. Sociology Press, Mill Valley, CA

Glaser B (1998) Doing grounded theory: issues and discussion. Sociology Press, Mill Valley, CA

Glaser BG, Strauss AL (1967) The discovery of grounded theory: strategies for qualitative research. Taylor & Francis, New York

Glassdoor (2018) U.S. cybersecurity data scientist job openings. Retrieved from https://www.glassdoor.com/Job/jobs.htm?suggestCount=0&suggestChosen=false&clickSource=searchBtn&typedKeyword=%22cyber%22+and+%22security%22+and++%22data+scientist%22&sc.keyword=%22cyber%22+and+%22security%22+and++%22data+scientist%22&locT=&locId=&jobType=

Goes PB (2014) Big data and IS research. MIS Q 38(3):iii–viii

Goffman E (1961) Encounters: two studies in the sociology of interaction. Bobbs-Merrill, Indianapolis, IN

Google Scholar (2019a) Google Scholar search for keywords "data science" and "cybersecurity"

Google Scholar (2019b) Google Scholar search for keywords "machine learning" and "cybersecurity"

Google Trends (2018) Google Trends search results

Gopalakrishnan C (2020) Security leaders consider AI-charged cyber-attacks inevitable: survey. SC Media

Gorelik A (2019) The enterprise data lake, Kindle edn. O'Reilly, Sebastopol, CA

Grahn K, Westerlund M, Pulkkis G (2017) Analytics for network security: a survey and taxonomy. In: Alsmadi IM, Karabatis G, AlEroud A (eds) Information fusion for cyber-security analytics. Springer, Cham

Grant RM (1996) Toward a knowledge-based theory of the firm. Strategic Manag J 17(Winter Special Issue):13

Grant RM (1997) The knowledge-based view of the firm: implications for management practice. Long Range Plan 30(3):4

Greenberg A (2017) How an entire nation became a test lab for Cyberwar. Wired. Retrieved from https://www.wired.com/story/russian-hackers-attack-ukraine/

Greengard S (2016) Cybersecurity gets smart. Commun ACM 59(5):2

Greenwood E (1957) Attributes of a profession. Social Work 2:11

Gregor S, Hevner AR (2013) Positioning and presenting design science research for maximum impact. MIS Q 37(2):18

Greig J (2019) Cybersecurity analysts overworked, undertrained and buckling under volume of alerts. TechRepublic

Grimmer J (2015) We are all social scientists now: how big data, machine learning, and causal inference work together. Polit Sci Politics 48:3

Groenland E (2009) ANNABEL toolbox - problem analysis. Nyenrode Business Universiteit Breukelen, Netherlands

Gupta BB, Sheng M (eds) (2019) Machine learning for computer and cyber security: principles, algorithms, and practices. Taylor & Francis, Boca Raton, FL

Guyon I, Elisseeff A (2003) An introduction to variable and feature selection. J Mach Learn Res 3(March):1157–1182

Haag S, Cummings M (2012) Management information systems for the information age, 9th edn. McGraw-Hill, New York

Haffar J (2015) Have you seen ASUM-DM? Retrieved from https://web.archive.org/web/20160308065035/https://developer.ibm.com/predictiveanalytics/2015/10/16/have-you-seen-asum-dm/

Hafner K (1998) Where wizards stay up late: the origins of the Internet. Simon & Schuster, New York

Halder S, Ozdemir S (2018) Hands-on machine learning for cybersecurity. Packt, Birmingham, UK

Harley E, Purdy S, Limiero M, Lu T, Mathews W (2018) CyGraph: big-data graph analysis for cybersecurity and mission resilience. Paper presented at the FLOCON, Tuscon, Arizona, US

Harlow LL (2005) The essence of multivariate thinking: basic themes and methods. Lawrence Erlbaum Associates, London

Harris H, Murphy S, Vaisman M (2013) Analyzing the analyzers: an introspective survey of data scientists and their work. O'Reilly Media, Sebastopol, CA

Hart C (2000) Doing a literature review. Sage, London

Hartwig F, Dearing BE (1979) Exploratory data analysis. Sage, London

Hastie T (2016) Statistical learning with big data. Retrieved from Palo Alto, CA, US: https://web.stanford.edu/~hastie/TALKS/SLBD_new.pdf

Hastie T, Tibshirani R, Friedman J (2009) The elements of statistical learning data mining, inference, and prediction, 2nd edn. Springer, New York

Hazen BT, Boone CA, Ezell JD, Jones-Farmer LA (2014) Data quality for data science, predictive analytics, and big data in supply chain management: an introduction to the problem and suggestions for research and applications. Int J Prod Econ 154:72–80. https://doi.org/10.1016/j.ijpe.2014.04.018

Hazen BT, Weigel FK, Ezell JD, Boehmke BC, Bradley RV (2017) Toward understanding outcomes associated with data quality improvement. Int J Prod Econ 193:737–747. https://doi.org/10.1016/j.ijpe.2017.08.027

Heard N, Adams N, Rubin-Delanchy P, Turcotte M (eds) (2018) Data science for cyber-security. World Scientific, London

Help Net Security (2019) More than 99% of cyberattacks rely on human interaction. Help Net Security

Henke N, Bughin J, Chui M, Manyika J, Saleh T, Wiseman B, Sethupathy G (2016) The age of analytics: competing in a data-driven world. Retrieved from https://www.mckinsey.com/business-functions/mckinsey-analytics/our-insights/the-age-of-analytics-competing-in-a-data-driven-world

Hern A (2019) Tim Berners-Lee on 30 years of the world wide web: 'We can get the web we want'. The Guardian

Heudecker N, Ronthal A (2018) How to avoid data lake failures

Hevner AR, March ST, Park J, Ram S (2004) Design science in information systems research. MIS Q 28(1):30

Hey T (2010) The big idea: the next scientific revolution. Harv Bus Rev 88(11). Retrieved from http://hbr.org/2010/11/the-big-idea-the-next-scientific-revolution/ar/1?conversatio nId=2653906

Hey T, Tansley S, Tolle KE (2012) The fourth paradigm data-intensive scientific discovery

Hiatt JM (2006) ADKAR: a model for change in business, government and our community. Prosci Research, Loveland, CO

Hillier FS, Lieberman GJ (2005) Introduction to operations research, 8th edn. McGraw-Hill, New York

Horling B, Lesser V (2005) A survey of multi-agent organizational paradigms. Knowl Eng Rev 19:35

Hostmann B (2012) Best practices in analytics: integrating analytical capabilities and process flows (G00219139)

Hsu C (2013) Information systems: the connection of people and resources for innovation. World Scientific, London

Hubbard DW, Seiersen R (2016) How to measure anything in cybersecurity risk. Wiley, Hoboken, NJ

Hull JJ, Li Y (1993) Word recognition and result interpretation using the vector space model for information retrieval. Paper presented at the 2nd annual symposium of the IEEE Computer Society on document analysis and information retrieval, Las Vegas, NV, US

IBM Security, Ponemon Institute (2019) Cost of a data breach report 2019. Retrieved from https://www.ibm.com/security/data-breach

ICS2 (2019) CISSP Domain Refresh FAQ. Retrieved from https://www.sans.org/courses/

IDC (2014) The digital universe of opportunities: rich data and the increasing value of the Internet of Things. Retrieved from https://www.emc.com/leadership/digital-universe/2014iview/executive-summary.htm

IIBA (2009) A guide to the business analysis body of knowledge (BABOK guide): version 2.0. International Institute of Business Analysis, Toronto, ON

INFORMS (2013) Certified Analytics Professional (CAP) examination candidate handbook. Retrieved from https://www.informs.org/content/download/266243/2511357/file/CAP_handbook_withforms.pdf

Institute for Advanced Analytics (2019) Graduate degree programs in analytics and data science. Retrieved from https://analytics.ncsu.edu/?page_id=4184

International Statistical Institute (2010) Declaration on professional ethics. Retrieved from Reykjavik, Iceland: https://www.isi-web.org/images/about/Declaration-EN2010.pdf

International Telecommunication Union (ITU-T) (2008) Recommendation ITU-T X.1205: overview of cybersecurity

Ishikawa K (1990) Introduction to quality control. 3A Corporation, Tokyo, Japan

ITIL (2019) ITIL: back to basics (people, process and technology). Retrieved from https://www.itilnews.com/index.php?pagename=ITIL__Back_to_basics_People_Process_and_Technology

Jacobs J, Rudis B (2014) Data-driven security: analysis, visualization and dashboards. Wiley, Hoboken, NJ

James G, Witten D, Hastie T, Tibshirani R (2013) An introduction to statistical learning with applications in R. Springer, New York

Jang-Jaccard J, Nepal S (2014) A survey of emerging threats in cybersecurity. J Comput Syst Sci 80:20

Jensen M, Meckling W (1992) Specific and general knowledge and organizational structure. In: Werin L, Wijkander H (eds) Contract economics. Blackwell, Oxford, pp 251–274

Jin X, Wah BW, Cheng X, Wang Y (2015) Significance and challenges of big data research. Big Data Res 2:59–64

Jockers ML (2014) Text analysis with R for students of literature, Kindle edn. Springer, New York

Johns Hopkins University (2014) Integrated Adaptive Cyber Defense (IACD). Retrieved from https://www.iacdautomate.org/

Johnson J (2019) Artificial intelligence & future warfare: implications for international security. Defense Secur Anal 35(2):22. https://doi.org/10.1080/14751798.2019.1600800

Jöreskog KG (1967) UMLFA, a computer program for unrestricted maximum likelihood factor analysis. Educational Testing Service, Princeton, NJ

Jöreskog KG (1977) Factor analysis by least squares and maximum likelihood methods. In: Enslein K, Ralston A, Wilf HS (eds) Statistical methods for digital computers. Wiley, New York

Jöreskog KG, Lawley DN (1968) New methods in maximum likelihood factor analysis. Br J Math Stat Psychol 21(1):9

Jowitt T (2020) Coronavirus: UK, US warn hackers are exploiting pandemic. Retrieved from https://securityboulevard.com/2020/03/coronavirus-ransomware/

Kahneman D (2011) Thinking, fast and slow. Farrar, Straus and Giroux, New York

Kahneman D, Klein G (2009) Conditions for intuitive expertise. Am Psychol 64(6):11

Kaisler SH, Espinosa JA, Armour F, Money WH (2014) Advanced analytics - issues and challenges in the global environment. Paper presented at the 47th Hawaii international conference on system science, Hawaii, US

Kambatla K, Kollias G, Kumar V, Grama A (2014) Trends in big data analytics. J Parallel Distrib Comput 74(7):2561–2573. https://doi.org/10.1016/j.jpdc.2014.01.003

Kaner M, Karni R (2004) A capability maturity model for knowledge-based decisionmaking. Inform Knowl Syst Manag 4:27

Keen PGW (1981) Information systems and organizational change. Commun ACM 24(1):9

Kelleher JD, Tierney B (2018) Data science. MIT Press, London

Keller B (2013) O.R. vs. analytics ... and now data science? Analytics 4

Kemp J (2007) In: Gartner_Hype_Cycle.svg.png (ed) Gartner research's Hype cycle. Gartner, Inc., Stamford, CT

Kent AD (2016) Cyber security data sources for dynamic network research. In: Adams N (ed) Dynamic networks and cyber-security. World Scientific, London

Khan S, Parkinson S (2018) Review into state of the art of vulnerability assessment using artificial intelligence. In: Parkinson S, Crampton A, Hill R (eds) Guide to vulnerability analysis for computer networks and systems: an artificial intelligence approach. Springer, Cham

Kilduff M, Tsai W (2003) Social networks and organizations. Sage, London

Kim J-O, Mueller CW (1978a) Factor analysis: statistical methods and practical issues. Sage, Thousand Oaks, CA

Kim J-O, Mueller CW (1978b) Introduction to factor analysis: what it is and how to do it. Sage, Thousand Oaks, CA

Kim J, Nie N, Verba S (1977) A note on factor analyzing dichotomous variables: the case of political participation. Polit Methodol 4:23

Kim H, Han Y, Kim S (2005) A curriculum design for E-commerce security. J Inf Syst Educ 16(1):9

Kirchhoff C, Upton D, Winnefeld J, Admiral JA (Producer) (2015) Defending your networks: lessons from the Pentagon [Webinar]. Retrieved from https://hbr.org/webinar/2015/10/defending-your-networks-lessons-from-the-pentagon

Kiron D, Shockley R (2011) Creating business value with analytics. MIT Sloan Manag Rev 53(1):10

Kiron D, Shockley R, Kruschwitz N, Finch G, Haydock M (2011) Analytics: the widening divide. MIT Sloan Manag Rev 53(2):1–22

Kiron D, Prentice PK, Ferguson RB (2014) The analytics mandate. MIT Sloan Manag Rev 55:1–25

Klein G, Moon B, Hoffman R (2006) Making sense of sensemaking 2: a macrocognitive model. IEEE Intell Syst 21(5):5

Klein G, Wiggins S, Dominguez CO (2010) Team sensemaking. Theor Issues Ergon Sci 11(4):16

Kleindorfer GB, Ganeshan R (1993) The philosophy of science and validation in simulation. Paper presented at the 1993 Winter simulation conference

Kleindorfer GB, O'Neill L, Ganeshan R (1998) Validation in simulation: various positions in the philosophy of science. Manag Sci 44(8):12

Kohnfelder L, Garg P (1999) The treats to our products. Retrieved from Redmond, Washington, US: https://adam.shostack.org/microsoft/The-Threats-To-Our-Products.docx

Kolini F, Janczewski L (2017) Cluster and topic modelling: a new approach for analysis of National cyber security strategies. Paper presented at the PACIS, Malaysia

Kotter JP, Cohen DS (2002) The heart of change. Harvard Business School Press, Boston, MA

Kreuter F, Peng RD (2014) Extracting information from big data: issues of measurement, inference and linkage. In: Lane J, Stodden V, Bender S, Nissenbaum H (eds) Privacy, big data, and the public good: frameworks for engagement. Cambridge University Press, Cambridge, UK

Krippendorff K (2019) Content analysis: an introduction to its methodology, 4th edn. Sage, London

Kuhn M (2016) Journals on computer security. University of Cambridge Department of Computer Science and Technology. Retrieved April 19, 2019 from https://www.cl.cam.ac.uk/research/security/journals.html

Landauer TK, Foltz PW, Laham D (1998) An introduction to latent semantic analysis. Discourse Process 25:25

Landwehr CE (2012) Cybersecurity: from engineering to science. The Next Wave 19(2):4. Retrieved from https://www.nsa.gov/Portals/70/documents/resources/everyone/digital-media-center/publications/the-next-wave/TNW-19-2.pdf

Laramore J (2017) Feature engineering and data preparation for analytics course notes (DMDP41_001). SAS Institute, Inc., Cary, NC

Latour B (2005) Reassembling the social: an introduction to actor-network-theory. Oxford University Press, Oxford

Laudon KC, Laudon JP (2017) Management information systems: managing the digital firm, 15th edn. Pearson, London

LaValle S, Hopkins MS, Lesser E, Shockley R, Kruschwitz N (2010) Analytics: the new path to value. MIT Sloan Manag Rev 22

LaValle S, Lesser E, Shockley R, Hopkins MS, Kruschwitz N (2011) Big data, analytics and the path from insights to value. MIT Sloan Manag Rev 52(2):13

Law AM (2007) Simulation modeling & anlaysis, 4th edn. McGraw-Hill, London

Law AM, Kelton DW (2000) Simulation modeling and analysis, 3rd edn. McGraw-Hill, New York

Lawley DN, Maxwell AE (1971) Factor analysis as a statistical method. Butterworths, London

Lazer D, Kennedy R, King G, Vespignani A (2014) The parable of Google flu: traps in big data analysis. Science 343(6176):3. https://doi.org/10.1126/science.1248506

Leavitt HJ (1964) Applied organization change in industry: structural, technical and human approaches. In: Cooper W, Leavitt HJ, Shelly MWI (eds) New perspectives in organization research. Wiley, New York, pp 55–71

Lee N, Lings I (2008) Doing business research: a guide to theory and practice. Sage, Thousand Oaks, CA

Lehto M (2015) Phenomena in the cyber world. In: Lehto M, Neittaanmaki P (eds) Intelligent systems, control and automation: science and engineering. Springer, London

Lehto M, Neittaanmaki P (eds) (2015) Cyber security: analytics, technology and automation, vol 78. Springer, London

Leiner BM, Cerf VG, Clark DD, Kahn RE, Kleinrock L, Lynch DC et al (1999) A brief history of the internet. ACM SIGCOMM Comput Commun Rev 39(5):22–31

Leodolter J (2013) Data mining and business analytics with R, Kindle edn. Wiley, Hoboken, NJ

Leuprecht C, Skillicorn DB, Tait VE (2016) Beyond the castle model of cyber-risk and cyber-security. Gov Inf Q 33(2):7

Levy Y, Ellis TJ (2006) A systems approach to conduct an effective literature review in support of information systems research. Inform Sci J 9:32

Lewin R (1993) Complexity: life at the edge of chaos. Phoenix, Chicago

Liberatore MJ, Luo W (2010) The analytics movement: implications for operations research. Interfaces 40(4):11

Lichfield G (2020) We're not going back to normal. MIT Technol Rev. Retrieved from https://www.technologyreview.com/2020/03/17/905264/coronavirus-pandemic-social-distancing-18-months/

Lim S, Saldanha TJV, Malladi S, Melville NP (2013) Theories used in information systems research: insights from complex network analysis. J Inf Technol Theory Appl 14(2):41

Linden A (2015) Hype cycle for advanced analytics and data science (G00277762). Retrieved from Gartner.com: https://www.gartner.com/en/documents/3087721

LinkedIn (2018) LinkedIn search results focused - Job Title refinement – 'cyber' + ('analytics' or 'data scientist'). Retrieved from https://www.linkedin.com/

LinkedIn (2019) About LinkedIn. Retrieved from https://about.linkedin.com

Liu J, Li J, Li W, Wu J (2016) Review article: Rethinking big data: a review on the data quality and usage issues. ISPRS J Photogramm Remote Sens 115:134–142. https://doi.org/10.1016/j.isprsjprs.2015.11.006

Loaiza FL, Birdwell JD, Kennedy GL, Visser D (2019) Utility of artificial intelligence and machine learning in cybersecurity (NS D-10694). Retrieved from https://www.ida.org/-/media/feature/publications/u/ut/utility-of-artificial-intelligence-and-machine-learning-in-cybersecurity/d-10694.ashx

Lockheed Martin (2012) The cyber kill chain. Retrieved from https://www.lockheedmartin.com/en-us/capabilities/cyber/cyber-kill-chain.html

Lohr S (2011) The internet gets physical, news analysis. The New York Times. Retrieved from https://www.nytimes.com/2011/12/18/sunday-review/the-internet-gets-physical.html

Longstaff T (2012) Barriers to achieving a science of cybersecurity. The Next Wave 19(4):2. Retrieved from https://www.nsa.gov/Portals/70/documents/resources/everyone/digital-media-center/publications/the-next-wave/TNW-19-2.pdf

Looijen M, Delen G (1992) Beheer van Informatievoorziening. Lansa, Amsterdam, Netherlands

Lorenzo-Seva U, Ferrando PJ (2012) TETRA-COM: a comprehensive SPSS program for estimating the tetrachoric correlation. Behav Res Methods 44(4):5

Lorenzo-Seva U, Ferrando PJ (2019a) Factor. Departament de Psicologia - Universitat Rovira i Virgili, Tarragona, Spain. Retrieved from http://psico.fcep.urv.es/utilitats/factor/index.html

Lorenzo-Seva U, Ferrando PJ (2019b) Not positive definite correlation matricies in exploratory item factor analysis: causes, consequences and a proposed solution. Struct Equat Model 28(1):138–147

Loshin D (2013) Big data analytics. Elsevier Science, Boston, MA

Loukides M (2011) What is data science? O'Reilly, Sebastopol, CA

Lovallo D, Sibony O (2010a) The case for behavioral strategy. McKinsey Q 1:16. Retrieved from https://www.mckinsey.com/business-functions/strategy-and-corporate-finance/our-insights/the-case-for-behavioral-strategy

Lovallo D, Sibony O (2010b) The case for behavioral strategy. McKinsey Q 16. Retrieved from https://www.mckinsey.com/business-functions/strategy-and-corporate-finance/our-insights/the-case-for-behavioral-strategy

Lowry PB, Willison R, Dinev T (2018) Why security and privacy research lies at the centre of the information systems (IS) artefact: proposing a bold research agenda. Eur J Inf Syst 26(6)

MacKay RJ, Oldford RW (2000) Scientific method, statistical method and the speed of light. Stat Sci 15(3):24

Magoulas R, King J (2016) 2016 Data science salary survey. O'Reilly, Sebastopol, CA

Maheshwari A (2015) Data analytics made accessible, Kindle edn. Amazon Media

Mahmood T, Afzal U (2013) Security analytics: big data analytics for cybersecurity: a review of trends, techniques and tools. Paper presented at the national conference on information assurance (NCIA), Rawalpindi, Pakistan

Maloof MA (ed) (2006) Machine learning and data mining for computer security: methods and applications. Springer, London

Manyika J, Chui M, Brown B, Bughin J, Dobbs R, Roxburgh C, Hung Byers A (2011) Big data: the next frontier for innovation, competition, and productivity. Retrieved from https://www.ebsco.com/

March S, Smith G (1995) Design and natural science research on information technology. Decis Support Syst 15:15

Marchette DJ (2001) Computer intrusion detection and network monitoring. Springer, New York

Mason J (1994) Linking qualitative and quantitative data analysis. In: Bryman A, Burgess RG (eds) Analyzing qualitative data. Routledge, London

Mastrogiacomo R (2017) The conflict between data science and cybersecurity. Information--management.com 1-1

Maxion R (2012) Making experiments dependable. The Next Wave 19(2):10. Retrieved from https://www.nsa.gov/Portals/70/documents/resources/everyone/digital-media-center/publications/the-next-wave/TNW-19-2.pdf

Maxion RA, Longstaff TA, McHugh J (2010) Why is there no science in cybersecurity science? Paper presented at the NSPW 2010, Concord, Massachusetts, US. https://www.nspw.org/2010/local

McDermott T, Rouse W, Goodman S, Loper M (2013) Multi-level modeling of complex socio-technical systems. Proc Comput Sci 16:9

McMullan T (2015) What does the panopticon mean the age of digital surveillance? The Guardian. Retrieved from https://www.theguardian.com/technology/2015/jul/23/panopticon-digital-surveillance-jeremy-bentham

McNeely C, Hahm J (2014) The big (data) bang: policy, prospects, and challenges. Rev Policy Res 31(4):6

McNulty E (2014) Understanding big data: the seven V's. Retrieved from https://dataconomy.com/2014/05/seven-vs-big-data/

Mehrotra KG, Mohan CK, Huang H (2018) Anomaly detection principles and algorithms. Springer, Cham

Merchant KA, van der Stede WA (2003) Management control systems: performance measurement, evaluation and incentives. Prentice Hall, London

Metcalf L, Casey W (eds) (2016) Cybersecurity and applied mathematics. Imperial College Press, London

Meushaw R, Landwehr CE (2012) NSA initiatives in cybersecurity science. The Next Wave 19(4):6. Retrieved from https://www.nsa.gov/Portals/70/documents/resources/everyone/digital-media-center/publications/the-next-wave/TNW-19-2.pdf

Microsoft (2017) What is the team data science process? Retrieved from https://docs.microsoft.com/en-us/azure/machine-learning/team-data-science-process/overview

Miles MB, Huberman A (1984) Qualitative data analysis: a sourcebook of new methods. Sage, Thousand Oaks, CA

Miller JH, Page SE (2007) Complex adaptive systems: an introduction to computational models of social life. Princeton University Press, Princeton, NJ

Miner G, Elder J, Fast A, Hill T, Nisbet R, Delen D (2012) Practical text mining and statistical analysis for non-structured text data applications, Kindle edn. Elsevier, Amsterdam

MITRE (2018) Adversarial tactics, techniques & common knowledge. Retrieved from https://attack.mitre.org/wiki/Main_Page

Mongeau SA (2013a) Business analytics model risk: framing model risk. Retrieved from https://sctr7.com/2013/06/13/business-analytics-model-risk-part-0-of-5-framing-model-risk-the-complexity-genie-and-challenge-of-deciding-on-decision-models/

Mongeau SA (2013b) Business analytics model risk: model scoping and complexity. Retrieved from https://sctr7.com/2013/06/13/business-analytics-model-risk-part-0-of-5-framing-model-risk-the-complexity-genie-and-challenge-of-deciding-on-decision-models/

Mongeau SA (2014) Emerging trends in data analytics. Retrieved from https://sctr7.com/2014/07/09/twelve-emerging-trends-in-data-analytics-part-1-of-4/

Mongeau SA (2015) Why is 'data scientist' a contentious title? Retrieved from https://sctr7.com/2015/09/30/why-is-data-scientist-a-contentious-title/

Mongeau SA (2019a) Cybersecurity research datasets. Retrieved from https://sctr7.com/2019/11/18/cybersecurity-research-datasets/

Mongeau SA (2019b) Wait… is data science even a thing? Retrieved from https://sctr7.com/2019/04/28/wait-is-data-science-even-a-thing/

Mongeau SA (2020a) Cybersecurity data science (CSDS) corpus. Retrieved from https://www.linkedin.com/pulse/cybersecurity-data-science-csds-corpus-scott-allen-mongeau/

Mongeau SA (Producer) (2020b) FloCon 2020: cybersecurity data science (CSDS): best practices in an emerging profession. 2020 FloCon Conference - Savannah, GA. Retrieved from http://www.sark7.com/docs/CSDS_FloCon_2020.01.09.pdf

Mongeau SA (Producer) (2020c) INFORMS security conference 2020: cybersecurity data science (CSDS): best practices in an emerging profession. 2020 INFORMS security conference. Retrieved from http://www.sark7.com/docs/CSDS_INFORMS_Security_Conference_2020.02.11.pdf

Morabito V (2015) Big data and analytics. Springer, Cham

Morgan MS (2012) The world in the model. Cambridge University Press, Cambridge, UK

Morgan S (2017a) 2017 Official annual cybercrime report. Retrieved from https://cybersecurityventures.com/hackerpocalypse-cybercrime-report-2016/

Morgan S (2017b) Cybersecurity jobs report. Retrieved from https://cybersecurityventures.com/jobs/

Morgan S (2017c) Ransomware damage report. Retrieved from https://cybersecurityventures.com/ransomware-damage-report-2017-part-2/

Morgan S (2018a) 2018 Cybersecurity market report. Retrieved from https://cybersecurityventures.com/cybersecurity-market-report/

Morgan S (2018b) 2018 Directory of M.S. in cybersecurity programs at universities in the US. Retrieved from https://cybersecurityventures.com/cybersecurity-university-masters-degree-programs/

Morgan S (2019a) 50 Cybersecurity titles that every job seeker should know about. Retrieved from https://cybersecurityventures.com/50-cybersecurity-titles-that-every-job-seeker-should-know-about/

Morgan S (2019b) Security awareness training for employees, and certification training programs. Retrieved from https://cybersecurityventures.com/cybersecurity-education/

Morrison M, Morgan MS (1999) Models as mediators: perspectives on natural and social science. In: Morrison M, Morgan MS (eds) Models as mediators: perspectives on natural and social science. Cambridge University Press, Cambridge

Mortelmans D (2019) Qualitative analysis: a grounded theory approach. Paper presented at the qualitative analysis PhD seminar, Breukelen, Netherlands

Morton A (2001) Kinds of models. Wiley, Hoboken, NJ

Muegge S, Craigen D (2015) A design science approach to constructing critical infrastructure and communicating cybersecurity risks. Technol Innov Manag Rev 5(6)

Munoz-Gonzalez L, Lupu EC (2018) The security of machine learning systems. In: Sikos LF (ed) AI in cybersecurity. Springer, Cham

Muthén B, Kaplan D (1992) A comparison of some methodologies for the factor analysis of non-normal Likert variables: a note on the size of the model. Br J Math Stat Psychol 45:11

Muzio D, Kirkpatrick I (2011) Introduction: professions and organizations – a conceptual framework. Curr Sociol 59(4):17

Myers MD, Newman M (2007) The qualitative interview in IS research: examining the craft. Inf Organ 17:24

Nakamura GC, Gerth JA, Bridges R, Goodall JR (2015) Developing an ontology for cyber security knowledge graphs. Paper presented at the 10th annual cyber and information security research conference, Oak Ridge, TN, US

Neil J (2019) Graph for security workshop: statistics graphs finding bad guys. [Video of workshop presentation]: YouTube

Nelson GS (2018) The analytics lifecycle toolkit: a practical guide for an effective analytics capability. Wiley, Hoboken, NJ

New J (2019) AI needs better data, not just more data. Retrieved from https://www.datainnovation.org/2019/03/ai-needs-better-data-not-just-more-data/

NIST (2017) National Initiative for Cybersecurity Education (NICE) cybersecurity workforce framework. (800-181). Retrieved from https://www.nist.gov/cyberframework

NIST (2018) NIST cybersecurity framework. Retrieved from https://www.nist.gov/cyberframework

NSA (2019) NSA academic criteria for cybersecurity degree programs. Retrieved from https://www.iad.gov/

Nutt PC (2002) Why decisions fail: avoiding the blunders and traps that lead to debacles. Berrett-Koehler, San Francisco, CA, USA

O'Neil C (2013) On being a data skeptic. O'Reilly, Sebastapol, CA

O'Neil C (2016) Weapons of math destruction. Crown, New York

O'Rourke N, Hatcher L (2013) A step-by-step approach to using SAS for factor analysis and structural equation modeling. SAS Press, Cary, NC

OASIS (2019) OASIS cyber threat intelligence (CTI) TC. Retrieved from https://www.oasis-open.org/committees/

Obrsta L, Chaseb P, Markeloffa R (2012) Developing an ontology of the cyber security domain. Paper presented at the semantic technology for intelligence, defense, and security: semantic technologies in cyber security, Fairfax, VA, US

Offer GJ, Contestabile M, Howey DA, Clague R, Brandon NP (2011) Techno-economic and behavioural analysis of battery electric, hydrogen fuel cell and hybrid vehicles in a future sustainable road transport system in the UK. Energy Policy 39(4):11

Office of the Director of National Intelligence (ODNI) (2018) Cyber threat framework. Retrieved from https://www.dni.gov/index.php/cyber-threat-framework

Oltsik J (2018) Goodbye SIEM, Hello SOAPA. Cybersecurity Snippets. Retrieved from https://www.csoonline.com/article/3145408/goodbye-siem-hello-soapa.html

Oltsik J (2019a) ESG data point of the week. ESG Global

Oltsik J (2019b) Looking for answers at black hat 2019: 5 important cybersecurity issues. CSO

Onwuegbuzie AJ, Leech NL, Collins KMT (2012) Qualitative analysis techniques for the review of the literature. Qual Rep 17(56):28

Open Data Science (2018) A technical look at how criminals use AI. Medium

Oreskes N, Belitz K (2001) Philosophical issues in model assessment. In: Anderson MG, Bates PD (eds) Model validation: perspectives in hydrological science. Wiley, Hoboken, NJ

Osbourne JW, Banjanovic ES (2016) Exploratory factor analysis with SAS. SAS Press, Cary, NC

Oxford Dictionary (2019) Definition of cybersphere. Retrieved from https://en.oxforddictionaries.com/definition/cybersphere

Palmer D (2019) Cybersecurity staff burnout risks leaving organizations vulnerable to cyberattacks. ZDNet

Parisi A (2019) Hands-on artificial intelligence for cybersecurity. Packt, Birmingham, UK

Park H, Cho S, Kwon H-C (2009) Cyber forensics ontology for cyber criminal investigation. Paper presented at the e-Forensics, Adelaide, Australia

Parkinson S, Crampton A, Hill R (eds) (2018) Guide to vulnerability analysis for computer networks and systems: an artificial intelligence approach. Springer, Cham

Pavlovic D (2012) On bugs and elephants: mining for science of security. The Next Wave 19(2):7. Retrieved from https://www.nsa.gov/Portals/70/documents/resources/everyone/digital-media-center/publications/the-next-wave/TNW-19-2.pdf

Pearlson KA, Saunders CS, Galletta DF (2016) Managing & using information systems: a strategic approach, 6th edn. Wiley, Hoboken, NJ

Pein C (2017) Live work work work die. Metropolitan Books, New York

Penrose E (1959) The theory of the growth of the firm. Wiley, New York

Pfleeger SL, Caputo DD (2012) Leveraging behavioral science to mitigate cyber security. Comput Secur 31:14

Phillips KA, Morrison KR, Andersen R, Aday LA (1998) Understanding the context of healthcare utilization: assessing environmental and provider-related variables in the behavioral model of utilization. Health Serv Res 33:25

Pidd M (2004) Computer simulation in management science. Wiley, Hoboken, NJ

Pigliucci M (2009) The end of theory in science? EMBO Rep 10(6):1. https://doi.org/10.1038/embor.2009.111

Podeswa H (2009) The business analyst's handbook. Cenage Learning, Boston, MA

Polkinghorne M (2015) Two blue hats student guide to analysing qualitative interview data using recursive abstraction, Kindle Ebook edn. Two Blue Hats Ltd., Dorset, UK

Ponemon Institute (2017) When seconds count: how security analytics improves cybersecurity defenses. Retrieved from https://www.sas.com/content/dam/SAS/en_us/doc/research2/ponemon-how-security-analytics-improves-cybersecurity-defenses-108679.pdf

Ponemon Institute (2018) Managing the risk of post-breach or "resident" attacks. Retrieved from https://www.illusivenetworks.com/resources/2018-ponemon-institute-research-report

Popov V (2003) Social network analysis in decision making: a literature review. PSIRU University of Greenwich, Greenwich, UK

Poppensieker T, Riemenschnitter R (2018) A new posture for cybersecurity in a networked world. Retrieved from https://www.mckinsey.com/business-functions/risk/our-insights/a-new-posture-for-cybersecurity-in-a-networked-world

Popper K (1963) Conjectures and refutations: the growth of scientific knowledge. Routledge, New York

Popper K (1992) The logic of scientific discovery. Routledge, London

Porter M (2008) The five competitive forces that shape strategy. Harv Bus Rev

Power DJ (2007) A brief history of decision support systems. Retrieved from http://DSSResources.COM/history/dsshistory.html

Power DJ (2008) Understanding data-driven decision support systems. Inf Syst Manag 25(2):6

Prell C (2012) Social network analysis: history, theory & methodology. Sage, London

Press G (2013a) A very short history of data science. Forbes. Retrieved from https://www.forbes.com/sites/gilpress/2013/05/28/a-very-short-history-of-data-science/

Press G (2013b) Data science: what's the half-life of a buzzword? Forbes

Provost F, Fawcett T (2013) Data science for business. O'Reilly, Sebastapol, CA

Puri R (2003) Bots & Botnet: an overview. Retrieved from https://www.sans.org/reading-room/whitepapers/malicious/paper/1299

Radware (2019) Understanding the Darknet and its impact on cybersecurity. Radware Blog. Retrieved from https://blog.radware.com/security/2019/02/understanding-the-darknet-and-its-impact-on-cybersecurity/

Ramirez JM (2017) Some criminal aspects of cybersecurity. In: Ramirez JM, Garcia-Segura LA (eds) Cyberspace: risks and benefits for society, security and development. Springer, New York

Razvan B, Ana-Ramona L, Catalin B, Tiberiu Marian G (2017) Sustaining employability: a process for introducing cloud computing, big data, social networks, mobile programming and cybersecurity into academic curricula. Sustainability 9(12):2235. https://doi.org/10.3390/su9122235

Refsdal A, Solhaug B, Stolen K (2015) Cyber-risk management. Springer, London

Reiner S (2020) CoronaVirus ransomware. Security Boulevard

Reinsel D, Rydning J (2018) Worldwide global storagesphere forecast, 2019–2023: meaningful shifts in how and where data is stored. Retrieved from https://www.emc.com/leadership/digital-universe/2014iview/executive-summary.htm

Rennie KM (1997) Exploratory and confirmatory rotation strategies in exploratory factor analysis. Paper presented at the annual meeting of the southwest educational research association, Austin, TX, US

Richards L, Morse JM (2007) README first for a User's guide to qualitative methods, 2nd edn. Sage, London

Riley S (2019a) Thoughts about AI in cybersecurity. IACD

Riley S (2019b) Thoughts on A.I. in cybersecurity. LinkedIn Pulse

Robinson S (2004) Simulation: the practice of model development and use. Wiley, Hoboken, NJ

Rogova GL (2019) Information quality in fusion-driven human-machine environments. In: Bosse E, Rogova G (eds) Information quality in information fusion and decision making, Kindle edn. Springer, New York

Rohanizadeh SS, Moghadam MB (2009) A proposed data mining methodology and its application to industrial procedures. J Ind Eng 4:13

Rollins JB (2015) Foundational methodology for data science. Retrieved from https://www.ibm.com/downloads/cas/WKK9DX51

Rose R (2016) Defining analytics: a conceptual framework. ORMS Today 43(3):36–41

Rosenzweig P (2013) Thinking about cybersecurity: from cyber crime to cyber warfare. The Great Courses

Ross A (2019) ML and AI in cyber security: real opportunities overshadowed by hype. Information Age

Roy R, Seitz B (2018) How to build a data-first culture for a digital transformation. Digital McKinsey

Ruvinsky A, Walker L, Abdullah W, Seale M, Bond WG, Leonard L et al (2019) An epistemological model for a data analysis process in support of verification and validation. In: Bosse E, Rogova G (eds) Information quality in information fusion and decision making, Kindle edn. Springer, New York

Sahinoglu M (2016) Cyber-risk informatics: engineering evaluation with data science. Wiley, Hoboken, NJ

Saldana J (2016) The coding manual for qualitative researchers. Sage, London

Sallam RL, Cearley DW (2012) Advanced analytics: predictive, collaborative and pervasive (G00230321). Gartner, Stamford, CT

Salmon M (2019) Artificial intelligence 'winter-time' continues in business school journals. LinkedIn

Samek W, Montavon G, Vedaldi A, Hansen LK, Müller KR (eds) (2019) Explainable AI: interpreting, explaining and visualizing deep learning. Springer, Cham

Sammon D, Nagle T, O'Raghallaigh P (2010) Assessing the theoretical strength within the literature review process: a tool for doctoral researchers. In: Respício A, Adam F, Phillips-Wren G, Teixeira C, Telhada J (eds) Bridging the socio-technical gap in DSS: challenges for the next decade, vol 212. IOS Press, Amsterdam, p 11

Samtani S, Chinn R, Chen H (2015) Exploring hacker assets in underground forums. Paper presented at the IEEE international conference security informatics (ISI)

Samuel S (2019) It's disturbingly easy to trick AI into doing something deadly. Vox

Sanders J (2019) Data breaches increased 54% in 2019 so far. TechRepubliic

Sanders C, Smith J (2013) Applied network security monitoring: collection, detection, and analysis (Bianco DJ, Ed). Elsevier, Amsterdam, Netherlands

SANS Institute (2015) 2015 Analytics and intelligence survey. Retrieved from https://www.sas.com/content/dam/SAS/en_us/doc/research2/sans-analytics-intelligence-survey-108031.pdf

SANS Institute (2016) Using analytics to predict future attacks and breaches. Retrieved from https://www.sas.com/content/dam/SAS/en_us/doc/whitepaper2/sans-using-analytics-to-predict-future-attacks-breaches-108130.pdf

SANS Institute (2019a) Cyber security courses. Retrieved from https://www.sans.org/courses/

SANS Institute (2019b) Security certification: GMON. Retrieved from https://www.giac.org/certification/continuous-monitoring-certification-gmon

Sapp C (2020) Assessing DevOps in artificial intelligence initiatives. Retrieved from https://www.gartner.com/en/documents/3981214

Sargent RG (2000) Verification, validation and accreditation of simulation models. Paper presented at the simulation conference, Orlando, FL

Sargent RG (2007) Verification and validation of simulation models. Paper presented at the 2007 winter simulation conference

Sargent RG (2013) Verification and validation of simulation models. J Simula 7:12

Sarkar C (2016) Beyond big data: from analytics to cognition – an interview with Thomas Davenport. Retrieved from http://www.marketingjournal.org/beyond-big-data-from-analytics-to-cognition-an-interview-with-thomas-davenport/

Sarker IH, Kayes ASM, Badsha S, Alqahtani H, Watters P, Ng A (2020) Cybersecurity data science: an overview from machine learning perspective. J Big Data 7(41)

SAS Institute (2018) Managing the analytics life cycle for decisions at scale: how to go from data to decisions as quickly as possible. Retrieved from SAS Institute Inc.: https://www.sas.com/content/dam/SAS/en_us/doc/whitepaper1/manage-analytical-life-cycle-continuous-innovation-106179.pdf

SAS Institute (2019a) Operationalizing analytics delivery approach. SAS Institute, Cary, NC

SAS Institute (2019b) The analytics life cycle. Retrieved from https://www.sas.com/en_us/software/platform/analytics-life-cycle.html

SAS Institute Inc (2016) Managing the analytics life cycle for decisions at scale. Retrieved from https://www.sas.com/content/dam/SAS/en_us/doc/whitepaper1/manage-analytical-life-cycle-continuous-innovation-106179.pdf

Saunders M, Lewis P, Thornhill A (2009) Research methods for business students, 5th edn. Prentice Hall, Harlow

Savas O, Deng J (eds) (2017) Big data analytics in cybersecurity. CRC Press, Boca Raton, FL

Saxe J, Sanders H (2018) Malware data science: attack detection and attribution. No Starch Press, Inc., San Francisco

Schatz D, Bashroush R, Wall J (2017) Towards a more representative definition of cyber security. J Digit Forensic Secur Law 12(2)

Schneider FB (2012) Blueprint for a science of cybersecurity. The Next Wave 19(2):11. Retrieved from https://www.nsa.gov/Portals/70/documents/resources/everyone/digital-media-center/publications/the-next-wave/TNW-19-2.pdf

Schneier B (2018) Click here to kill everybody: security and survival in a hyper-connected world. W. W. Norton & Company Ltd., London

Schutt R, O'Neil C (2014) Doing data science, Kindle edn. O'Reilly Media, Inc., Sebastopol, CA

Schwartz MJ (2018) Criminals' cryptocurrency addiction continues. Bank Info Security. Retrieved from https://www.bankinfosecurity.com/criminals-cryptocurrency-addiction-continues-a-11594

Schwartz MJ (2019a) How cybercriminals continue to innovate. Bank Info Security

Schwartz MJ (2019b) Malware most foul: Emotet, Trickbot, Cryptocurrency Miners. Bank Info Security

Security Brief Magazine (2016) Analyze this! Who's implementing security analytics now? Security Brief Magazine. Retrieved from https://www.sas.com/en_th/whitepapers/analyze-this-108217.html

Shanks G, Sharma R, Seddon P, Reynolds P (2010) The impact of strategy and maturity on business analytics and firm performance: a review and research agenda. Paper presented at the ACIS 2010, Brisbane, Australia

Shannon RE (1975) Systems simulation: the art and science. Prentice-Hall, Englewood Cliffs, NJ

Sharda R, Delen D, Turban E (2014) Business intelligence: a managerial perspective on analytics. Pearson Education Limited, London

Shearer C (2000) The CRISP-DM model: the new blueprint for data mining. J Data Warehousing 5(4):5

Shmueli G (2010) To explain or to predict? Stat Sci 25(3):289–310

Shortt SED (1983) Physicians, science, and status: issues in the professionalization of Anglo-American medicine in the nineteenth century. Med Hist 27:17

Shostack A (2012) The evolution of information security. The Next Wave 19(2):6. Retrieved from https://www.nsa.gov/Portals/70/documents/resources/everyone/digital-media-center/publications/the-next-wave/TNW-19-2.pdf

Shostack A (2014) Threat modeling: designing for security. Wiley, Indiana

Shuai B, Li H, Li M, Zhang Q, Tang C (2013) Automatic classification for vulnerability based on machine learning. Paper presented at the IEEE international conference on information and automation (ICIA)

Sikos LF (2018a) OWL ontologies in cybersecurity: conceptual modeling of cyber-knowledge. In: Sikos LF (ed) AI in cybersecurity. Springer, Cham, p 17

Sikos LF (ed) (2018b) AI in cybersecurity, vol 151. Springer, Cham

Sikos LF, Choo K-KR (eds) (2020) Data science in cybersecurity and cyberthreat intelligence. Springer, Cham

Sikos LF, Philip D, Howard C, Voigt S, Strumptner M, Mayer W (2018) Knowledge representation of network semantics for reasoning-powered cyber-situational awareness. In: Sikos LF (ed) AI in cybersecurity. Springer, Cham, p 17

Simon HA (1960) The new science of management decision. Harper, New York

Simon HA (1996) Sciences of the artificial, 3rd edn. MIT Press, Cambridge, MA

Singh S (1999) The code book: the science and secrecy from ancient Egypt to quantum cryptography. Anchor Books, New York

Singh S, Singh N (2012) Big data analytics. Paper presented at the international conference on communication, information & computing technology, Mumbai, India

Sivarajah U, Kamal MM, Irani Z, Weerakkody V (2017) Critical analysis of big data challenges and analytical methods. J Bus Res 70:263–286. https://doi.org/10.1016/j.jbusres.2016.08.001

Skiena SS (2017) The data science design manual. Springer, Cham

Skyrius R, Kazakeviciene G, Bujauskas V (2013) The relationship between management decision support and business intelligence: developing awareness. In: Rocha A, Correia A, Wilson T, Stroetmann K (eds) Advances in information systems and technologies, vol 206. Springer, Berlin, DE

Smith G (2018) The AI dellusion. Oxford University Press, Oxford, UK

Sodhi MS, Tang CS (2010) Introduction: a long view of research and practice in operations research and management science. In: Hillier FS (ed) International series in operations research & management science, vol 148. Springer, New York, p 297

Sousa KJ, Oz E (2014) Management information systems, 7th edn. Cengage Learning, London

Spadafora A (2019) US utility firms hit by state-sponsored spear-phishing attack. TechRadar

Spiegelhalter D (2019) The art of statistics. Basic Books, New York

Spring JM, Fallon J, Galyardt A, Horneman A, Metcalf L, Stoner E (2019) Machine learning in cybersecurity: a guide (REF 270912). Retrieved from Pittsburgh, Pennsylvania, US: https://resources.sei.cmu.edu/asset_files/TechnicalReport/2019_005_001_633597.pdf

Stallings W (2019) Effective cybersecurity: a guide to using best practices and standards. Addison-Wesley, London

Stamp M (2017) Introduction to machine learning with applications in information security. CRC Press, London

Stamp M (2018) A survey of machine learning algorithms and their application in information security. In: Parkinson S, Crampton A, Hill R (eds) Guide to vulnerability analysis for computer networks and systems: an artificial intelligence approach. Springer, New York, p 384

Strauss AL (1987) Qualitative analysis for social scientists. Cambridge University Press, Cambridge, UK

Strawn G (2016) Data scientist. IT Professional 3. Retrieved from https://ieeexplore.ieee.org/document/7478500

Strout N (2019) The 3 major security threats to AI. C4ISRNET

Subrahmanian VS, Ovelgonne M, Dumitras T, Prakash A (2015) The global cyber-vulnerability report. Springer, Cham

Svolba G (2017) Applying data science: business case studies using SAS. SAS Institute Inc., Cary, NC

Taleb NN (2014) Antifragile: things that gain from disorder, Kindle edn. Random House, New York

Tardiff MF, Bonheyo GT, Cort KA, Edgar TW, Hess NJ, Hutton III WJ, et al (2016) Applying the scientific method to cybersecurity research. Paper presented at the IEEE symposium on technologies for homeland security (HST), Waltham, MA, US

Tatnall A (2003) Actor-network theory as a socio-technical approach to information systems research. In: Clarke S, Coakes E, Hunter MG, Wenn A (eds) Socio-technical and human cognition elements of information systems. Information Science Publishing, London

The Economist (2018) The joys of data hygine: Europe's tough new data-protection law. The Economist

The Economist (2019a) Internet policy: first, do no harms. The Economist

The Economist (2019b) Spooky. The Economist. Retrieved from https://www.economist.com/business/2019/12/12/offering-software-for-snooping-to-governments-is-a-booming-business

The Economist (2019c) Western firms should not sell spyware to tyrants. The Economist. Retrieved from https://www.economist.com/leaders/2019/12/12/western-firms-should-not-sell-spyware-to-tyrants

Thomason M (2018) Market forecast: worldwide software license, maintenance, and subscription forecast, 2018–2022 (US40752216)

Trayan L (2019) A glimpse into the future: science fiction or fact? Paper presented at the ISC2 secure summit EMEA, Den Haag, Netherlands

Trustwave (2015) 2015 Trustwave global security report. Retrieved from https://www.trustwave.com/Company/Newsroom/News/New-Trustwave-Report-Reveals-Criminals-Receive-1,425-Percent-Return-on-Investment-from-Malware-Attacks/

Tsukerman E (Producer) (2019a) Cybersecurity data science. [Online Course]. Retrieved from https://www.udemy.com/cybersecurity-data-science/

Tsukerman E (2019b) Machine learning for cybersecurity cookbook. Packt, Birmingham, UK

Tukey JW (1962) The future of data analysis. Ann Math Stat 33(1):67

Tukey JW (1977) Exploratory data analysis. Addison-Wesley, Reading, MA

Tulley JR (2008) Is there techne in my logos? On the origins and evolution of the ideographic term—technology. Int J Technol Knowl Soc 4(1):11

Tuor A, Kaplan S, Hutchinson B, Nichols N, Robinson S (2017) Deep learning for unsupervised insider threat detection in structured cybersecurity data streams

Tushman ML, Fombrun C (1979) Social network analysis for organizations. Acad Manag Rev 4(4):12

UBM (2016) Close the detection deficit with security analytics

Vacca JR (ed) (2017) Cloud computing security: foundations and challenges, Kindle edn. CRC Press, Boca Raton, FL

Vaishnavi VK, Kuechler W (2015) Design science research methods and patterns. CRC Press, Boca Raton, FL

van den Berg J, van Zoggel J, Snels M, van Leeuwen M, Boeke S, van de Koppen L, et al (2014) On (the emergence of) cyber security science and its challenges for cyber security education, Talin, ES

van den Hoonaard WC (1996) Working with sensitizing concepts: analytical field research. Sage, London

van der Aalst W (2016) Process mining: data science in action, 2nd edn. Springer, London

Van der Krogt T (1981) Professionalisering En Collectieve Macht: Een Conceptueel Kader. Vuga, 's-Gravenhage, Netherlands

Van der Krogt T (2015) Professionals and their work. Retrieved from https://profqual.wordpress. com/2015/07/21/theoretical-perspectives-on-professionalization-1-trait-and-functionalist-approach/

van Smeden M, Lash TL, Groenwold RHH (2019) Reflection on modern methods: five myths about measurement error in epidemiological research. Int J Epidemiol 1(10):10

Varian HR (2014) Big data: new tricks for econometrics. J Econ Perspect 28(2):25

Vasilomanolakis E, Karuppayah S, Kikiras P, Mühlhäuser M (2015) A honeypot-driven cyber incident monitor: lessons learned and steps ahead. Paper presented at the 8th international conference on security of information and networks (SIN)

Verizon (2019) 2019 Data breach investigations report. Retrieved from verizon.com

Verma RM, Marchette D (2020) Cybersecurity analytics. CRC Press, Boca Raton, FL

Verma R, Kantarcioglu M, Marchette D, Leiss E, Solorio T (2015) Security analytics: essential data analytics knowledge for cybersecurity professionals and students. IEEE Secur Privacy 13(6):5

Verschuren P and Doorewaard H (2010) Designing a research project, 2nd edn. Eleven International Publishing, The Hague

Verschuren P, Doorewaard H (2010) Designing a research project, 2nd edn. Eleven International Publishing, The Hague

Violino B (2019a) Major infrastructure investments needed to thwart growing cyber risks. Information Management

Violino B (2019b) Most firms want AI in their data security arsenal, but aren't sure why. Information Management

Vollmer HM, Mills DL (eds) (1966) Professionalization. Prentice-Hall, Englewwod Cliffs, NJ

von Bertalanffy L (1968) General system theory. George Braziller, New York

von Bertalanffy L (1981) A systems view of man. Westview Press, Boulder

Voulgaris Z (2017) Data science. Technics Publications, New Jersey

Waldron K (2019) Resources for measuring cybersecurity: a partial annotated bibliography. Retrieved from Washington D.C.: https://www.rstreet.org/wp-content/uploads/2019/10/Final-Cyberbibliography-2019.pdf

Walker M, Burton B (2015) Hype cycle for emerging technologies, 2015. Retrieved from Gartner. com: https://www.gartner.com/en/documents/3100227

Walsham G (1997) Actor-network theory and IS research: current status and future prospects. In: Lee A, Liebenau J, DeGross J (eds) Information systems and qualitative research. Chapman and Hall, London

Weick KE (1995) Sensemaking in organizations. Sage, London

Weick KE (2001) Making sense of the organization. Blackwell Publishing Ltd., Oxford, UK

Weick KE (2009) Making sense of the organization: the impermanent organization, vol 2. Wiley, West Sussex, UK

Weisberg M (2013) Simulation and similarity: using models to understand the world. Oxford University Press, Oxford

Wernerfelt B (1984) A resource-based view of the firm. Strateg Manag J 5(2):9

Wheeler E (2011) Security risk management: building an information security risk management program from the ground up (Swick K, ed), Kindle edn. Elsevier, Amsterdam

Wieringa RJ (2014) Design science methodology for information systems and software engineering. Springer, New York

Wikipedia (2019) List of computer security certifications. Retrieved from https://en.wikipedia.org/wiki/List_of_computer_security_certifications

Williamson OE (1975) Markets and hierarchies. Free Press, New York

Willis D, Burton B (2015) Gartner's hype cycles for 2015: five megatrends shift the computing landscape. Retrieved from Gartner.com: https://www.gartner.com/en/documents/3111522

Wilson JR (2002) Responsible authorship and peer review. Sci Eng Ethics 8(2):19

Winn Z (2020) A human-machine collaboration to defend against cyberattacks. Technology.org

Wirth R, Hipp J (2000) CRISP-DM: towards a standard process model for data mining. Paper presented at the 4th international conference on the practical applications of knowledge discovery and data mining

Wolff J (2014) Cybersecurity as metaphor: policy and defense implications of computer security metaphors. Paper presented at the research conference on communications, information and internet policy (TPRC), Pennsylvania State University, Pennsylvania, US

Wong PK, Yang Z, Vong CM, Zhong J (2014) Real-time fault diagnosis for gas turbine generator systems using extreme learning machine. Neurocomputing 128:249–257. https://doi.org/10.1016/j.neucom.2013.03.059

World Economic Forum (2015) Partnering for cyber resilience: towards the quantification of cyber threats (REF 301214). Retrieved from Geneva, Switzerland: http://www3.weforum.org/docs/WEF_IT_PartneringCyberResilience_Guidelines_2012.pdf

World Economic Forum (2016) The future of jobs: employment, skill and workforce strategy for the fourth industrial revolution (REF 010116). Retrieved from Geneva, Switzerland: http://www3.weforum.org/docs/WEF_Future_of_Jobs.pdf

World Economic Forum (2018) The future of jobs report 2018. Retrieved from Geneva, Switzerland: http://www3.weforum.org/docs/WEF_Future_of_Jobs_2018.pdf

Wu J (2020) ModelOps is the key to enterprise AI. Forbes. Retrieved from https://www.forbes.com/sites/cognitiveworld/2020/03/31/modelops-is-the-key-to-enterprise-ai/#22ba320a6f5a

Wu WW, Lin S-H, Cheng Y-Y, Liou C-H, Wu J-Y, Lin Y-H, Wu FH (2006) Changes in MIS research status and themes from 1989 to 2000. Int J Inf Syst Change Manag 1(1):3–35. https://doi.org/10.1504/ijiscm.2006.008285

Yin Z, Liu W, Chawla S (2019) Adversarial attack, defense, and applications with deep learning frameworks. In: Alazab M, Tang M (eds) Deep learning applications for cyber security. Springer, Cham

Yu Z, Tsai JJP (2011) Instrusion detection: a machine learning approach, vol 3. Imperial College Press, London

Zaidi E (2019) Market guide for data preparation tools. Gartner, Stamford, CT

Zicari RV (2014) Big data: challenges and opportunities. In: Akerkar R (ed) Big data computing. CRC Press, London

Zicari RV, Rosselli M, Ivanov T, Korfiatis N, Tolle K, Niemann R, Reichenbach C (2016) Setting up a big data project: challenges, opportunities, technologies and optimization. In: Emrouznejad A (ed) Big data optimization: recent developments & challenges. Springer, Cham, p 17

Zinatullin L (2019) Artificial intelligence and cybersecurity: attacking and defending. Information Management

Zrahia A (2019) Should we let the machines take over? AI implementations in the cyber space. Paper presented at the ISC2 secure summit EMEA, Den Haag, Netherlands

Printed in the United States
by Baker & Taylor Publisher Services